FIRE PROTECTION EQUIPMENT AND SYSTEMS

Ronald R. Spadafora

PEARSON

Boston Columbus Indianapolis New York San Francisco Upper Saddle River
Amsterdam Cape Town Dubai London Madrid Milan Munich Paris Montreal Toronto
Delhi Mexico City São Paulo Sydney Hong Kong Seoul Singapore Taipei Tokyo

Publisher: Julie Levin Alexander
Publisher's Assistant: Regina Bruno
Product Manager: Sladjana Repic Bruno
Program Manager: Monica Moosang
Development Editor: Jean Sirois
Editorial Assistant: Daria Galbo Ballard
Director of Marketing: David Gesell
Executive Marketing Manager: Brian Hoehl
Marketing Specialist: Michael Sirinides
Project Management Lead: Cynthia Zonneveld
Project Manager: Julie Boddorf
Full-Service Project Manager: Peggy Kellar, Aptara®, Inc.
Editorial Media Manager: Richard Bretan
Media Project Manager: Ellen Martino
Creative Director: Diane Ernsberger
Cover Designer: Cenveo Publisher Services
Cover Image: Darren Greenwood/Design Pics/Corbis
Manufacturing Buyer: Mary Ann Gloriande
Composition: Aptara®, Inc.
Printer/Binder: Courier/Kendallville
Cover Printer: Courier/Kendallville

Credits and acknowledgments borrowed from other sources and reproduced, with permission, in this textbook appear on appropriate pages with text.

Copyright © 2015 by Pearson Education, Inc. All rights reserved. Manufactured in the United States of America. This publication is protected by Copyright, and permission should be obtained from the publisher prior to any prohibited reproduction, storage in a retrieval system, or transmission in any form or by any means, electronic, mechanical, photocopying, recording, or likewise. To obtain permission(s) to use material from this work, please submit a written request to Pearson Education, Inc., Permissions Department, One Lake Street, Upper Saddle River, New Jersey 07458, or you may fax your request to 201-236-3290.

Notice: The authors and the publisher of this volume have taken care that the information and technical recommendations contained herein are based on research and expert consultation, and are accurate and compatible with the standards generally accepted at the time of publication. Nevertheless, as new information becomes available, changes in clinical and technical practices become necessary. The reader is advised to carefully consult manufacturers' instructions and information material for all supplies and equipment before use, and to consult with a health care professional as necessary. This advice is especially important when using new supplies or equipment for clinical purposes. The authors and publisher disclaim all responsibility for any liability, loss, injury, or damage incurred as a consequence, directly or indirectly, of the use and application of any of the contents of this volume.

Many of the designations by manufacturers and sellers to distinguish their products are claimed as trademarks. Where those designations appear in this book, and the publisher was aware of a trademark claim, the designations have been printed in initial caps or all caps.

Library of Congress Cataloging-in-Publication Data

Spadafora, Ronald R.
  Fire protection equipment and systems/Ronald R. Spadafora.
      pages cm
  ISBN 978-0-13-502828-5—ISBN 0-13-502828-0
  1. Fire prevention—Equipment and supplies. 2. Fire departments—Equipment and supplies. I. Title.
    TH9360.S68 2015
    628.9'22—dc23

2014017685

10 9 8 7 6 5 4 3 2 1

PEARSON

ISBN-13: 978-0-13-502828-5
ISBN-10:   0-13-502828-0

DEDICATION

For Fred and Nick (my loyal older brothers) and my good friend Jack Tobin.

CONTENTS

Preface xix
Acknowledgments xxiii
About the Author xxv
Fire and Emergency Services Higher Education (FESHE) Grid xxvi

Chapter 1 Concepts of Fire 1

Introduction 1

Overview 2

Fire 2

Fuel (Solids, Liquids, and Gases) 5
  Solids 5
  Liquids 6
  Gases 7

Heat (Thermal Energy) 8
  Heat and Temperature 8
  Heat Units 8
  Heat Release Rate (HRR) 9
  Temperature Units 10
  Heat Transfer 10

Oxidizer 11

Chemical Chain Reaction 11

Phases of Fire 11
  Incipient 12
  Growth 12
  Fully Developed 13
  Decay 13

Classification of Fire 13
  Class A Fires 13
  Class B Fires 13
  Class C Fires 13
  Class D Fires 14
  Class K Fires 14

Portable Fire Extinguishers 14
  Labeling and Marking System 14
  Nonmagnetic Fire Extinguishers 15
  Fire Rating 15
  Fire Hazard Categories 16
  Number, Size, and Placement 16
  Maintenance, Inspection, and Tests 17
  When to Use a Fire Extinguisher 18

Summary 20
Review Questions 20
Endnotes 21

Chapter 2 Water Supply Systems 22

Introduction 22

Overview 23

Types of Water Supply Systems 23

Water Supply Sources and Design Criteria 23

Dry Hydrant Advantages 26

Insurance Services Office (ISO) 27
- Fire Alarm and Communications Systems 28
- Fire Department 28
- Water Supply 28

Distribution Network 29

Types of Water Distribution Systems 30

Dual Water Distribution Systems 32

Fire Hydrants 32
- Valves 33
- Capacity 34
- Spacing 34
- Signage 34
- Color Coding 34
- Inspection and Maintenance 35
- Flow Testing 35

Summary 36
Review Questions 36
Endnotes 36

Chapter 3 Water, Wetting Agent, and Wet Chemical 37

Introduction 37

Water 38
- Chemical Properties 38
- Physical Properties 39
- Advantageous Properties of Water 39
- Disadvantageous Properties of Water 40

Wetting Agent 40
- Applications 40

Heat-Encapsulating Technology 41
- Applications 41

Wet Chemical Extinguishing Agent   42
    Physical Properties   42

Saponification   42

National Association of Fire Equipment Distributors   44

Underwriters Laboratories Inc.   44

Extinguishing Systems   45
    Local Application   45

Inspection, Testing, and Maintenance   48

Portable Fire Extinguishers   49

Summary   50

Review Questions   50

Endnotes   50

Chapter 4 Sprinkler and Water Spray Fire Suppression Systems   51

Introduction   51

Sprinkler Systems   52
    Performance Statistics   52

Types of Fire Sprinkler Systems   54
    Wet Pipe Automatic Sprinkler Systems   54
    Residential Sprinkler Systems   56
    Quick-Response Sprinklers   57

Dry Pipe Automatic Sprinkler System   57
    Air Supply   57
    Dry Pipe Differential Valve Design   58
    Low-Differential Dry Pipe Valves   58
    Quick-Opening Devices   58
    Accelerator   59
    Applications   59

Nonautomatic Dry Pipe Systems   59

Deluge Sprinkler System   60

Preaction Sprinkler System   61

Inspection, Testing, and Maintenance   62
    Inspection   62
    Testing   62
    Maintenance   62

System Components   62
    Fire Pumps   62
    Fire Pump Inspection, Testing, and Maintenance   64
    Alternate Power   64
    Fire Department Connection (FDC)   64

        Valves   64
        Gate and Control Valves   65
        Valve Security   67
        Piping Arrangement   67
        Sprinkler Heads   68
        Components   68
        Sidewall Sprinklers   70

In-Rack Sprinkler System   71

Early-Suppression, Fast-Response Sprinklers   72

Summary   73

Review Questions   73

Endnotes   73

Additional References   74

Chapter 5 Standpipe Fire Suppression Systems   75

Introduction   75

Overview   76

Types of Standpipe Systems   76
        Combined (Dual)   76
        Automatic Wet   76
        Manual Wet   77
        Automatic Dry   77
        Manual Dry   77
        Semiautomatic Dry   78

Classes of Standpipe Systems   78
        Class I   78
        Class II   80
        Class III   81

Firefighting Water Supply Operations   81

Water Supply   82
        Public Waterworks Connection   82
        Fire Pumps   82
        Gravity Tanks   83
        Pressure Tanks   84

Fire Department Connection   86

Valves   86
        Hose Outlet Valves   86
        Check Valves   86
        Swing Check Valves   86
        Alarm Check Valves   86
        Gate Valves (Nonrising Stem)   86
        Outside Screw and Yoke Gate Valves   87
        Drain Valve   87

Hose Stations 87
Standpipe Kit 87
Alarm Devices and Supervisory Switches 87
Water Flow Alarms 87
Supervisory (Tamper) Switches 87
Inspection, Testing, and Maintenance 88
    Inspection 88
    Testing 88
    Maintenance 89

Summary 90
Review Questions 90
Endnotes 90
Additional References 91

Chapter 6 Foam and Foam Fire Suppression Systems 92

Introduction 92
Overview 93
Physical Characteristics 93
General Principals of Foam Extinguishing Systems 93
Chemical and Mechanical Foam 94
Protein and Synthetic Foam 95
Low-Expansion Foam 98
    Fire and Spill Applications 98
High-Expansion Foam 99
    Fire and Spill Applications 100
    Firefighter Safety Concerns 101
Class A Foam 102
    Generating Process 102
    Firefighting Operations 102
Compressed-Air Foam System 102
    Generating Process 103
    Firefighting Operations 103
    Use of a CAFS in Fixed Systems 103
Class B Foam 105
    Generating Process 105
    Firefighting Operations 105
Fixed Foam Systems 106
    Foam Concentrate Storage Tanks 107
    Bladder Tanks 107

        Proportioning Devices and Eductors   107
        Pumps   108
        Foam Chambers and Foam Makers   108
        Sprinkler Foam-Water Heads and Nozzles   109
        Foam Generators   109
        Foam Monitors   109
        Fire Detection and Control Equipment   110

Fixed Foam System Applications   110
        Subsurface Injection for Permanent Roof Liquid Storage Tanks   110
        Surface Application for Permanent Roof Liquid Storage Tanks   110
        Seal Protection for Floating Roof Liquid Storage Tanks   111
        Aircraft Hangars   111

Truck Loading Racks   113

Summary   115

Review Questions   115

Endnotes   115

Additional References   116

Chapter 7 Carbon Dioxide   117

Introduction   117

Overview   118

Properties of Carbon Dioxide   118
        Physical Properties   119
        Environmental Properties   119

Phases   119
        Solid   120
        Liquid   121
        Vapor   122

Fire Extinguishment Applications   122

Local-Application Extinguishing Systems   123
        Dip Tanks   123

Total-Flooding Extinguishing Systems   123

Integrity of Enclosure   123

Minimum Carbon Dioxide Concentrations for Extinguishment   125
        Surface Fires and Deep-Seated Fires   125
        Extended Discharge System   125

Hand Hose-Line Extinguishing Systems   125

Portable Fire Extinguishers   127

Examining the Risks   128

Firefighting Operations   130

Low-Pressure and High-Pressure Storage Systems   132
    Low Pressure   132
    High Pressure   134

Summary   136

Review Questions   136

Endnotes   136

Additional References   137

Chapter 8 Inert Gas and INERGEN Fire Protection Systems   138

Introduction   138

Overview   140

Argon   140
    Physical Properties   141

Environmental Issues   141
    Clean agent   142

Liquid Argon   142
    Why Inert Atmospheres in Confined Spaces Are Dangerous   143
    Firefighting Procedures   143

Nitrogen   143
    Firefighting Procedures   146

Firefighter Safety Concerns   146

INERGEN   147
    Physiology of INERGEN   148
    Design and Operation   149
    Applications   149
    Firefighting Procedures   151

Summary   152

Review Questions   152

Endnotes   152

Additional References   153

Chapter 9 Dry Chemical Fire Suppression Systems   154

Introduction   154

Overview   155

Physical Properties   155

Varieties   156
    Sodium Bicarbonate   156
    Foam-Compatible Sodium Bicarbonate   156

        MET-L-KYL   156
        Monoammonium Phosphate   157
        Potassium Bicarbonate   157
        Potassium Chloride   158
        Urea-Potassium Bicarbonate   158

Extinguishing Properties   158
        Interruption of the Chemical Chain-Breaking Reaction   159

System Components   159

Sequence of Operation   161

Total-Flooding and Local-Application System Methods   162

Hand Hose-Line Method   164

Portable Fire Extinguishers   164

Application Techniques   165
        Flammable Liquid Spill Fire   166
        Three-Dimensional (Dip Tank), Flammable Liquid, Gravity-Fed Fire   166
        Flammable Liquid Pressure Fire   166
        Flammable Gas Pressure Fire   166
        Liquefied Petroleum Gas (Broken Flange) Fire (In Conjunction with Water Streams)   166

FDNY Purple-K Units   167

Summary   168
Review Questions   168
Endnotes   168
Additional References   168

Chapter 10 Combustible Metal Suppression Agents   170

Introduction   170

Overview   171

Dust Explosion Pentagon   172
        Fuel   172
        Ignition Source   173
        Oxidizer   173
        Confinement   173
        Dispersion   173

Physical Properties   174
        Aluminum   174
        Beryllium   175
        Caesium   176
        Calcium   176
        Hafnium   176
        Iron   176
        Lithium   177
        Magnesium   177

　　　　Manganese   177
　　　　Phosphorous   178
　　　　Plutonium   179
　　　　Potassium   179
　　　　Rubidium   179
　　　　Silicon   179
　　　　Sodium   179
　　　　Strontium   179
　　　　Tantalum   180
　　　　Thallium   180
　　　　Thorium   180
　　　　Tin   180
　　　　Titanium   180
　　　　Uranium   182
　　　　Zinc   182
　　　　Zirconium   182

Extinguishing Agent Applications   182

Categories   183

Alkali Metals   183

Alkaline Earth, Transitional, and Other Combustible Metals   185

Explosions   186

Explosion Prevention and Protection Systems   186

Detection Devices and Fire Alarm Control Units   188

Alternatives and Enhancements to Explosion Suppression Systems   189

Firefighting Operations and Emergency Procedures   191

Summary   193

Review Questions   193

Endnotes   193

Additional References   194

Chapter 11 Halon Fire Suppression Systems   195

Introduction   195

Overview   196

Montreal Protocol   196

Physical and Chemical Properties   197
　　　　US Army Corps of Engineers Numbering System   198

Types and Applications   198

Halon 1211   199
　　　　Local-Application Systems   200
　　　　System Components   200
　　　　Portable Fire Extinguishers   200
　　　　Essential Applications   201

Clean Agent Substitute for Halon 1211   201

Halon 1301   202

Total-Flooding Systems   202

    System Components   203
    Activation Process   204
    Emergency Procedures for Employees   205
    Clean Agent Substitute for Halon 1301   205

Decommissioning   205

Summary   207

Review Questions   207

Endnotes   207

Additional References   207

Chapter 12 Clean Agent Fire Suppression Systems   208

Introduction   208

Overview   209

Significant New Alternatives Policy   209

Industry Standards   210

Halocarbons   211

    Novec 1230   211
    FM-200®   211
    Additional Halocarbon Agents   212

Halocarbon Total-Flooding System Design   215

Protection under Raised Floors   215

Deep-Seated Smoldering Fires   216

Powdered Aerosols   216

    Dispersed Systems   217
    Condensed Systems   217

Powdered Aerosol System Design   218

    Propelled Extinguishing Agent Technology Total-Flooding Systems   218
    Propelled Launcher Total-Flooding and Local-Application Systems   219
    Pneumatic Impulse Total-Flooding and Local-Application Systems   219
    Heat-Actuated Total-Flooding Systems   219

Firefighter Operations   219

National Fallen Firefighters Foundation: 16 Firefighter Life Safety Initiatives Program   221

Portable Fire Extinguishers   221

Summary   222

Review Questions   222

Endnotes   222

Chapter 13 Water Mist Fire Suppression Systems 224

Introduction 224

Overview 224

Extinguishing Properties 225

Droplet Size 225

Water Mist Advantages over Conventional Sprinkler Systems 227

Water Mist Disadvantages over Conventional Sprinkler Systems 227

Pressure Classifications 228
- Low Pressure 228
- Medium Pressure 228
- High Pressure 228

System Components 229
- Water Supply 229
- Pumps/Cylinders 230
- Distribution Piping 230
- Nozzles 230

Fire Detection and Alarm 231

Actuation and Service Disconnect 232

Types of Systems 232
- Wet Pipe System 232
- Dry Pipe System 232
- Deluge System 232
- Preaction System 232

Water Supply Methods 232
- Single Fluid 233
- Twin Fluid 233

Applications 233
- Total-Flooding Application Systems 234
- Local-Application Systems 234
- Zoned-Application Systems 234
- Portable Handheld Fire Extinguishers 234
- Water Mist with Foam Additives 235
- Occupant Safety 235

Summary 236

Review Questions 236

Endnotes 236

Additional References 237

Chapter 14 Fire Alarm and Detection Systems 238

Introduction 238

Overview 239

Fire Alarm System Basics 239
Fire Alarm Control Unit 240
Annunciator Panel 241
Conventional Fire Alarm System 241
Addressable Fire Alarm System 241
Analog Addressable (Intelligent) Fire Alarm System 241
Primary and Secondary Power Supplies 242
- Primary (Main) Power Supply 242
- Secondary (Backup) Power Supply 242

Initiating Devices 242
Manual (Pull Station) Alarm Boxes 242
Smoke Detectors 243
- Spot Photoelectric Light-Scattering Smoke Detectors 244
- Spot Photoelectric Light-Obscuration Smoke Detectors 244
- Spot Ionization Smoke Detectors 245
- Air-Aspirating Smoke Detection Systems 246
- Linear-Beam Smoke Detector Systems 248
- Video-Imaging Smoke Detection Systems 248

Heat Detectors 249
- Spot Fixed-Temperature Heat Detectors 250
- Spot Rate-of-Rise Heat Detectors 250
- Spot Rate-Compensation Heat Detectors 251
- Coaxial Conductor Line Heat Detectors 251
- Paired-Wire Line Heat Detector 252

Flame (Radiant) Detectors 253
Gas-Sensing Detectors 254
Pressure-Sensing Detectors 255
Notification Appliances 256
Fire Alarm Connection Stations 256
- Central Station 257
- Proprietary Supervisory Station 257
- Remote Supervisory Station 257

Summary 258
Review Questions 258
Endnotes 258
Additional Reference 259

Chapter 15 Smoke Control Systems 260

Introduction 260
Overview 261

NFPA Smoke Control and Management Strategies 261

Smoke and Smoke Inhalation 261

Toxic Gases 262
- Hydrogen Cyanide 262

Nondedicated Smoke Control Systems 263
- Single-Zone HVAC Systems 263
- Central HVAC Systems 264

Dedicated Smoke Control Systems 265
- Stair Pressurization Systems 265
- Elevator Shaft Pressurization Systems 266
- Smoke Shaft Systems 268
- Atrium Systems 268

Firefighting Planning, Strategy, and Tactics for a Fire in an Atrium 271

Principles of Smoke Control 271
- Positive Stack Effect 272
- Positive-Pressure Ventilation for Firefighting Purposes 272
- Negative Stack Effect 273

Passive Protection 273
- Smoke Dampers 274
- Fire Dampers 275
- Combination Smoke/Fire Dampers 275
- Smoke Curtains 275
- Smoke-Proof Enclosure 276

Summary 277

Review Questions 277

Endnotes 277

Additional References 278

Index 279

PREFACE

This book is written for young minds interested in understanding fire protection and life safety systems, and it will provide a commonsense approach to learning. I have been in the FDNY for 36 years and taught firefighters and college students aspiring to become uniformed civil servants. This has allowed me to recognize what first responders should know about fire. In this book, I have tried to incorporate my own experiences as a student, instructor, firefighter, fire marshal, fire officer, and fire chief to describe the concepts of fire and the traditional and modern extinguishing systems developed to fight it. Years of conducting building and fire prevention inspections, as well as participating in site-planning exercises and familiarization drills, have allowed me to enhance my knowledge of fire extinguishing agents, water supply and distribution systems, fire alarms and detectors, and smoke management systems and relate this knowledge to the work performed by first responders at fires and emergencies. These areas are the main topics of this book.

The purpose of this book is to enhance the safety of first responders when operating at the scene of an incident. Through a greater knowledge of fire hazards, life safety systems, and fire protection features, efficient strategies and tactics can be formulated to reduce the time required to extinguish fires and mitigate emergencies in the built environment. In addition, architects, fire protection engineers, fire code officials, and other design professionals will be provided with a glimpse of how fire chiefs and firefighters think. Designers traditionally are tasked with the requirements of a building's owner and stakeholders when constructing the structural layout and building management systems. This interaction of ideas will allow designers to tailor their work to also meet the operational needs of the fire service.

Fire science is the study of the nature of fires, fire protection, and firefighting techniques. Each chapter in this book provides the reader with important information in each of these three areas. Fire occurs in defined phases, and the fire's phase is important to recognize on the fire scene for enhanced firefighter safety. Fire protection professionals must also understand these transitions when designing systems and equipment to protect building occupants and property. Fire is also classified according to the combustibles involved. This classification corresponds to the type of extinguishing agent needed and the application design process. Extinguishing agents vary; Chapter 1 reviews the nature of these agents and focuses on the use of portable fire extinguishers to fight fire in its Incipient or beginning phase.

Water delivery systems in municipalities transport potable water from a water treatment facility to town or city residents for use as drinking water. They are also used to supply water to designated locations to provide the public with an effective level of fire protection. Chapter 2 examines water supply systems and their design features. It explains how water is distributed through a system of piping for fire department use. Layouts and components (valves, piping, and hydrants) are examined to better understand the workings of this valuable fire service infrastructure. Chapter 3 looks at the properties of water and water-based extinguishing agents. The pros and cons of using water on fires correlates to its application design. The replacement of dry chemical extinguishing agent with wet chemical extinguishing agent (which contains 40% to 60% water) in pre-engineered commercial cooking equipment in the 1990s is one of the most important fire protection advancements in modern times.

Chapters 4 and 5 cover sprinkler and standpipe water-based fire protection systems. Fire sprinkler systems are used extensively worldwide and have an outstanding success rate in controlling fires. They commonly have a water supply and distribution piping designed to flow water out of sprinkler heads over the fire area. Standpipe systems consist of a network of piping designed to feed hose outlet valves for manual firefighting. They have water supply features similar to a sprinkler system. Standpipe systems are especially important in high-rise and large-footprint buildings, where the fire floor can be beyond the reach of aerial streams and where long, manual hose stretches from the fire apparatus to the fire is impractical. Modern fire protection design combines sprinkler and standpipe systems, allowing the standpipe riser to perform a dual function in supplying water to the sprinkler heads.

Firefighting foam and the fire protection systems that use it are discussed in Chapter 6. This water-based extinguishing agent is an aggregate of air-filled bubbles formed from an aqueous solution. Its use in fixed (permanently installed) fire suppression systems and by the fire service in extinguishing structural fires and mitigating fuel spills is reviewed. The cohesive foam blanket formed by the bubbles extinguishes fire by excluding air, cooling fuel, suppressing vaporization, and preventing reignition.

Chapters 7 and 8 cover carbon dioxide (CO_2), inert gas, and INERGEN, which is a mixture of inert gas (nitrogen and argon) and carbon dioxide. These non-water-based fire extinguishing agents suppress fire by displacing oxygen and reducing its concentration to a percentage at which the fire can no longer be sustained. They perform this task without creating a residue that can damage delicate electronics and electrical equipment. Fixed systems consist of the agent, storage containers or cylinders, fire detection system, agent release valves, piping, and nozzles or applicators. In addition, because of the nonreactive properties of inert gases they are often useful to prevent undesirable reactions and fire in the chemical and oil industries. When used in this manner, first responders arriving at emergencies involving workers in distress within confined spaces must monitor the air for oxygen concentration levels prior to rescue efforts.

The first mention of local-application and total-flooding systems is found in these two chapters. Local-application systems have fixed piping that apply an extinguishing agent via nozzles directly onto a fire or onto an area immediately surrounding the hazard. There are no physical barriers surrounding the fire area. The term "local application" does not refer to the use of manually operated wheeled or portable fire extinguishers. Total-flooding extinguishing systems have nozzles that are arranged to discharge into an enclosure around the hazard. The protected, three-dimensional space must enable the design concentration of the agent (volume percent of the agent in air) to build up and remain for the prescribed period of time for effective extinguishment of fire. Both systems may be operated either automatically by detection or manually. Local-application and total-flooding systems are also installed for dry chemicals, halons, clean agents, and water mist.

In Chapter 9, dry chemical extinguishing agent and its suppression applications are discussed. It is composed of tiny particles, usually of sodium bicarbonate, monoammonium phosphate, potassium bicarbonate, potassium chloride, or urea-potassium bicarbonate. Added particulate material enhances flow capabilities and provides resistance to moisture absorption. The principle way that dry chemical extinguishes fire is via the interruption of the chain reaction sequence. Typical local-application and total-flooding fire protection applications of dry chemical agent include petroleum and chemical loading racks, refinery processing equipment, and offshore platforms; marine tanker decks, machinery spaces, and loading docks; and paint spray booths, dip and quench tanks, automobile service stations, warehouses, hazardous materials storage buildings, and flammable liquid storage areas.

When all of the elements in the dust explosion pentagon are in place, rapid combustion can occur. When a dust cloud is ignited within a confined area or a building, it burns very rapidly and may explode. The safety of building occupants is threatened by the ensuing fires, additional explosions, flying debris, and collapsing building structural elements. Chapter 10 explores managing explosion risks by identifying the hazards and implementing extinguishment and prevention measures. Alternative techniques to explosion-suppression systems are also reviewed.

Chapter 11 examines halon gas and its extinguishing system applications. Halon suppresses fire in a manner similar to dry chemical agents but leaves no residue after the fire is extinguished. In halon local-application and total-flooding systems, pressurized tanks holding the gas are connected to a series of pipes for discharge through nozzles. When a fire detection system is activated, the gas is released. Under the Clean Air Act, however, the United States banned the production and import of halon as of January 1, 1994, in compliance with the Montreal Protocol on Substances that Deplete the Ozone Layer (Montreal Protocol). Recycled halon and inventories produced before January 1, 1994, are now the only sources of supply. Today, the installation of new halon systems is only allowed for "essential applications." These systems, in general, are used for military aircraft fire protection. Older existing systems are legal, although decommissioning and/or replacing them with more environmentally friendly fire extinguishing agents is encouraged.

Clean agent fire-suppression criteria include low toxicity, environmentally friendly, fast acting and effective, noncorrosive, electrically nonconductive, and leaves no residue. Clean agents have zero ozone-depletion potential (ODP), low (or no) atmospheric lifetimes, and are listed by the US Environmental Protection Agency (EPA) with no usage restrictions. They have been produced as replacement agents for the halons. Chapter 12 reviews how some of these non-water-based clean agents (Novec 1230, FM-200, and powdered aerosols) extinguish fire and their methods of application. Chapter 13 deals with water mist technology. Water mist is a water-based fire extinguishing replacement agent for the halons. A major feature of water mist is the minute droplets generated by its nozzles. Small drops of water, with a greater surface area to mass ratio, are more effective in extinguishing fire than large droplets. They more readily evaporate and change to steam, resulting in the dilution of surrounding oxygen and fuel vapors.

A fire alarm system is a set of electrical components and equipment that work together to detect the products of combustion and alert people through visual and audio appliances when a fire has occurred. These alarms may be activated from various types of detectors and water flow sensors that are automatic or from a manual fire alarm pull station. Chapter 14 also covers fire alarm connection stations that automatically notify building occupants and the fire department. Early detection helps to save lives and gives firefighters the ability to arrive on the scene in a timely manner to control and suppress the fire. The most common type of fire alarm connecting stations include central stations, proprietary supervisory stations, and remote supervisory stations.

Chapter 15 discusses smoke control systems and how they save lives by keeping exits and exit corridors free of smoke to facilitate evacuation. They also prevent or delay fire from developing further and spreading rapidly. Firefighting operations are facilitated, and there is a reduced risk of damage to the building. Smoke control management strategies include controlling the airflow into a smoke-filled room and using exhaust fans (negative air pressure technique) and forcing air into neighboring areas and turning off exhaust fans (positive air pressure technique). Smoke control equipment can be either non-dedicated or dedicated. Non-dedicated systems use heating, ventilation, and air-conditioning (HVAC) equipment components to create differential pressure zones for smoke control. Dedicated equipment systems are designed to be used only for smoke control. They include stair pressurization, elevator shaft pressurization, smoke shafts, and atrium systems. Passive smoke control components are also reviewed. They may include smoke and fire dampers, smoke curtains, and smoke-proof enclosures.

Photographs taken on location will hopefully add depth to the subject matter discussed in each chapter of the book. Real-world case studies attempt to further enlighten the reader about the tragic consequences possible for uninformed professionals. Incidents reviewed include loss-of-life fires, explosions, hazardous material spills, improper use of firefighting equipment, and the inadvertent release of fire extinguishing and inerting agents. Use the information provided to enhance your value to the community, whether it be as a first responder or a public safety official. Good luck, and stay safe in all future endeavors.

–Chief Ron

ACKNOWLEDGMENTS

To the late Sid Aconsky, Engineer/President, Acotech Services. A valued mentor in fire protection engineering. Thank you and rest in peace.

To Stephen G. Smith, Senior Acquisitions Editor. Thank you for your confidence in me on this book project.

To Monica Moosang, Program Manager, and Jean Sirois, Development Editor, who were with me throughout the entire editing process of the manuscript. Thank you both for providing professional guidance and encouragement.

Reviewers

The publisher and authors would like to thank the reviewers for their feedback and suggestions, which helped guide the development of this text:

Andrew S. Caldwell, P.E.
Fire Protection Engineer
NFS, Inc.
Erwin, TN

Richard A. Marinucci, Fire Chief
Northville Township Fire Department
Northville Township, MI

Chief Mark Martin, MBA
Stark State College
Perry Township Fire Department
North Canton, OH

Jack J. Murphy, Chairman
Fire Safety Directors Association of New York City for High-Rise Buildings

David Walsh
Program Chair—Fire Science
Dutchess Community College
Poughkeepsie, NY

ABOUT THE AUTHOR

Assistant Chief Ronald R. Spadafora is a 36-year veteran in the Fire Department of New York (FDNY) and is the Chief of Logistics in the Bureau of Operations. On 9/11, he responded to the World Trade Center (WTC) and supervised both rescue and fire suppression efforts at the North Tower and WTC 7. He was named the WTC Chief of Safety in October 2001 for the entire Recovery Operation that ended in June 2002. On August 14–15, 2003, he headed the Logistics Section for the FDNY during the New York City Blackout. In the aftermath of Hurricane Katrina in September 2005, he was designated Deputy Incident Commander of the FDNY Incident Management Team and sent to Louisiana to assist the New Orleans Fire Department in the protection of  the city. In October–November 2013, Chief Spadafora once again played a major role in coordinating logistics for the FDNY during and in the months following Hurricane Sandy. Chief Spadafora has taught fire science at John Jay College (CUNY) for 25 years and is currently teaching emergency management at Metropolitan College of New York at both the graduate and undergraduate level. He is also the senior instructor for Fire Technology Inc., an organization devoted to preparing FDNY members for promotional exams. Since 2008, Chief Spadafora has been a visiting instructor and advisor on urban firefighting and incident management for the Working on Fire (WoF) program of South Africa. Chief Spadafora holds an MPS degree in Criminal Justice from Long Island University (C.W. Post Center), a BS degree in Fire Science from John Jay College, and a BA degree in Health Education from Queens College (at CUNY). He has written dozens of articles in publications such as *WNYF* (*With New York Firefighters*), *Fire Engineering*, *Size Up*, and the *American Journal of Industrial Medicine*.

FIRE AND EMERGENCY SERVICES HIGHER EDUCATION (FESHE) GRID

The following grid outlines Fire Protection Systems course requirements developed as part of the FESHE Model Curriculum and where specific content can be located within this text.

Course Objectives

Students will

1. Identify and describe various types and uses of fire protection systems.
2. Describe the basic elements of a public water supply system as it relates to fire protection.

COURSE REQUIREMENTS	1	2	3	4	5	6	7	8	9	10	11	12	13	14	15
Explain the benefits of fire protection systems in various types of structures.	X		X	X		X	X	X	X	X	X	X	X	X	X
Describe the basic elements of a public water supply system including sources, distribution networks, piping, and hydrants.		X		X	X										
Explain why water is a commonly used extinguishing agent.	X	X	X	X	X	X							X		
Identify the different types and components of sprinkler, standpipe, and foam systems.				X	X								X		
Review residential and commercial sprinkler legislation.				X											
Identify the different types of non-water-based fire-suppression systems.	X						X	X	X	X	X	X			
Explain the basic components of a fire alarm system.				X			X	X	X	X	X	X	X	X	X
Identify the different types of detectors and explain how they detect fire.				X			X	X	X	X	X	X	X	X	X
Describe the hazards of smoke and list the four factors that can influence smoke movement in a building.									X				X	X	X
Discuss the appropriate application of fire protection systems.			X	X	X	X	X	X	X	X	X	X	X	X	X
Explain the operation and appropriate application for the different types of portable fire protection systems.	X		X			X	X		X	X	X	X	X		

xxvi

CHAPTER 1

Concepts of Fire

Source: Ronald R. Spadafora

KEY TERMS

Wildland, *p. 4*
Flashover, *p. 9*
Heat flux, *p. 9*

Incipient, *p. 12*
Growth, *p. 12*

Fully Developed, *p. 13*
Decay, *p. 13*

OBJECTIVES

After reading this chapter, the reader should be able to:

- Understand the basic terminology of fire science
- Describe the components of the fire triangle
- Explain the chemical chain reaction process denoted in the fire tetrahedron
- List the four phases of fire
- List the classifications of fire
- Explain the role of portable fire extinguishers in fire protection

Introduction

The curriculum of fire science involves the study of how fires are started and how they spread. Students learn about fire prevention and protection as well as activation and suppression systems. Water, the traditional fire quencher, and its supply, storage, and delivery systems are standard topics covered. Extinguishing agents other than water are also examined, as is the equipment that discharges these agents. Different techniques for suppressing and isolating fires are also reviewed. The products of combustion and their effect on humans and human behavior is yet another area of analysis. Smoke control is also extremely important in fire safety design.

A basic understanding of the characteristics and behavior of fire is important to firefighters and professionals in the fire engineering and protection fields. This chapter will review important concepts related to the science of fire. Understanding fire requires a good working knowledge of both chemistry and physics. This information is designed to give the reader a vital overview of fire science fundamentals. It will facilitate learning the material presented in the forthcoming chapters pertaining to fire safety.

Overview

Effective fire control and extinguishment requires an understanding of the chemical and physical nature of fire. This includes information describing the major components necessary to start and sustain fire. Terminology that describes the composition and characteristics of fuels as well as the sources of heat energy will provide a keen insight into common fire hazards and corresponding fire protection systems.

The burning process occurs in defined phases. It is important that firefighters recognize these phases to better understand the combustion process in order to develop successful and safe operational firefighting strategies and tactics. Understanding these phases is also beneficial to fire protection professionals when they design equipment to protect lives and property.

Fires are classified according to the nature of combustibles involved. The classification of fire corresponds to the type of extinguishing agent needed and the application design process. Extinguishing agents vary in scope from multipurpose to specific. This chapter focuses on the utilization of portable fire extinguishers to fight fire in the beginning or incipient phase. Portable fire extinguishers are the first line of defense and an important tool for both firefighters and building occupants. Information provided in this chapter will help clarify the fire extinguisher rating and selection process and facilitate understanding of the utilization, inspection, testing, and maintenance of portable fire extinguishers. The first sections ahead provide definitions of vital terms, grouped into major fire science categories.

Fire

Fire: Fire is a rapid oxidation process, which is a chemical reaction resulting in the evolution of light and heat in varying intensities.[1] Fires are created when combustible or flammable fuel ignites. This takes place when a sufficient quantity of an oxidizer (oxygen gas or another oxygen-rich compound) is exposed to a source of heat or ambient temperature and is able to sustain a rate of rapid oxidation that produces a chain reaction. Fire cannot exist without all of these elements in place and in the right proportions.

Backdraft: A backdraft is a phenomenon that occurs when a smoldering fire that has consumed most of the available oxygen inside an enclosure suddenly explodes when additional oxygen is introduced (commonly when a window or door is opened).

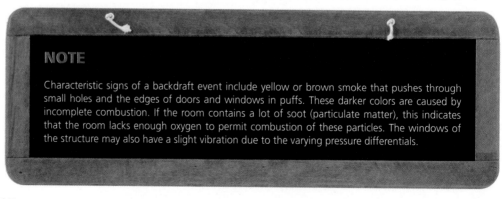

NOTE

Characteristic signs of a backdraft event include yellow or brown smoke that pushes through small holes and the edges of doors and windows in puffs. These darker colors are caused by incomplete combustion. If the room contains a lot of soot (particulate matter), this indicates that the room lacks enough oxygen to permit combustion of these particles. The windows of the structure may also have a slight vibration due to the varying pressure differentials.

BLEVE: A *boiling liquid expanding vapor explosion* is caused by the rupture of a vessel or container that holds a pressurized liquid above its boiling point. The storage vessel is often weakened from the heat of a fire, but a BLEVE can also be created through container corrosion over time or mechanical damage by an impact that occurred in an accident.

Chemical explosion: A rapid combustion reaction classified as a detonation or deflagration, depending on the rate of propagation.

Combustion: Combustion is rapid oxidation reaction that can produce fire. The oxygen in the air (21%) is generally the oxidizer, chemically reacting with the fuel. Combustion reactions are almost always exothermic reactions. Combustion is commonly called fire. If combustion is confined and a rapid pressure rise occurs, it is called an explosion.

Deflagration: A deflagration is a term describing the propagation of combustion at a velocity less than the speed of sound. Deflagrations are typically associated with natural gas or propane releases, gasoline and hydrocarbon vapors, finely divided fuels, and certain reactive chemicals. They can occur either immediately before or immediately after a fire.

Detonation: A detonation is combustion propagating at a velocity faster than the speed of sound. Detonations are commonly associated with high explosives (dynamite, blasting agents, or munitions). Detonations generate pressure waves many times greater than deflagrations, resulting at times in structural damage and collapse.

Endothermic reaction: An endothermic reaction is a process that absorbs energy in the form of heat. New substances formed by the reaction contain more heat energy than was in the reacting materials. Examples of the endothermic process include melting ice cubes and evaporating liquid water.

Exothermic reaction: An exothermic reaction is a process that generates heat. New substances formed by the reaction have less heat energy than was in the reacting materials. Examples of the exothermic process include freezing water and condensing water vapor.

Explosion: An explosion is a rapid expansion of gases (fuel and oxygen) that have premixed prior to ignition. Common explosions encountered by firefighters are chemical and mechanical explosions.

Fire tetrahedron: The fire tetrahedron is a model that expands on the one-dimensional fire triangle. The fire tetrahedron shows the interrelationship between the three components of the fire triangle visually and further clarifies the definition of combustion by adding a fourth component (uninhibited chemical chain reaction). The fire tetrahedron helps to describe "ignition," which the fire triangle does not.

Once a fire has started, the resulting exothermic chain reaction sustains the fire and allows it to continue until at least one of the elements of the fire is eliminated. Foam can be highly effective in denying the fire the oxygen it needs. Water can be used to lower the temperature of the fuel below its ignition point. Halons, dry chemical agents, and dry powders interact with free radicals (unstable fuel particles), thereby keeping them from reacting with oxygen to enhance the combustion process (see Figure 1.1).

Fire triangle: The fire triangle is a model used to help in the understanding of the three major elements necessary for ignition: heat (thermal energy), fuel, and oxidizer (oxygen). Fire will be extinguished by removing any one of the elements in the fire triangle. A fire naturally occurs when the elements are present and combined in the right mixture.[2]

Without the right amount of heat, a fire cannot begin, nor can it continue. Heat can be removed or absorbed by the application of an extinguishing agent

FIGURE 1.1 Fire tetrahedron

Wildland ■ An area in which development is essentially nonexistent, except for roads, railroads, power lines, and similar transportation facilities. Structures, if any, are widely scattered.

that reduces the amount of heat available to the fire reaction. Water is often used for this purpose. Other types of suppression agents (gases and powders) used in sufficient quantities at the point at which combustion is taking place reduce the amount of heat available for the fire reaction.

A fire cannot be sustained without fuel. Fuel can be removed naturally or deliberately by mechanically or chemically removing the fuel from the fire. Fuel separation is an important factor in **wildland** fire suppression. Tactics using the fuel-separation method include back burning as well as the use of heavy equipment (bulldozers) and hand tools. These techniques involve the deliberate setting of fires or the uprooting of foliage in order to remove the fuel in the path of a quickly moving forest fire. When the fire reaches the burnt-out or uprooted area, a firebreak is created. There is little fuel left for the fire to feed on. Subsequently, the fire burns itself out. Removing the fuel decreases the heat generated.[3]

In general, without sufficient oxygen a fire cannot be created and cannot propagate. With a decreased oxygen concentration, the combustion process slows. Oxygen can be denied to a fire by using inert gases to displace or purge ambient oxygen in the air, foam to smother the fire, or tools and equipment that cover the fire (for example, a fire blanket; see Figure 1.2).

Flashover: A flashover is a phenomenon whereby an enclosure or room suddenly becomes engulfed in flames from floor to ceiling. This is a result of the fire's heat energy raising the temperature of the contents of the space to their ignition temperature.

Pyrolysis: Pyrolysis is a chemical decomposition of organic material at elevated temperatures. It does not involve reactions with oxygen, but it can take place in its presence. It is one of the processes involved in charring wood, starting at 390°F

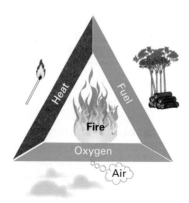

FIGURE 1.2 Fire triangle

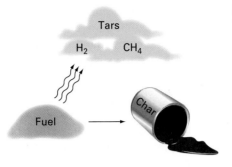

FIGURE 1.3 Pyrolysis

to 570°F (200°C to 300°C). In general, pyrolysis of organic substances produces gas and liquid products and leaves a solid residue (char; see Figure 1.3).

Spontaneous combustion: Spontaneous combustion is the ignition of a substance as a result of internal oxidation processes without the application of an external heat source. Many substances undergo a slow oxidation that releases heat. If the heat cannot escape, the temperature of the substance rises until ignition occurs.

EXAMPLE

Hay, straw, or mulch with relatively low ignition temperature can release heat due to oxidation in conjunction with a little moisture and bacterial fermentation. This heat is unable to escape, because the material involved is a good thermal insulator. The temperature continues to rise until it goes above the ignition point of the material. Combustion is initiated if adequate oxygen and fuel are present to maintain the reaction.

Fuel (Solids, Liquids, and Gases)

Any materials that burn can be considered fuel. Most common fuels contain carbon, hydrogen, and oxygen. Examples include wood, paper, propane gas, methane gas, and plastics. Combustible matter may be in a solid, liquid, or gaseous state.

SOLIDS

Solids are materials with defined volume, size, and shape at a given temperature. Examples include wood and wood products (paper and cardboard), carbon-containing materials (coal and charcoal), plastics (organic polymers and epoxies), textiles (cotton, wool, and rayon), and combustible metals (magnesium, lithium, and aluminum).

Wood and Wood Products

Wood and wood products are the most common solids encountered by firefighters. They are considered Class A–type materials and require water or water solutions to cool them below their ignition temperature and extinguish them. The average ignition temperature of wood is approximately 400°F (204°C). Major components are carbon, hydrogen, and oxygen.

Factors Affecting the Ignition and Combustibility of Wood and Wood Products

Many factors influence the ignition and combustibility of wood and wood products, the most important of which include the following:

- *Physical form (size, form, shape, and mass):* In general, the greater the mass in relation to surface area, the more heat energy will be required to ignite it, and the slower the rate of burning will be once ignited.

- *Autoignition temperature:* This occurs through self-heating of the reactants of the combustible material. It is the minimum temperature a material must be heated to for it to ignite and be self-sustaining without an external heat source. Autoignition temperature is commonly referred to as ignition temperature.
- *Piloted-ignition temperature:* This occurs with the assistance of an external (flame or spark) heat source. It is generally lower than autoignition temperature.

Plastics

Plastics are another common combustible solid. Most plastics contain organic polymers and are petroleum based (hydrocarbons). The vast majority of these polymers are based on chains of carbon atoms either alone or with oxygen, sulfur, or nitrogen. They can be soft or hard and can be electrically conductive or nonconductive (insulators). Manufactured plastics usually contain additives (colorants, stabilizers, and lubricants), which change the chemical nature and combustibility of the original plastic. Pyrolysis occurs less readily in plastics than in wood, and therefore plastics tend to have a higher ignition temperature than wood and wood products.

Synthetic Fiber

Synthetic fiber (rayon, nylon, and polyester) is material woven from manmade fiber (plastic, hydrocarbon, metal, and glass). Burning characteristics of synthetic fiber include decomposition, burning, and melting. Synthetic fiber can, however, be made flame retardant, and various kinds of synthetic fiber (Nomex, Polybenzimidazole [PBI], and Kevlar) are used in the production of flame- and thermal-resistant clothing for firefighters. Materials used in the production of firefighter bunker gear are high-temperature resistant synthetic fibers. *NFPA 1971: Standard on Protective Ensembles for Structural Fire Fighting and Proximity Fire Fighting,* is followed by the fire service.[4]

Combustible Metals

Combustible metals are elements that will combine with oxygen, reach their ignition temperature, and burn. Metals do not, however, undergo pyrolysis to produce combustible vapors when heated. They burn on their surface with no flaming combustion. Metals that do burn produce an abundance of heat energy. When water is applied and the water molecule separates, steam and hydrogen explosions can occur. For this reason, water, unless in large amounts and readily available, is not recommended as an extinguishing agent on combustible metals. Specific extinguishing agents (graphite, sodium chloride, and copper powder) have been developed to cover the surface of the burning metal and exclude oxygen. Combustible metals are classified as Class D–type materials.

LIQUIDS

Liquids are materials in the stage of matter between solids and gases. A liquid has definite volume but takes the shape of the container it is being stored in. Liquids that produce vapors that burn can be divided into two categories: combustible liquids (kerosene, diesel, and heavy fuel oils) and flammable liquids (gasoline, methyl alcohol, and acetone). Liquids can present other hazards to firefighters besides fire (corrosiveness and toxicity). In general, liquids that burn are classified as Class B-type materials. Vegetable oils, however, used in cooking and the preparation of foods, are classified as Class K materials. Some key characteristics concerning liquids that burn are flash point, fire point, boiling point, specific gravity, solubility, and viscosity. These are reviewed ahead.

- *Flash point:* The flash point is the minimum temperature of a liquid in degrees Fahrenheit at which it emits vapors to form an ignitable mixture with air near the surface but will not sustain combustion. For firefighters, the flash point is one of the most important properties of liquids that burn. Liquids with low flash points pose the greatest danger. The degree of hazard will be determined by the flash point of the liquid because it is the vapors of the liquid that burn, not the liquid itself. In general, flammable liquids have a flash point below 100°F (40°C), and combustible liquids have a flash point at or above 100°F (40°C).

- *Fire point:* The fire point is the lowest temperature at which a liquid will ignite and continue to burn in a self-sustaining fashion. The fire point will therefore be a higher temperature than the flash point.
- *Boiling point:* The boiling point is the temperature of the liquid at which it will liberate the most vapors. It is the temperature at which the vapor pressure of the liquid equals atmospheric pressure. It is impossible to raise the temperature of a liquid above its boiling point unless it is under pressure.
- *Specific gravity:* The specific gravity of a liquid is the ratio of the weight of the liquid to the weight of an equal volume of water. The specific gravity of water is 1. Hence, nonsoluble liquids with a specific gravity of less than 1 (gasoline at 0.8, for example) will float on water. Nonsoluble liquids with a specific gravity more than 1 (sulfuric acid at 1.8, for example) will sink in water.
- *Solubility:* The solubility of a liquid is the percentage by weight of the liquid that will dissolve in water. The solubility of a liquid ranges from negligible (less than one-tenth of 1%) to complete (100%).
- *Viscosity:* Viscosity is a measure of a liquid's flow (through an opening or into a container) in relation to time. Thick liquids (molasses, asphalt, and wax) are on the borderline between liquids and solids and are considered viscous.

GASES

Gases are the third stage of matter. The volume of a given amount of gas is dependent on its temperature and the surrounding pressure. An important concept for firefighters to understand regarding gases and vapors being emitted from a liquid is vapor density.

- *Vapor density:* Vapor density is the relative density of the gas/vapor as compared to air. The vapor density of air is 1. Gases with a vapor density of less than 1, in general, are less difficult for firefighters to dissipate. Gases are classified as Class B–type materials.

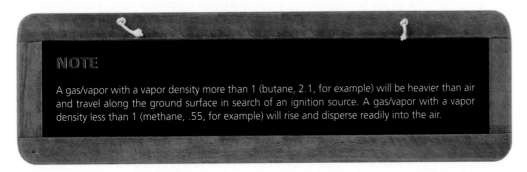

NOTE

A gas/vapor with a vapor density more than 1 (butane, 2.1, for example) will be heavier than air and travel along the ground surface in search of an ignition source. A gas/vapor with a vapor density less than 1 (methane, .55, for example) will rise and disperse readily into the air.

Chemical Properties of Gases

Gases can be classified according to their chemical properties as flammable (burn in air), inert (will not burn in air or in any concentration of oxygen and will not support combustion), oxidizer (will not burn in air or in any concentration of oxygen but will support combustion), toxic (poisonous or irritating when inhaled), and reactive (can rearrange chemically when exposed to heat or shock and explode, or can react with other materials and ignite).

- *Flammable:* A gas that will burn in normal concentrations of oxygen in air is a flammable gas. When discussing flammable gases (or flammable vapors boiling off a liquid) mixing with air, the concept of flammable range must be understood. The flammable range is defined as the ratio of gas/vapor in air that is between the upper and lower flammable limits. The upper flammable limit is the maximum ratio of flammable gases/vapors above which ignition will not occur because it is too rich a mixture. The lower

FIGURE 1.4 Flammable gases

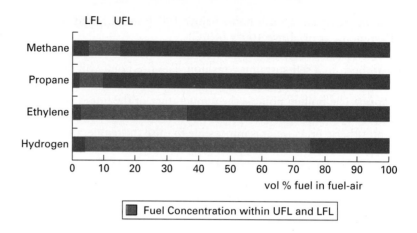

flammable limit is the minimum ratio of flammable gases/vapors in air below which ignition will not occur because it is too lean a mixture. Therefore, gases with wide flammable ranges are more dangerous than gases with narrow flammable ranges. Flammable gases with wide flammable ranges include acetylene and hydrogen. Gases with narrow flammable ranges include butane and propane (see Figure 1.4).

- *Inert:* An inert gas is a nonflammable gas that will not undergo chemical reactions under a set of given conditions and will not support combustion. They are odorless and colorless. Inert gases are often used to avoid undesirable chemical reactions (oxidation, for example). The noble gases and nitrogen are included in this category. The noble gases commonly used with fire protection systems include helium and argon.
- *Oxidizer:* A nonflammable gas that will support combustion is known as an oxidizer. Examples include oxygen and chlorine.
- *Toxic:* Gases that cause harm to living tissue via chemical activity are called toxic gases. They can endanger the lives and health of all those who inhale or come into skin contact with them. Examples include hydrogen cyanide (HCN), carbon monoxide (CO), and ammonia.
- *Reactive:* Gases that react internally and with other materials are reactive gases. They can be heat and shock sensitive and also react with organic and inorganic substances, leading to combustion. Examples include fluorine and vinyl chloride.

Heat (Thermal Energy)

Heat is a form of energy associated with the motion of atoms or molecules. Heat is capable of being transmitted via conduction, convection, and radiation.

HEAT AND TEMPERATURE

Heat and temperature are two distinct, but closely related, concepts. Heat, as stated previously, is thermal energy in the process of transfer or conversion across a boundary of one region of matter to another. Temperature, on the other hand, is a measure of how fast molecules are moving within a substance. It is an indicator of the level at which the heat energy exists.

HEAT UNITS

Heat is a form of energy and is measured in energy units: British thermal units, calories, or joules.

- *British thermal unit:* A British thermal unit (Btu) is the amount of heat energy required to raise the temperature of 1 pound (lb.) of water (measured at 60°F [15°C]

TABLE 1.1	Heat of Combustion of Some Common Fuels	
FUEL	HEAT OF COMBUSTION	
Paper	6,000 Btu/lb.	13.9 kJ/g
Wood	8,000 Btu/lb.	18.6 kJ/g
Coal (lignite)	8,000 Btu/lb.	18.6 kJ/g
Ethanol	12,000 Btu/lb.	27.9 kJ/g
Coal (anthracite)	14,000 Btu/lb.	32.5 kJ/g
Kerosene	20,000 Btu/lb.	46.5 kJ/g
Gasoline	20,000 Btu/lb.	46.5 kJ/g
Paraffin wax	20,000 Btu/lb.	46.5 kJ/g
Butane	21,000 Btu/lb.	48.8 kJ/g
Propane	22,000 Btu/lb.	51.1 kJ/g
Methane	24,000 Btu/lb.	55.8 kJ/g
Hydrogen	61,000 Btu/lb.	141.8 kJ/g

Note: Approximate values in Btu/lb. and kilojoules/gram (kJ/g)

at sea level) by 1°F. This information is valuable to fire protection engineers who need to calculate the amount of water required when designing and installing fire extinguishment systems and equipment.

Table 1.1 provides a listing of some common combustibles and their equated heat of combustion in Btu/lb. as well as kilojoules (kJ)/gram (g). Heat of combustion is energy released as heat when a compound undergoes complete combustion with oxygen under standard conditions. The chemical reaction is commonly a hydrocarbon reacting with oxygen to form carbon dioxide, water, and heat.

- *Calorie:* A calorie is the amount of heat energy required to raise the temperature of 1 gram of water (measured at 15°C at sea level) by 1°C.
- *Joule:* The joule is the heat energy unit in the International System of Units (SI)*.

HEAT RELEASE RATE (HRR)

Heat release rate (HRR) pertains to how rapidly a material burns. It is measured in joules per second (J/s) or watt (W). In general, HRR is considered by many as the most important variable when describing a material's fire hazard. This is due to the fact that most other fire signature (smoke, heat, light) and fire characteristic (generation of toxic gases) variables escalate as the HRR increases. This is not the case for other variables. A high HRR relates directly to life safety, because high room temperatures correlate to faster **flashover** times and an increase in the products of combustion. Elevated HRR causes soaring temperatures and high **heat flux** conditions, which may prove lethal to occupants.[5]

Roughly, a burning candle will have an HRR of 80 W, whereas a burning wastepaper basket will have an HRR of 100 kW. A burning pool of gasoline 10.7 ft.^{2} (1 m^2) in size will have an HRR of 2.5 MW. (*Note:* 1 kW = 1,000 W and 1 MW = 1,000 kW.) The increase in the HRR depends on fuel characteristics and available air supply. In addition, in a typical fire scenario as the fire grows in magnitude the heat release rate increases to a peak value. This is referred to as the peak HRR.[6]

Flashover ■ The near-simultaneous ignition of most of the directly exposed combustible material in an enclosed area. When certain organic materials are heated, they undergo thermal decomposition and release flammable gases. Flashover occurs when most of the exposed surfaces in a room are heated to their autoignition temperature and emit flammable gases. Flashover normally occurs at 930° F (500°C) or 1,100°F (600°C) for ordinary combustibles.

Heat flux ■ Heat flux is the rate of heat energy transfer through a given surface. It is measured in heat rate per unit area (W/m^2). The SI derived unit of heat rate is joules per second (J/s).

---
*The International System of Units (SI) is the modern form of the metric system. It is the most widely used system of measurement in the world.

TEMPERATURE UNITS

Temperature units can be used to compare the difference in heat energy levels between two materials. A temperature is a numerical measure of hot or cold. It may be calibrated in any of various temperature scales: Celsius, Fahrenheit, Rankine, or Kelvin.

- *Fahrenheit (F) degree:* There are 180 increments between the temperature of melting ice (32°F) and boiling water (212°F) on the Fahrenheit (F) temperature scale. Therefore, a degree on the Fahrenheit scale is $\frac{1}{180}$ of the interval between the freezing point and the boiling point. On the Celsius scale, the freezing and boiling points of water are 100 degrees apart. A temperature interval of 1°F is equal to an interval of five-ninths degrees Celsius. The Fahrenheit and Celsius scales intersect at −40° (−40°F and −40°C represent the same temperature).
- *Celsius (C) degree:* The Celsius (C) degree is a metric unit of temperature measurement. The Celsius temperature scale uses the freezing point of water as 0 degrees and the boiling point of water as 100 degrees. This unit is approved by the SI.
- *Rankine (R) degree:* The Rankine (R) degree is a traditional unit of absolute temperature. The temperature units for Rankine and Fahrenheit are equal (1 degree Rankine represents the same temperature difference as 1 degree Fahrenheit), but the zero points differ. The zero point on the Rankine scale is set at absolute zero (−457.6°F on the Fahrenheit scale), the hypothetical point at which all molecular movement ceases.
- *Kelvin (K) degree:* The Kelvin (K) degree is equal to the Celsius degree, but the Kelvin scale has its zero point set at absolute zero (−273.15°C on the Celsius scale). This unit is approved by the SI.

HEAT TRANSFER

Heat can be transferred to other materials through various mechanisms. Although these mechanisms have distinct characteristics, they often occur simultaneously in the same system. Conduction, convection, and radiation are common ways heat is transferred.

- *Conduction:* Conduction is the transfer of heat energy through a medium (usually a solid). Heat causes molecules within the material to move at a faster rate and transmit their energy to neighboring molecules. Touching a hot stove and getting burned is an example of conduction. The heat of conduction can also be transferred from one material to another via direct contact in the same fashion as internal molecular movement. Direct flame contact is the transfer of heat energy via direct flame impingement or autoexposure, such as occurs with a flame traveling upward and outward from a roof, window, or doorway to a neighboring building.

    The amount of heat transferred and rate of travel is dependent upon the thermal conductivity of the material. Dense materials (metals) are good conductors of heat energy. Fibrous materials (wood, paper, cloth) and air are poor conductors. In a fire situation, heat can be conducted via steel columns and girders to abutting wood components and other combustible materials, causing them to smolder and eventually ignite.
- *Convection:* Convection is the transfer of heat energy through a circulating medium (liquids and gases). During firefighting operations, hot air expands and rises as do the products of incomplete combustion. Fire spread by convection is mostly in an upward and outward direction through corridors, stairwells, atria, elevator shafts, and voids from floor to floor. An old-style radiator is an example of convection in a room; it emits warm air at the top while drawing in cool air at the bottom.
- *Radiation:* Radiation is the transfer of heat via infrared/ultraviolet waves or rays. These heat waves travel in a straight line through space at the speed of light in all

directions and are not affected by the wind. Heat felt on your face coming from the sun on a clear day is a good example of radiation. Objects exposed to radiated heat will absorb and reflect a certain amount of heat energy, depending on certain factors. The darker and duller the object, the more heat it will absorb, and the greater the chance it will reach its ignition temperature and burst into flames.

Light-colored, shiny objects tend to reflect radiated heat and absorb less energy and are less likely to reach their ignition temperature. Due to the intense radiant heat generated by burning jet fuels, aircraft rescue and firefighting (ARFF) personnel wear personal protective equipment (PPE), called a fire proximity suit, that is coated with a silvered material to reflect heat away from their bodies. Radiated heat waves will travel through space until they are absorbed by an opaque object. These waves will pass through air, glass, transparent plastics, and water. Large amounts of radiated heat can travel 50 to 100 ft. (15 to 30 m) to ignite nearby structures.

Oxidizer

Oxygen gas is the most common oxidizer that firefighters deal with during fire operations. The atmosphere consists of 21% oxygen, 78% nitrogen, and 1% other elements. An oxygen-enriched atmosphere (greater than 21% oxygen) will enhance the rate and intensity of burning. Conversely, an oxygen-deprived atmosphere (less than 15% oxygen) will not be able to sustain combustion. Key terms to review include oxidation and oxidizer.

- *Oxidation:* Oxidation is a chemical reaction between an oxidizer and fuel. An oxidation reaction involves the loss of at least one electron when two or more substances interact. The opposite of oxidation is reduction, which is the addition of at least one electron when substances come in contact with each other.
- *Oxidizer:* In general, a substance containing oxygen that will chemically react with fuel to start and/or feed a fire is considered an oxidizer. Examples include oxygen in the air, fluorine gas, hydrogen peroxide, ozone, nitric acid, chlorine gas, bromine, and iodine.

Chemical Chain Reaction

The chemical chain reaction process that occurs during flaming combustion is the fourth component of fire, which was added to the fire triangle model to form the fire tetrahedron model. It depicts self-sustaining combustion with an ample amount of fuel and oxygen chemically interacting. As fuel burns, it generates radiant heat. Heat directed back onto the burning substance denoted as radiated feedback helps to raise more fuel to its ignition temperature and generate more vapors to mix with air and form a combustible mixture. Additional oxygen is then drawn into the zone of chemical reaction. This scenario is known as entrainment. Oxygen enrichment via entrainment also increases the heat being generated by the burning material.

Phases of Fire

There are four phases of fire: Incipient, Growth, Fully Developed, and Decay (see Figure 1.5). Each phase has its own unique characteristics and dangers to firefighters and should be understood thoroughly to enhance safety during firefighting operations inside buildings. These phases are part of the Standard Time/Temperature Curve. The curve helps to visually perceive the HRR and temperatures attained during a fire.

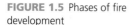

FIGURE 1.5 Phases of fire development

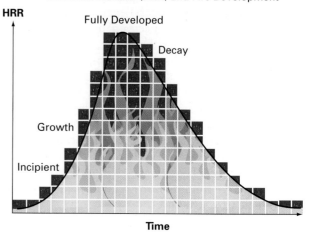

INCIPIENT

Incipient ■ The earliest phase of a fire; fires in this phase are often small and can be readily extinguished with the proper type and size of portable fire extinguisher.

The **Incipient** phase is normally represented by a small fire that can be readily extinguished with the proper type and size of portable fire extinguisher. Detecting a fire in this phase provides your best chance at suppression or escape. Most fires extinguished by firefighters are in this beginning phase. In this phase, the oxygen content in the area is still within the normal range (21%). There is limited heat being generated, and the levels of smoke production and flammable carbon monoxide gas are slowly increasing. Physical destruction from fire is limited to the immediate surrounding area.

GROWTH

Growth ■ This second phase of a fire incorporates the building's fuel or fire load.

This second phase of a fire incorporates the building's fuel or fire load. There are many factors affecting the **Growth** phase. Some of these factors include the location of fire origin, nearby combustibles, ventilation, and ceiling height. It is in this phase during fire operations that there is a possibility for backdraft and flashover to occur. In certain situations, the introduction of fresh air by firefighters entering the area of fire can cause flammable gases to react violently and explode (backdraft), leading to serious injury while also increasing the intensity of the fire (see Figure 1.6). During this phase, there is

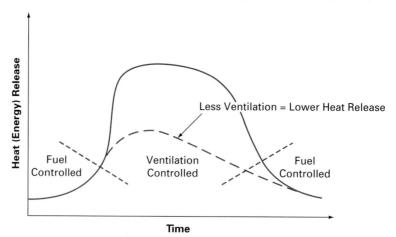

FIGURE 1.6 The rate of fire development is limited by the availability of atmospheric oxygen provided by the room of fire origin inside a structure or building. When the fire is burning in a ventilation-controlled state, any increase in the supply of oxygen to the fire will result in an increase in HRR. Firefighters forcing doors and breaking windows to access the fire building can increase ventilation

also the possibility of fire gases and the contents of the enclosure in which the fire is located reaching their ignition temperatures, causing the entire area's contents to become suddenly engulfed in fire (flashover) and greatly increasing the HRR and temperature of the fire of the fire.

FULLY DEVELOPED

When the Growth phase is at its apex, a fire is considered in its **Fully Developed** phase. This is the hottest phase of a fire. As fire spreads throughout an area, more heat and smoke are generated and travel in an upward direction toward the ceiling. During the Fully Developed phase, oxygen content in the area drops from 21% to approximately 15%, causing the volume of flames to eventually decrease, while smoke production continues to increase. When the oxygen level falls below 15%, flame generation ceases, and the fire enters the next and last phase, the Decay phase.

Fully Developed ■ When the Growth phase is at its apex, a fire is considered in its Fully Developed phase. This is the hottest phase of a fire.

DECAY

Commonly the longest phase of a fire, the **Decay** phase is characterized by a significant decrease in oxygen. Combustibles in the room have been largely consumed by the fire and are no longer actively burning. These combustibles, however, are still emitting large amounts of smoke and flammable gases. Two hazardous conditions can occur during this phase: The existence of smoldering combustible materials that can potentially reignite if not totally extinguished is one concern; and if fresh air (oxygen) is introduced into the fire area at this time, a backdraft situation is possible.

Decay ■ Usually the longest phase of a fire, the Decay phase is characterized by a significant decrease in oxygen.

Classification of Fire

Fires are classified according to the nature of the combustibles (or fuels) involved. The classification of any particular fire is of great importance, because it determines the manner in which the fire must be extinguished as well as what type of fire suppression agent to use. Fires are classified as Class A, Class B, Class C, Class D, or Class K.

CLASS A FIRES

Class A fires can involve any material that has a burning ember or leaves an ash. Common examples of Class A fires are wood, paper, cardboard, plastics, textiles, natural fibers, and rubber. The preferred method for extinguishing Class A fires is to remove the heat. Water is the most common agent used on Class A fires, but other extinguishing agents, such as multipurpose dry chemical, inert gases, clean agents, water mist, and foam, can also be used effectively.

CLASS B FIRES

Class B fires include combustible and flammable liquids as well as flammable gases. These fires require the use of dry chemical extinguishing agents, carbon dioxide (CO_2), inert gases, clean agents, water mist, or foam.

CLASS C FIRES

Class C fires involve live (energized) electrical equipment. These fires require an extinguishing agent that is nonconductive. Electricity is an energy source and an ignition source, but by itself it will not burn. Instead, the energized electrical equipment may serve as a source of ignition for a Class A fire or a Class B fire. Water mist, dry chemical, CO_2, inert gases, and clean agents should be used on these types of fires. It should be noted that when electrical equipment is de-energized extinguishing agents for Class A and Class B fires may be used.

CLASS D FIRES

Class D fires are in materials such as combustible metals or combustible metal alloys. They involve extremely hot temperatures and highly reactive fuels. Examples of combustible metals include magnesium, lithium, sodium, potassium, sodium potassium alloys, zirconium, uranium, and aluminum. There is no one type of extinguishing agent for all kinds of combustible metals. Some of the most common extinguishing agents include sodium chloride (table salt), copper-based dry powder, finely powdered graphite (preferred on lithium fires), and very dry sand. These materials must act as a heat-absorbing medium as well as a smothering agent without reacting with the burning metal. Generally, the extinguishing agents and methods used on Class A, Class B, and Class C fires will not be successful on Class D fires, nor will the agents and methods used for Class D fires work on any other classification of fire.

CLASS K FIRES

Class K fires involve the cooking medium and greases, fats, and vegetable oils used today with more efficient cooking appliances. Vegetable oils have higher ignition temperatures and burn hotter than animal-based, saturated fat cooking oils and require a wet chemical extinguishing agent with superior cooling capabilities when compared to dry chemical.

Portable Fire Extinguishers

In general, fires start out small and can be controlled and suppressed quickly if the proper portable fire extinguisher is available and used effectively. In the United States, portable fire extinguishers are approved or certified by Factory Mutual Insurance Company (FM) and Listed by Underwriters Laboratories, Inc. (UL). UL "lists" certified products and components in their own product directories. These documents are then referenced by the authority having jurisdiction (AHJ) to verify that the portable fire extinguishers have the appropriate fire extinguishing ratings.

LABELING AND MARKING SYSTEM

Fire extinguishers are labeled so that users can quickly identify the classes of fire on which the extinguisher will be effective. *NFPA 10: Standard for Portable Fire Extinguishers* recommends a pictorial concept marking system that combines the uses and nonuses of fire extinguishers on a single label. Fire extinguishers suitable for more than one class of fire should be identified by multiple symbols placed in a horizontal sequence[7] (see Figure 1.7).

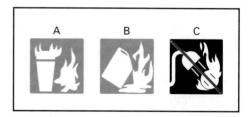

FIGURE 1.7 This pictogram denotes that a portable fire extinguisher is designed to fight both Class A and Class B fires but not Class C fires. *Source:* "Portable Fire Extinguishers Classifications" from Safety Emporium. Used by permission of Safety Emporium.

---

*The *tesla* (symbol T) is the SI-derived unit used to measure magnetic fields.

NONMAGNETIC FIRE EXTINGUISHERS

Nonmagnetic fire extinguishers are used in hazard areas that require a reliable and safe handheld fire suppression tool. They offer a solution to fighting fires in telecommunications and hospital equipment. Extinguishers of this type are generally tested in a 3-tesla* Magnetic Resonance Imaging (MRI) system and found to pose no magnetic safety hazard. An "MR Conditional" rating as a result of such testing characterizes the extinguisher as safe to operate in MRI environments.

MRI equipment uses magnetic waves that could cause harm to firefighters using ferrous tools and equipment. This necessitates state-of-the-art fire technology that is designed to be ideal for hazard areas that require a reliable and safe nonmagnetic fire extinguisher. These extinguishers have a completely nonmagnetic stainless steel shell, valve, hose, and nozzle. The extinguishers can safely be mounted within the MRI room.[8]

FIRE RATING

Also located on the fire extinguisher label is the UL fire rating. The fire rating of a portable fire extinguisher provides a guide to its extinguishing capability. Class A and Class B fire extinguishers may have their classification on a nameplate located on the shell of the extinguisher. The fire rating consists of a numeral followed by a letter. The number indicates the approximate fire extinguishing capacity of the extinguisher for the class of fire that is identified by the letter.

EXAMPLE

A 10A-rated portable fire extinguisher would have approximately five times as much fire suppression capability as a 2A extinguisher. A 20B-rated portable fire extinguisher would have approximately four times as much fire suppression capability as a 5B extinguisher.

The determination of numerical ratings also entails the use of test fires with experienced operators at testing facilities. Class A tests utilize wood cribbing, a wooden wall, and other materials. A 1A fire rating represents the firefighting equivalent of 1.25 gallons (4.7 L) of water. Firefighters normally carry several pressurized-water portable fire extinguishers on their apparatus that have a capacity of 2.5 gallons (9.5 L). These extinguishers, therefore, have a fire rating of 2A.

EXAMPLE

Class B fire-rating tests are based upon the effectiveness of an extinguisher on a heptane fire within square metal pans of the appropriate size. The area of fire that the portable fire extinguisher can put out is measured. The fire rating for Class B portable fire extinguishers also indicates the fire extinguishing capability of the extinguisher when used by an inexperienced operator. It has been established in NFPA 10 that an inexperienced operator or amateur is capable of putting out only 40% of the fire area extinguished by an expert or professional firefighter. For a portable fire extinguisher to receive a 10B fire rating, an expert in the testing lab must extinguish 25 ft.^{2} (2.3 m^2) of fire. The inexperienced operator would therefore be expected to extinguish just 10 ft.^{2} (0.9 m^2), 40% of what the expert or professional firefighter could accomplish.

EXAMPLE

Class C portable fire extinguishers carry only the letter symbol and have no numerical rating, because when the power is removed from the equipment or wiring what remains is essentially a Class A or Class B fire. Class D extinguishers also have no numerical rating. Fire extinguishers and extinguishing Class D agents for use on Class D hazards shall be of the type specifically designed and approved for use on the appropriate metal fire. This information should be on the extinguisher label or nameplate. Class K portable fire extinguishers also have no numerical fire ratings. Their size and capability must be commensurate with NFPA 10.

FIRE HAZARD CATEGORIES

In accordance with NFPA 10, occupancies are typically classified as being light (low) hazard, ordinary (moderate) hazard, or extra (high) hazard. These categories correlate to the amount of Class A and Class B materials being stored inside these locations. A brief description of these three NFPA fire hazards follows.

Light Hazard

Light hazard areas are locations where the quantity and combustibility of Class A combustibles and Class B flammables is minimal. In these areas, expected fires have relatively low HRR. Light hazard areas are typically offices, classrooms, and meeting rooms.

Ordinary Hazard

Ordinary hazard areas are locations where the quantity and combustibility of Class A combustible materials and Class B flammables is moderate. Fires with moderate HRR are expected in these occupancies. Ordinary hazard locations are commonly malls, light manufacturing spaces, parking garages, workshops, and maintenance/service areas.

Extra Hazard

Extra hazard areas are locations where the quantity and combustibility of Class A combustible material is high or where large quantities of Class B flammables are present. It is anticipated that these occupancies will experience rapidly developing fires due to materials with high HRR being stored. Locations may include aircraft and boat servicing areas, spray paint booths, and flammable liquid storage areas.[7]

NUMBER, SIZE, AND PLACEMENT

Determining the proper type, size, and placement of portable fire extinguishers is based primarily on NFPA 10. The determination is based upon the fire hazards presented, projected intensity of the fire, travel distance from any point inside a building to the nearest fire extinguisher, and the accessibility of the fire (see Table 1.2).

The requirements for travel distance for Class A and Class D fire extinguishers from any point to the nearest fire extinguisher must not exceed 75 ft. (22 m). Maximum travel distance to a Class B fire extinguisher is based upon the fire hazard it is protecting as well as its rating. In general, 5B, 10B, and 40B extinguishers protecting light, ordinary, and extra hazard occupancies, respectively, must have a maximum travel distance of 30 ft. (9 m); 10B, 20B, and 80B extinguishers protecting the same hazard occupancies, respectively, must have a maximum travel distance of 50 ft. (15 m) (see Table 1.3). Because once Class C equipment is de-energized a Class A and/or Class B fire exists, portable fire extinguishers used for Class C fires must meet travel distance requirements for the existing Class A or Class B hazards. Class K extinguishers shall be located no more than 30 ft. (9 m) from the hazard.

TABLE 1.2 Fire Extinguisher Size and Placement for Class A Hazards

CRITERIA	LIGHT (LOW) HAZARD OCCUPANCY	ORDINARY (MODERATE) HAZARD OCCUPANCY	EXTRA (HIGH) HAZARD OCCUPANCY
Minimum rated single extinguisher	2-A	2-A	4-A
Maximum floor area per unit of A	3,000 ft.^{2}	1,500 ft.^{2}	1,000 ft.^{2}
Maximum floor area for extinguisher	11,250 ft.^{2}	11,250 ft.^{2}	11,250 ft.^{2}
Maximum travel distance to extinguisher	75 ft.	75 ft.	75 ft.

Note: For SI units, 1 ft. = 0.305 m; 1 ft.^{2} = 0.0929 m^2

TABLE 1.3 Fire Extinguisher Size and Placement for Class B Hazards

TYPE OF HAZARD	BASIC MINIMUM EXTINGUISHER RATING	MAXIMUM TRAVEL DISTANCE TO EXTINGUISHERS FEET	MAXIMUM TRAVEL DISTANCE TO EXTINGUISHERS METERS
Light (low)	5-B	30	9.14
	10-B	50	15.25
Ordinary (moderate)	10-B	30	9.14
	20-B	50	15.25
Extra (high)	40-B	30	9.14
	80-B	50	15.25

MAINTENANCE, INSPECTION, AND TESTS

Maintenance

Portable fire extinguishers are maintained and inspected in accordance with NFPA 10. Fire extinguishers should be maintained at regular intervals (at least once a year) or when specifically indicated by an inspection. Maintenance is a comprehensive check of the extinguisher. It is intended to give maximum assurance that an extinguisher will operate effectively and safely. It includes a thorough examination and any necessary repair, recharging, or replacement. It will normally reveal the need for hydrostatic testing of an extinguisher. In a business setting, the maintenance of a portable fire extinguisher is the responsibility of the employer. The onus is on this person to ensure that accurate records are kept for a period of at least one year. In addition, the employer shall record maintenance work performed on every extinguisher on its individual, attached maintenance record card.

Inspection

An inspection is a fast check to provide assurance that a fire extinguisher is available and operational. The value of any inspection lies in the frequency and thoroughness with which it is conducted. The frequency will vary based upon the needs of the situation. Inspections should always be conducted when extinguishers are initially placed in service.

In general, inspections of portable fire extinguishers are required every 30 days to ensure that the unit is pressurized and unobstructed. This inspection may be discontinued so long as the fire extinguisher has a monitoring system that electronically ensures the extinguisher's physical presence, internal pressure, and whether an obstruction exists that could prevent its access and usage. In addition, an annual inspection by a qualified technician is deemed necessary.

Visual inspections of portable fire extinguishers should check that:

- Fire extinguishers are in their assigned place
- Fire extinguishers are not blocked or hidden
- Fire extinguishers are mounted in accordance with NFPA 10
- Pressure gauges show adequate pressure
- Where required, fire extinguishers are weighed to determine if leakage has occurred
- Pin and seals are not missing
- There are no obvious signs of physical damage (dents and leaks) to shell and components
- Nozzles are free of blockage

Tests

Portable fire extinguishers are required to be hydrostatically pressure tested using water or another type of fluid at certain intervals to help prevent unwanted failure or rupture of the cylinders, according to NFPA 10. Hydrostatic testing includes both an internal and external examination of the cylinder. Generally, water, foam, wet chemical, dry chemical (stainless steel shell), and carbon dioxide (CO_2) extinguishers are tested every five years, whereas dry chemical (aluminum shell) and halogenated extinguishers are tested every 12 years. Hydrostatic testing must be performed by trained personnel with proper test equipment and facilities.

OBSOLETE PORTABLE FIRE EXTINGUISHERS

There are portable fire extinguishers that should be removed from service because they are obsolete.[7] This should be done primarily to ensure life safety and secondarily for property protection. The following types of portable fire extinguishers are considered by NFPA 10 to be obsolete and should be removed from service:

1. Soda acid
2. Chemical foam (excluding film-forming agents)
3. Vaporizing liquid (e.g., carbon tetrachloride)
4. Cartridge-operated water
5. Cartridge-operated loaded stream
6. Copper or brass shell (excluding pump tanks) joined by soft solder or rivets
7. Carbon dioxide extinguishers with metal horns
8. Solid charge–type AFFF extinguishers (paper cartridge)
9. Pressurized-water fire extinguishers manufactured prior to 1971
10. Any extinguisher that needs to be inverted to operate
11. Any stored pressure extinguisher manufactured prior to 1955
12. Any extinguishers with 4B, 6B, 8B, 12B, and 16B fire ratings
13. Stored-pressure water extinguishers with fiberglass shells (pre-1976)

WHEN TO USE A FIRE EXTINGUISHER

Portable fire extinguishers are valuable for immediate use on small fires. They contain a limited amount of extinguishing agent and need to be efficiently used. When cooking oil in a frying pan, for example, initially catches fire it may be proper to turn off the burner and put a metal cover over the pan. The use of a portable fire extinguisher at this time to control the flames lapping out of the pan may also be appropriate. If the fire has spread to

FIGURE 1.8 P.A.S.S.: Pull, Aim, Squeeze, Sweep

kitchen cabinets, however, these actions will not be adequate. In general, fire extinguishers should be used under the following circumstances:

- In conjunction with alerting occupants and the fire department
- When the fire is small (wastebasket)
- When smoke produced is at a minimum
- Only if an escape route is an option and the fire is not between the user and the escape route
- When the user is physically capable of operating the extinguisher

(See Figure 1.8.)

Chapter 1 Concepts of Fire

CHAPTER REVIEW

Summary

Effective fire control and extinguishment requires a basic understanding of the chemical and physical nature of fire. This includes information describing sources of heat energy, composition/characteristics of fuels, and conditions necessary to sustain the combustion process. Fire occurs in clearly defined phases. By recognizing the different phases of fire, a firefighter and fire protection engineer can better understand the methods required to suppress it. The extinguishment of fire is based on an interruption of one or more of the essential elements in the combustion process. With flaming combustion, the fire may be extinguished by reducing temperature, eliminating fuel or oxygen, or by stopping the chemical chain reaction. Fires are classified according to the nature of the combustibles (or fuels) involved. Knowing these classifications is essential in determining what extinguishing agent to use and what method to employ. Portable fire extinguishers are valuable for immediate use on small fires. They contain a limited amount of extinguishing material, however, and operators need to learn how to use them effectively and efficiently.

Review Questions

1. What component of the fire triangle triggers a backdraft?
2. Name the fourth component of the fire tetrahedron that expands on the one-dimensional fire triangle.
3. What is a decomposition reaction in a solid material—usually brought on by the introduction of heat—that is not fast enough to be self-sustaining called?
4. How does the physical form (mass) of a wooden object affect its ignition and combustibility characteristics?
5. Why is water, unless in large amounts, not recommended as an extinguishing agent on combustible metals?
6. For firefighters, what is the most important property of liquids that burn?
7. Define the term British thermal unit (Btu).
8. In a fire situation, heat conducted via steel girders to abutting wood joists is an example of what type of heat transfer?
9. What type of heat transfer does personal firefighting gear coated with reflective, silvered material look to protect against?
10. There are four phases of fire. In what phase do most fires get extinguished by firefighters?
11. Name the phase of fire that is characterized by a significant decrease in oxygen and fuel.
12. What is the preferred method for extinguishing Class A fires?
13. How do Class D extinguishing agents extinguish fire?
14. Fires involving the cooking medium are recognized as what class of fire?
15. Fire extinguishers are labeled so that users can quickly identify the class of fire on which the extinguisher will be effective. What type of marking system does the latest NFPA 10 standard recommend?
16. What type of portable fire extinguishers are safe to use in MRI rooms?
17. What fire rating would a pressurized-water portable fire extinguisher have if it was filled to a capacity of five gallons (18.9 L)?
18. In accordance with NFPA 10, occupancies are typically classified as being light hazard, ordinary hazard, or extra hazard. A parking garage would fit into which one of these fire hazard categories?
19. What is the maximum travel distance requirement for Class A portable fire extinguishers?
20. Why is hydrostatic pressure testing of portable fire extinguishers a requirement?

Endnotes

1. National Fire Protection Association (NFPA), *NFPA 921: Guide for Fire and Explosion Investigations* (Quincy, MA: NFPA, 2011).
2. Frank L. Fire, *The Common Sense Approach to Hazardous Materials*, 3rd ed. (Tulsa, OK: PennWell Corporation, 2009).
3. Stephen J. Pyne, Patricia L. Andrews, and Richard D. Laven, *Introduction to Wildland Fire*, 2nd ed. (New York: John Wiley and Sons, 1996).
4. National Fire Protection Association (NFPA), *NFPA 1971: Standard on Protective Ensembles for Structural Fire Fighting and Proximity Fire Fighting* (Quincy, MA: NFPA, 2013).
5. Vytenis Babrauskas and Stephen J. Grayson, ed., *Heat Release in Fires* (London: E & FN Spon Publishers, 1995).
6. Bjorn Karlsson and James G. Quintiere, "A Qualitative Description of Enclosure Fires," *Enclosure Fire Dynamics* (New York: CRC Press, 1999), 11–24.
7. National Fire Protection Association (NFPA), *NFPA 10: Standard for Portable Fire Extinguishers* (Quincy, MA: NFPA, 2013).
8. Stewart C. Bushong, *Magnetic Resonance Imaging: Physical and Biological Principles*, 3rd ed. (Saint Louis, MO: Mosby, 2003).

CHAPTER 2

Water Supply Systems

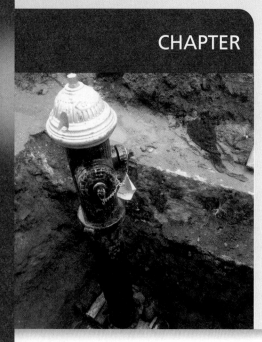

Source: Ronald R. Spadafora

KEY TERMS

Gravity system, p. 23
Pumped system, p. 23
Combination system, p. 23
Dry hydrant, p. 26
Transmission main, p. 29

Distribution main, p. 29
Fire line, p. 29
Feeder main, p. 30
Friction loss, p. 31
Dead end main, p. 31

Potable water, p. 32
Nonpotable water, p. 32
Greywater, p. 32
Blackwater, p. 32
Gate valves, p. 33

OBJECTIVES

After reading this chapter, the reader should be able to:

- List water supply alternatives
- Understand the role of the Insurance Services Office relating to water supply
- Explain water distribution system networks
- Describe fire hydrant features, uses, and maintenance/inspection procedures

Introduction

A water supply system is engineered with hydraulic components that commonly include a water collection location, aqueducts, water purification facilities, storage areas, pumping stations, and a distribution piping system to provide water to the consumer as well as the fire service (fire hydrants) (see Figure 2.1). These systems are typically owned and maintained by municipalities or other public entities, although private water supply systems may also provide water under contract to a municipality. Water supply and distribution networks are an important component of a community's master plan. Location, current water demand, future community growth, pressure, pipe size, and firefighting flows are some of the factors city planners and civil engineers must consider.

Overview

Although there are numerous published minimum standards, design criteria for water systems should be based upon the expected service needs over the life of a particular system. Storage and water delivery capacities should include maximum domestic (private) consumption combined with peak anticipated fire operational needs. Most municipal water systems do provide this dual service for domestic use and firefighting. The determination for volume of water stored, therefore, is based on a number of factors.

Fire hydrants are installed on water distribution piping in most urban and suburban areas where municipal water supply service is available for firefighting use. Hydrants strategically installed to protect large and complex facilities may qualify for insurance reductions, because the fire service should be able to extinguish or control a fire on the insured property. Regularly scheduled inspections, testing, and maintenance of fire hydrants is essential to ensure optimal efficiency and operability.

Types of Water Supply Systems

Gravity and pump systems are used to flow water to the supply network (see Figure 2.2). Some communities use a combination of gravity and a pump to move the water supply. In a **gravity system**, the water supply (reservoir or water tower) is at a higher elevation than the community it serves with both domestic and fire-protection water. Conversely, the water supply for a **pumped system** is at a lower elevation than the location it serves. This type of water supply system therefore requires a pump at the water source. A water supply system that uses both gravity and pumps is known as a **combination system**.

Water Supply Sources and Design Criteria

Natural sources, such as rivers and lakes, are prime sources of surface water. Water is withdrawn from these bodies of water through intakes. Standard intakes are pipes extending from the shore into deep water. Intakes for large municipal supplies may consist of

Gravity system ■ A supply system having its water supply at a higher elevation than the community it serves with both domestic and fire protection water.

Pumped system ■ A supply system having its water supply at a lower elevation than the location it serves. This type of water supply system requires a pump at the water source.

Combination system ■ A supply system that is a combination of both gravity and direct pumping systems. It is the most common type of municipal water supply system.

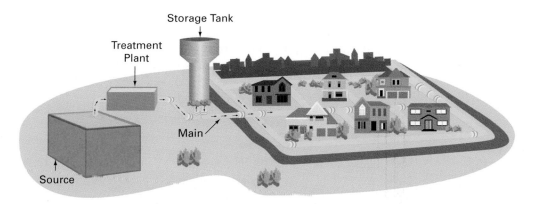

FIGURE 2.1 Water supply distribution system denoting supply source, treatment facility, storage tank, and delivery piping grid

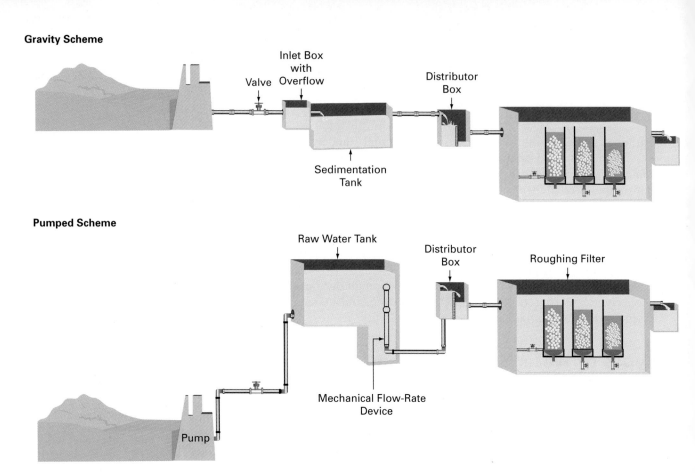

FIGURE 2.2 Water source/supply networks. (Top) Gravity system with water source at a higher elevation than the community it services. (Bottom) Pumped system with water source located below the location it services.

large conduits or tunnels. Manmade sources include water towers, cisterns, and water tender apparatus (see Table 2.1 for common water sources).

Reservoirs are used to supplement the main source of water supply and transmission system during peak demands. They also can provide water during a temporary failure of the supply system. Ground storage reservoirs built at high elevation can supplement the supply system by gravity. Circular steel tanks and basins built of earth embankments, concrete, or rock masonry are used. Elevated storage tanks (water towers) are also common. Storage tanks should be built high enough to maintain adequate pressure in the distribution system at all times. Elevated tanks are normally of steel plate or concrete (see

TABLE 2.1	Common Sources of Water
Oceans	Cisterns
Lakes	Swimming pools
Rivers	Irrigation ditches
Streams	Portable folding tanks
Ponds	Water tender apparatus
Reservoirs	Booster tanks on fire engines
Water towers	

FIGURE 2.3 Water tower. *Source:* Ronald R. Spadafora

Figure 2.3). Elevation-based systems are reliable and not dependent on pumps that could fail or be shut down as a result of an electrical power outage.

The major role of pumping facilities is to maintain appropriate water levels in reservoirs. Pumps are generally required wherever the water supply is not high enough to provide gravity flow and adequate pressure in the distribution system. Booster pumps are also installed on pipelines to increase the pressure and discharge. Pumping stations usually include two or more pumps, each of sufficient capacity to meet demands when one unit is down for repairs or maintenance. The station must also include piping and valves arranged so that a break can be isolated quickly.

To conserve energy and save money, pumping should take place during late night and early morning hours, when electricity demand is light. Automatic standby generators are commonly provided or should be readily available in the event of a long-term power failure. Connections should also be installed so that a portable pump can be used in the event that the stationary pumps fail.

Firefighters require access to large volumes of water in order to control and extinguish fire. Minimum water volumes and flow rates are recommended by the National Fire Protection Association (NFPA). For example, *NFPA 1142: Standard on Water Supplies for Suburban and Rural Fire Fighting* identifies a method of calculating the minimum requirements for alternative water supplies for structural firefighting purposes in areas in which the authority having jurisdiction (AHJ) determines that adequate and reliable water supply systems for these purposes do not exist. Fire departments can effectively use flowing and standing bodies of water as a significant tool for effective fire suppression. Often, fire departments access such supplies in conjunction with a water tender/tanker (see Figure 2.4) in a carefully orchestrated program of shuttling water from a supply point to the fire site.[1]

Small community water supplies, however, may not be adequate for fighting fires. Most rural water supplies can provide only enough water for an initial attack to try to

FIGURE 2.4 Water tender apparatus. *Source:* Ronald R. Spadafora

control a fire until adequate water from other sources can be brought to the site. Using potable water sources to fight a fire will likely disrupt customer service.

Small rural communities have developed several water reserve options for fire protection. Swimming pools, for example, can serve as a viable water source. Natural bodies of water can also be developed to provide water for fire protection. Many water bodies have steep banks, however, making access by fire apparatus in an emergency difficult. Water providers should think proactively and map the locations of each water supply in the community. An adequately wide, all-weather roadway must lead to each water source. A **dry hydrant** that has valve stems and threads matching local fire department equipment should also be installed at each water source. A community should have multiple sources of water throughout the area rather than rely on only one central source.[2]

Dry hydrant ■ A non-pressurized water-delivery system that, when properly installed in a natural or manmade body of water, will provide a ready source of water for fire departments to use.

Dry Hydrant Advantages

Available water in area streams, ponds, and cisterns helps a fire department only if the water is readily accessible. Installation of dry hydrants into nearby and developed water supplies eliminates the inefficiency and complexity of long-distance water-shuttle operations. In areas without water mains and domestic fire hydrants, the dry hydrant concept can provide a simple, cost-effective solution to the need for rapid access to water sources. A dry hydrant consists of an arrangement of piping in which one end is in the water and the other end extends to dry land and is available for connection to a pumper. Multiple lengths of hard-suction hose extending to the water are not needed. Fewer firefighters are necessary to make a hookup to the dry hydrant compared to the manpower required during a conventional direct-drafting operation. Communities can also preserve more of their treated water supplies, because dry hydrants use untreated water[3] (see Figure 2.5).

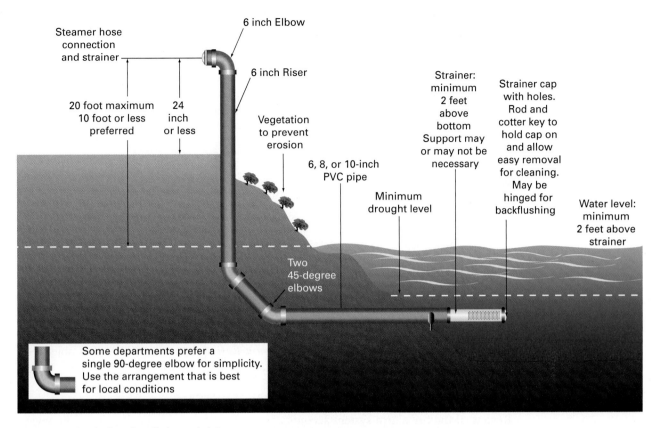

FIGURE 2.5 Dry-hydrant installation and piping

 CASE STUDY

Seawater suction connections are located along the San Francisco Bay, where they are easily accessed by engine apparatus during drafting operations. They facilitate the use of this natural water supply by the San Francisco Fire Department (SFFD). A pump operator attaches a 10 ft. (3 m) hard-suction hose from their outlets into the engine's suction inlet for drafting. A designated fire lane has been established alongside these pier-side drafting locations that allows fire engines easy access. To minimize the corrosive effect saltwater has on apparatus fire pumps, firefighters should flow freshwater through the whole system and all discharges upon conclusion of operations. This will lengthen the life of the pumping system.

In addition, the manmade Stow Lake inside Golden Gate Park provides nonpotable water to surrounding lake hydrants. Because these are nonpotable fire hydrants, the SFFD paints the bonnets blue.

Insurance Services Office (ISO)

The Insurance Services Office, Inc. (ISO) was founded in 1971 through a merger of smaller underwriting service organizations. ISO has developed into an international organization with huge databases of information about hundreds of millions of individual insurance policies along with a large volume of public records pertaining to fraud, real property, employment screening, and motor vehicles. ISO is a provider of data, underwriting, risk management, and legal/regulatory services, with a priority on community fire-protection efforts and building code effectiveness evaluation. Because a community's investment in fire mitigation is a reliable predictor of future fire losses, insurance companies rely upon ISO's public protection classification (PPC™) program to help establish fair premiums for fire insurance (lower premiums are set in communities with better fire protection). As part of the PPC program, ISO conducts evaluations of water supply for fire suppression.

PPC PROGRAM

ISO provides reliable and current information about a community's fire protection services through the use of its Public Protection Classification (PPC™) program. ISO's PPC evaluates the capacity of the local fire department to respond to and fight structure fires. This evaluation provides important information for understanding risk associated with a specific property. The PPC program provides grades for communities in the 48 contiguous states. A Class 1 PPC grade generally represents superior property fire protection. A Class 10 PPC grade indicates that the area's fire-suppression program does not meet ISO's minimum criteria.

Using a manual called the *Fire Suppression Rating Schedule* (FSRS), ISO objectively evaluates three major areas, discussed ahead.

FIRE ALARM AND COMMUNICATIONS SYSTEMS

Review of the fire alarm system accounts for 10% of the total classification. The review focuses on the community's facilities and support for handling and dispatching fire alarms.

FIRE DEPARTMENT

Review of the fire department accounts for 50% of the total classification. ISO focuses on a fire department's first-alarm response and initial attack to minimize potential loss. Review items include engine companies, ladder or service companies, distribution of fire stations and fire companies, equipment carried on apparatus, pumping capacity, reserve apparatus, department personnel, and training.

WATER SUPPLY

Review of the water supply system accounts for 40% of the total classification. ISO examines how a community uses its water supply to determine the adequacy for fire-suppression purposes. Also evaluated in this area are hydrant size, type, and installation as well as the inspection frequency and condition of fire hydrants. Some key factors that examiners evaluate when determining the water supply component of the FSRS include:

- Fire flow tests, observed at strategic locations in the community to determine the rate of flow the water mains provide
- Examination of the condition and maintenance of fire hydrants
- Review of the distribution of fire hydrants from representative locations[4]

To be recognized within the ISO PPC program, the water supply location must be capable of withstanding a 50-year drought, and the water shuttle operation has to provide a minimum of 250 gallons per minute (gpm; 946 L/min.), uninterrupted, for a period of two hours. ISO credits pumps at their effective capacities when delivering at normal operating pressures. Filters, softeners, or other devices in suction or discharge lines may limit the effective capacity.[5]

FIGURE 2.6 Pipe corrosion (curb valve enclosure).
Source: Ronald R. Spadafora

Distribution Network

Traditionally, water distribution systems were installed to meet both domestic consumption as well as fire protection needs. This single-system arrangement requires pipes and storage tanks to be larger than those for drinking water, which can allow the water to stagnate and thus cause degradation of drinking water quality. New pipes added to distribution systems as development occurs result in a wide variation in pipe sizes, materials, methods of construction, and age within individual distribution systems. Deterioration can occur due to corrosion, materials erosion, and external pressures. This can also lead to breaches in pipes and main breaks (see Figure 2.6).

A good distribution system provides adequate water pressure to the consumer for a specific rate of flow. Pressures should be high enough to adequately meet firefighting needs without being excessive; too much pressure can cause leakages and damage to piping. The distribution system must also be watertight to maintain water purity. In addition, maintenance of the distribution system should be easy and economical. Water must still remain available during breakdown periods of sections of pipeline.

Water distribution systems consist of an interconnected series of pipes and components that convey water to meet the needs for municipalities, businesses, industries, and other facilities. Public water systems depend on distribution systems to provide an uninterrupted supply of pressurized, safe drinking water to all consumers. The **transmission mains** (generally greater than 24 in. [610 mm]) of the distribution system carry large amounts of water from the treatment plant to the public.

Tapped off from transmission mains are **distribution mains** (varying in diameter from 24 in. [610 mm] to 6 in. [152 mm]) that supply a number of smaller areas of the distribution system. **Fire lines** (mains at least 6 in. [152 mm]) in diameter that flow water into fire hydrants and building fire protection systems (see Figure 2.7) are fed by the distribution mains. Distribution systems represent the vast majority of physical infrastructure for water supplies. The American Water Works Association (AWWA) *M31 Standard*: *Distribution System Requirements for Fire Protection* provides guidance on designing, operating, and maintaining water distribution systems as they relate to fire protection and fire-suppression activities.[6]

Transmission main ■ The primary piping for transporting the major quantity of water from the treatment plant to the community.

Distribution main ■ Intermediate piping receiving water from the transmission main and delivering it to fire lines.

Fire line ■ Piping that receives water from the distribution main. It supplies fire hydrants as well as individual structures with water for fire protection systems.

FIGURE 2.7 Fire line connected to dry hydrant (note the curb valve operating nut at the lower right).
Source: Ronald R. Spadafora

Types of Water Distribution Systems

Adequate and reliable distribution of water to points of use is a product of sound engineering practices and an understanding of water distribution principles to support firefighting operations. Distribution systems include gridiron and branch design patterns (see Figures 2.8 and 2.9).

The most reliable means to provide water for firefighting is by designing redundancy into the system. This can be accomplished by connecting transmission mains and distribution mains together to form grids that allow water to flow from different directions. In the gridiron or grid system, the piping is laid out in checkerboard fashion. Piping decreases in size as the distance increases from the source of supply.

Loop piping may be installed in conjunction to the grid pattern to enhance water pressure and volume in locations requiring high demand (industrial, commercial, and business areas). This loop piping is supplied by **feeder mains** (commonly 16 in. [406 mm] in diameter) originating directly from a pumping station.

There are several advantages gained by laying out water mains in this pattern. A grid with mains interconnecting at roadway intersections and other regular intervals will still allow water to be distributed through the system if a single section fails. The damaged section can be isolated, and the remainder of the system will still flow water. In addition, a

Feeder main ■ Piping connected directly to the pumping station designed to bolster the water supply to remote distribution areas.

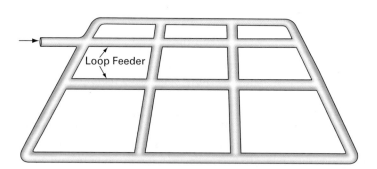

FIGURE 2.8 Grid distribution pattern with feeder main loops

30 Chapter 2 Water Supply Systems

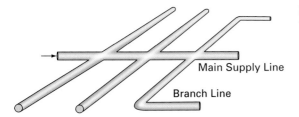

FIGURE 2.9 Branch (tree) distribution pattern

grid system supplies water to fire hydrants from multiple directions. This has benefits during periods of peak fire flow demand. There will be less impact from **friction loss** in water mains because several mains will be sharing the supply. Moreover, hydrants will not be supplied by **dead end mains**, so discharges will remain more stable when multiple hydrants are in use simultaneously.

Street valves should be installed at every junction for all mains tapping off from those junctions. This design allows workers to isolate any single section of main to be taken out of service for repair without disrupting water service beyond the affected section of pipe. Interconnected grid systems also compensate for a disruption in a section of primary feeder piping. Valve components of the system provide a bypass around the out-of-service area.[7]

Branch design is a simple method of water distribution. Calculations are easy to make, and the required dimensions of the pipes are economical. This method of distribution requires a comparatively small number of shutoff valves. Similar to the branching of a tree, a branch distribution pattern consists of a single main (trunk) line, reducing in size to submains and branches with increasing distance from the trunk's source of supply. There is no water distribution to consumers directly from the trunk line. Submains are connected to the trunk line, and they are located along the main roads. Branches are connected to the submains, and they are situated along the streets. Service connections are provided to the consumers from the branches. Fire lines come off of the trunk line to feed fire hydrants and water-based fire protection systems (see Figure 2.10). The American

Friction loss ■ Loss of water pressure resulting from water traveling away from its source through piping.

Dead end main ■ A water main not being fed at both ends.

FIGURE 2.10 In NYC, the number on the hydrant barrel indicates the size (diameter) of the fire line that connects to the hydrant. *Source:* Ronald R. Spadafora

Water Works Association (AWWA) *M17 Standard: Installation, Field Testing, and Maintenance of Fire Hydrants* provides procedures for fire hydrant design, installation, testing, and maintenance practices.[8]

Branch design, however, is not commonly followed in modern waterworks practice. One reason for this is that water available for fighting fires will be limited because it is being supplied by only one water main. In addition, the pressure at the end of the line may become undesirably low as additional areas are connected to the water supply system. Other drawbacks include the fact that the area of the system receiving water from a pipe undergoing repair cannot be supplied until the work is finished. Branch systems also have a large number of dead ends in which water does not circulate and remains static.

Sediments accumulate due to stagnation at the dead ends, and bacterial growth may occur. Drain valves are provided at dead ends to overcome this problem. These valves are opened periodically to clear the dead end piping of stagnant water, but in doing so a large amount of water is wasted. Another drawback of the branch system is that it is difficult to maintain chlorine at the dead ends of the pipe.

Dual Water Distribution Systems

Potable water ■ Water safe for humans to drink.

Nonpotable water ■ Water not treated to drinking water standards that is not meant for human consumption.

Greywater ■ Wastewater generated from domestic activities such as laundry and bathing.

Blackwater ■ Wastewater generated from toilets, kitchen sinks, and dishwashers.

Dual water distribution systems supply **potable water** through one distribution network and **nonpotable water** through another. The two systems work independently of each other within the same service area. Using dual systems can boost public water supplies; they lessen the burden on drinking water systems because they do not have to provide water treated to drinking water standards for activities such as firefighting.

Two separate, underground, piped water systems characterize dual water systems. The potable water system conveys drinking water. It does not have fire hydrants. The potable-water system operates like any other standard, potable-water distribution system. It requires a water source, treatment plant, storage facility, and distribution system.

The nonpotable water system is often referred to as a secondary system. It distributes water such as seawater or household **greywater**. Water used from bathroom sinks, showers, tubs, and washing machines is defined as greywater. Water used from toilet flushing, kitchen sinks and dishwashers, however, is known as **blackwater**. It can contain fecal matter and urine. Blackwater is not used for firefighting. Nonpotable water systems may include storage, treatment, a pumping system, distribution pipes, valves, hydrants, and standpipes. Pump operators should use a screen or strainer when using nonpotable water to keep particulates out that may be present because less treatment of the water is required. In urban areas, requirements for disinfection and additional filtration is normally stricter than in rural areas.

Pipes in either system can be made of PVC, ductile iron, or high-density polyethylene. Cross-connections and contamination of the potable water system by the nonpotable water system is a constant concern. Installers can reduce the risk by using proper layout procedures and color-coded pipe.

Fire Hydrants

There are two principle types of hydrants: dry barrel and wet barrel. Dry-barrel hydrants are pressurized and drained via a main valve located in the base of the hydrant. This type of hydrant does not have pressurized water up to its outlets. When the main valve is opened, the barrel is pressurized. When it is closed, the barrel drains. There are no valves at the outlets. The main valve is located below the frost line to protect the hydrant from freezing. Dry-barrel hydrants are especially suited for cold climates.

A major disadvantage to dry-barrel hydrants is that either the hydrant must be shut down to make a second fire hose connection or a manually installed valve must be placed on the second discharge outlet during the connection of the first fire hose. Dry-barrel

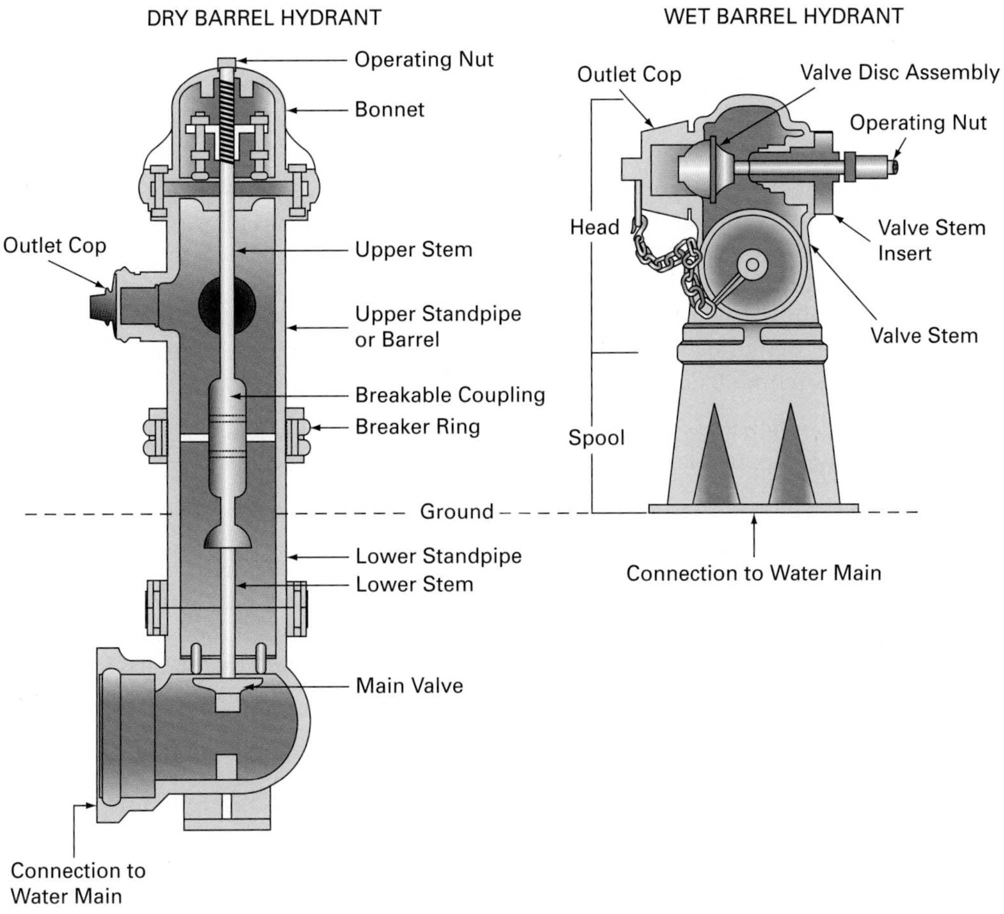

FIGURE 2.11 There are two types of pressurized fire hydrants: wet barrel and dry barrel. Wet-barrel hydrants are connected directly to the pressurized water source, and the upper section is always filled with water. Each outlet on the hydrant has its own valve with a stem that protrudes out the side of the barrel. A dry-barrel hydrant is separated from the pressurized water source by a main valve in the lower section of the hydrant below ground. The upper section does not fill with water until its main valve is opened by means of a long stem that extends up through the top of the hydrant. There are no valves on the outlets.

hydrants are manufactured in accordance with AWWA Standard C-502, *Standard for Dry-Barrel Fire Hydrants*.[9] The wet-barrel hydrant has water in the barrel up to each of its outlet valves and is used exclusively in warm-weather areas of the country, where freezing is not an issue. Wet-barrel hydrants are simpler in construction than dry barrel hydrants, and all their mechanical parts are above ground for easy accessibility. The outlet valves operate independently, so there is no need to shut down the hydrant when adding additional hose lines (see Figure 2.11). Wet-barrel hydrants are manufactured in accordance with AWWA Standard C-503, *Standard for Wet-Barrel Fire Hydrants*.[10]

Hydrants should have a minimum of two outlets, one of which should be at least 4 in. (102 mm) for pumper suction. During major emergencies, fire companies may be requested from multiple jurisdictions to provide assistance. Hydrant outlet threads, therefore, should meet the regional standard for compatibility.

VALVES

Gate valves installed on both sides of fire lines facilitate servicing and limit the number of hydrants being placed out of service. In addition, a sectional control valve should be

Gate valve ▪ A fluid-control device that allows or prohibits the flow of hydraulic fluid. It opens by lifting a round or rectangular gate/wedge out of the path of the liquid.

installed in the water distribution system so that no more than two hydrants will be out of service at any one time. Gate valves for the fire service lines may serve as a sectional control valve. Each hydrant should also have an individual shutoff valve.

CAPACITY

Hydrants in residential areas are commonly designed to deliver no less than 1,000 gpm (3,785 L/min.). In high-hazard areas, however, this volume should increase based on the required fire flows of the buildings being protected. The minimum working pressure rating of any fire hydrant should be 150 psi (103 kPa).* Newly installed hydrants must keep pace with the supply demands of modern fire apparatus and fire protection suppression systems. Hydrant design planning should take into consideration fire engines currently in service as well as more modern fire engines.

> The pascal (symbol: Pa) is the SI-derived unit of pressure. The kilopascal (1 kPa = 1000 Pa) is a common multiple unit of the pascal (6894.75729 Pa is equal to 1 psi).

SPACING

The water supply engineer should strive to have all hydrants of the fire protection system within adequate distance to structures in need of protection from fire. Standard practice calls for the installation of hydrants every 500 ft. (152 m). Consideration must be given, however, to apparatus accessibility, obstructions, vulnerability, and other factors for which positioning adjustments are warranted.

SIGNAGE

In areas of the country that do not experience snowfall, yellow or blue reflectors are used to allow rapid identification. In regions with snow cover, flags or tall, narrow posts painted with a highly visible color may be employed so that hydrants can be located even if covered over.

COLOR CODING

Fire hydrants should be immediately recognizable to firefighters as well as to the general public. *NFPA 291: Recommended Practice for Fire Flow Testing and Marking of Hydrants* specifies that fire hydrants are to be painted chrome yellow. Other highly visible colors that have been used include white, bright red, chrome silver, and lime yellow. Standard colors should be adopted uniformly throughout the region.

NFPA also recognizes that there are often functional differences in service provided by municipal and private hydrant systems. Therefore, NFPA specifies that nonmunicipal hydrants be painted a color that distinguishes them from municipal hydrants.[11]

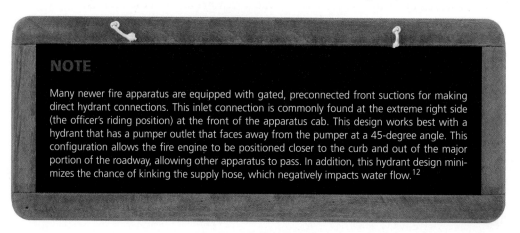

NOTE

Many newer fire apparatus are equipped with gated, preconnected front suctions for making direct hydrant connections. This inlet connection is commonly found at the extreme right side (the officer's riding position) at the front of the apparatus cab. This design works best with a hydrant that has a pumper outlet that faces away from the pumper at a 45-degree angle. This configuration allows the fire engine to be positioned closer to the curb and out of the major portion of the roadway, allowing other apparatus to pass. In addition, this hydrant design minimizes the chance of kinking the supply hose, which negatively impacts water flow.[12]

INSPECTION AND MAINTENANCE

Fire hydrants require regular inspections and maintenance. These inspections are normally performed by the local municipality water department and/or fire department. Private hydrants are usually installed to protect large properties and complexes of buildings. The inspection and maintenance of these hydrants may be contracted out to companies in the fire protection business unless they are under the municipality jurisdiction.

Some fire hydrant manufacturers recommend lubricating the head mechanism and restoring the head gaskets and O-rings annually. Others have incorporated proprietary features to provide long-term lubrication. Food-grade, nonpetroleum lubrication is recommended so as to avoid contamination of the distribution system.

FLOW TESTING

Hydrants are opened during inspections to remove built-up sediment in the water lines. The hydrant should be fully opened by turning the valve stem to its stopping point. If the hydrant is not fully open, water will flow out of drain holes and may undermine the drain field around the hydrant. Water should be allowed to flow out of the hydrant until it becomes clear and no sediment, rocks, or foreign objects are flowing from the hydrant. The direction of the flow must be controlled so that damage is not done to property in the water's path.

While performing flow testing, observe the hydrant for leaks or other component failures. It should be understood, however, that operational flow-testing of fire hydrants can be expensive. The costs of this testing can include manpower wages, the cost of the water flowed, and the equipment necessary to perform the task of measuring quantity and pressure.

CHAPTER REVIEW

Summary

Water supply systems consist of an interconnected series of components that convey drinking water to a population and hopefully also meet fire protection needs for both municipalities and rural areas. Spanning approximately one million miles in the United States, distribution systems represent the vast majority of physical infrastructure for water supplies.

Review Questions

1. Why is elevation as the means for developing proper water pressure in water mains and hydrants a reliable design feature?
2. What is the function of a dry hydrant?
3. Insurance companies rely upon ISO's PPC program to help establish fair premiums for fire insurance in various communities. What section of the program constitutes 40% of the available PPC score?
4. How is redundancy incorporated into the firefighting design of a water distribution system?
5. What function do street valves have when installed at water main junctions?
6. Why are branch distribution systems not commonly installed in modern waterworks practice?
7. Dual water systems feature what two types of distribution systems?
8. For what areas of the country are wet-barrel hydrants ideally suited?
9. How is a dry-barrel hydrant separated from its pressurized water source?
10. Describe the procedure used by firefighters during hydrant inspection to remove built-up sediment in the water lines.

Endnotes

1. National Fire Protection Association (NFPA), *NFPA 1142: Standard on Water Supplies for Suburban and Rural Fire Fighting* (Quincy, MA: NFPA, 2012).
2. Ibid.
3. William F. Eckman, *The Fire Department Water Supply Handbook* (Saddle Brook, NJ: PennWell Publishing Company, 1994).
4. Harry R. Carter, and Erwin Rausch, *Management in the Fire Service*, 4th ed. (Sudbury, MA: Jones & Bartlett Publishers, 2007).
5. Ibid.
6. American Water Works Association (AWWA), *M31 Standard: Distribution System Requirements for Fire Protection* (Denver, CO: AWWA, 2008).
7. Harry E. Hickey, *Water Supply Systems and Evaluation Methods—Volume 1: Water Supply System Concepts* (Washington, DC: US Fire Administration, October 2008).
8. American Water Works Association (AWWA), *M17 Standard: Installation, Field Testing, and Maintenance of Fire Hydrants* (Denver, CO: AWWA, 2006).
9. American Water Works Association (AWWA), *C-502-05 Standard for Dry-Barrel Fire Hydrants* (Denver, CO: AWWA, 2005).
10. American Water Works Association (AWWA), *C-503-05 Standard for Wet-Barrel Fire Hydrants* (Denver, CO: AWWA, 2006).
11. National Fire Protection Association (NFPA), *NFPA 291: Recommended Practice for Fire Flow Testing and Marking of Hydrants* (Quincy, MA: NFPA, 2013).
12. Ibid.

CHAPTER 3

Water, Wetting Agent, and Wet Chemical

Source: Ronald R. Spadafora

KEY TERMS

Covalent bond, *p. 38*
Specific heat, *p. 39*
Water curtain, *p. 40*
Micelle, *p. 41*
Hydrophobic, *p. 41*
Hydrophilic, *p. 41*
Saponification, *p. 42*

OBJECTIVES

After reading this chapter, the reader should be able to:

- Examine the chemical and physical properties of water
- Review the advantages and disadvantages of water as a fire extinguishing agent
- List the physical properties of wet chemical extinguishing agent
- Understand and describe the saponification process
- Evaluate NAFED and UL standards and testing requirements
- Explain the methods of application of agents
- Discuss inspection procedures for pre-engineered fire extinguishing systems

Introduction

Water is the most widely used and readily available fire extinguishing agent. It is effective and relatively inexpensive. Water is commonly safe, nontoxic, noncorrosive, and stable. It is available from municipal water supply and distribution systems as well as natural sources (lakes, rivers, and streams). In addition, water is transportable and can be pumped from its source to a fire. Although water is considered the extinguishing agent of choice for the majority of fires, water can exacerbate the hazard of some types of fires. For these fires, alternative extinguishing agents must be used.

Wetting agents are chemicals added to water to reduce its surface tension as well as enhance water's ability to penetrate and spread. Wetting agents evolved from foam technology and are commonly applied as foams. They are effective on Class A fires and some Class B fires. Wetting agents also use encapsulating technology to enhance water's ability to absorb heat.

Wet chemical agents are solutions of water mixed with potassium-based substances for use on range hood and stove fires, for which hot cooking oils, fats, and grease are the primary hazard. Used in both pre-engineered, fixed systems and portable fire extinguishers, wet chemical agents have replaced dry chemical agents in commercial kitchen fire suppression applications.

Water

Firefighters extinguish most fires using water. It is commonly available in large quantities at or near the location of the fire. There are several ways that water can be delivered onto the fire. These include hose lines stretched from fire department apparatus, hose lines stretched from a standpipe system located inside a building, and firefighting appliances (deck pipes, ladder nozzles, distributors, cockloft nozzles, cellar pipes, subcellar pipes, and portable fire extinguishers). Moreover, water is applied using fixed fire-suppression systems (sprinkler or water mist; see Figure 3.1).

Water extinguishes fire in many ways. It absorbs heat and thereby lowers the temperature of the burning material to below its ignition temperature. Water also smothers a fire by changing to steam during fire operations. Emulsification is another way that water is utilized to suppress fire. When applied on insoluble liquids (ones that will not dissolve in water), such as oils, the agitation upon application can produce a froth. Moreover, dilution (adding water to reduce the concentration of a burning soluble liquid and thereby raising its flashpoint) is yet another method.

CHEMICAL PROPERTIES

The chemical formula for water is H_2O. In a water molecule, there is a single bond between the oxygen and hydrogen atoms. Each of the **covalent bonds** contains two electrons, one

> **Covalent bond** ■ A chemical bond between two nonmetal atoms. One example is water, in which hydrogen (H) and oxygen (O) bond together to make H_2O. Each of the covalent bonds contains two electrons that are shared.

FIGURE 3.1 Portable 2.5-gallon water fire extinguisher. *Source:* Ronald R. Spadafora

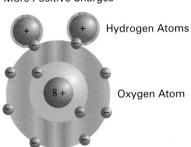

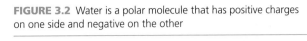

FIGURE 3.2 Water is a polar molecule that has positive charges on one side and negative on the other

from a hydrogen atom and one from the oxygen atom. Both atoms share the electrons. The configuration of this bonding results in a charge separation in the water molecule. There is a positive charge on the side where the hydrogen atoms are located and a negative charge on the opposite side, where the oxygen atom is located. This is also known as polar covalent bonding. Opposite electrical charges attract, causing water molecules to cling together (see Figure 3.2).

PHYSICAL PROPERTIES

Water is unique in that it is the only natural substance that is found in all three states—liquid, solid (ice), and gas (steam)—at normal temperatures found on Earth. An unusual property of water is that in its solid form it is less dense than in its liquid form, allowing it to float. Water has a high surface tension, which gives it a "tacky" texture. It tends to gather together in drops rather than spread out in a thin film (see Table 3.1).

ADVANTAGEOUS PROPERTIES OF WATER

There are several advantages to water as an extinguishing agent. Some of the characteristics that make it so valuable include the following:

- It is a stable liquid at ordinary temperatures.
- It can be transported via water tanker, engine apparatus booster tank, and containers as well as stored through infrastructure installation designs.
- It has a high specific heat.

The heat capacity of a substance is given in terms of its mass. **Specific heat** is the amount of heat per unit mass required to raise the temperature by one degree Celsius. The specific heat of water is 1 calorie/gram °C.

This characteristic enables it to absorb heat without having its temperature rise too quickly. All solids, liquids, and gases have specific heats. There are only two liquids having higher specific heats than water: ammonia and ether. Some additional advantages of water as an extinguishing agent include the following:

- *Latent heat of fusion:* The melting of 1 lb. (0.45 kg) of ice into water at 32°F (0°C) absorbs 143.4 Btu (151 kJ).

Specific heat ■ The heat capacity of a substance per unit mass. For example, it takes 1 calorie to raise the temperature of 1 gram of water by 1°C. One calorie is equal to 4.186 joules.

TABLE 3.1	Physical Properties of Water
Density: 62.4 lb./ft.³ (990 kg/m³) at 32°F (0°C)	
Melting point: 32°F (0°C)	
Boiling point: 212°F (100°C)	

Chapter 3 Water, Wetting Agent, and Wet Chemical

- *Latent heat of vaporization:* The latent heat of vaporization is the conversion of 1 lb. (0.45 kg) of water into steam at a constant temperature with the absorption of 970 Btu (1,023 kJ).
- *Conversion to steam:* The conversion of liquid water to steam also increases its volume approximately 1,600 times, which displaces an equal volume of air, thereby reducing the volume of oxygen available for the oxidation reaction.

DISADVANTAGEOUS PROPERTIES OF WATER

Water also has disadvantages. Some of the negative aspects of water use on fires include the following:

- It conducts electricity.
- It has low viscosity, allowing it to run off of smoldering material readily.
- It has high surface tension, causing it to have poor penetration qualities.
- It is transparent to radiated heat, allowing fires of large magnitude to ignite nearby buildings from a distance despite the establishment of a **water curtain**.
- It freezes at a relatively high temperature.
- It displaces flammable liquids.
- It reacts violently with combustible metals.

Water curtain ■ A water stream in the shape of a fan of water droplets that forms a "shield" against fire, in an attempt to reduce the risk of radiated heat igniting a nearby structure.

Wetting Agent

A wetting agent is a chemical compound (aerosol, ether, and glycol) that reduces the surface tension of a liquid. This ability of the wetting agent allows water to penetrate more readily into burning solid (Class A) materials as well as spread more easily across their surface. Typically, the reduction in surface tension is achieved by adding the wetting agent to water in an admixture rate of 0.1% to 0.5%, depending upon the wetting agent used. A proportioner system is required to mix the wetting agent with water.

Wetting agents are similar to foams with regard to increasing the wetting effectiveness of water, but they do not have the foaming capabilities. They can contain a surfactant or emulsifying ingredient that allows them to mix with hydrocarbon (Class B) fuels. In general, wetting agents require diluting Class B fuels with approximately 6% of solution (wetting agent and water) by volume.

APPLICATIONS

By enhancing water's ability to penetrate the source of the fire, extinguishment with a wetting agent requires significantly less water. This leads to less water damage and a reduced need for diking and reclamation. Wetting agents are especially valuable when used on fires burning within densely packed paper, cardboard, wood piles, and hay.

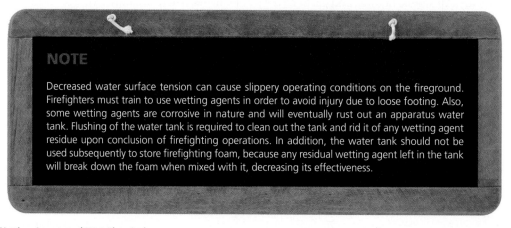

NOTE

Decreased water surface tension can cause slippery operating conditions on the fireground. Firefighters must train to use wetting agents in order to avoid injury due to loose footing. Also, some wetting agents are corrosive in nature and will eventually rust out an apparatus water tank. Flushing of the water tank is required to clean out the tank and rid it of any wetting agent residue upon conclusion of firefighting operations. In addition, the water tank should not be used subsequently to store firefighting foam, because any residual wetting agent left in the tank will break down the foam when mixed with it, decreasing its effectiveness.

Wetting agent solution use on Class B fires involving flammable and combustible liquids is limited to fuels not soluble in water. In addition, wetting agent solution has the same limitations as water with respect to extinguishing fires involving chemicals that react violently with water and extinguishing fires involving coolants and lubricants within live electrical equipment. For health, safety, and environmental reasons, wetting agent solutions must only be used in concentrations specified by their Listing.[1]

There are a number of methods of add wetting agent concentrate to water. The concentrate can be premixed in apparatus water tanks to form a solution, or the wetting agent concentrate can be brought in contact with water via a proportioning device. Wetting agent concentrate that complies with *NFPA 18: Standard on Wetting Agents* is permitted for use with standard firefighting equipment, provided the equipment is designed primarily to use water or foam as an extinguishment medium.[2]

Heat-Encapsulating Technology

Encapsulating agents are based on the high-performance, three-dimensional firefighting capabilities of the **micelle**. The micelle is made up of a water-insoluble (attracted to fats and oils) component, often illustrated as a **hydrophobic** tail, and a water-soluble component, commonly depicted as a **hydrophilic** head. It therefore possesses both oil-soluble and water-soluble characteristics. The insoluble component may extend out of water into oil, whereas the water-soluble component remains in the water. This alignment modifies the surface properties of water at the water/oil interface (see Figures 3.3 and 3.4).

Encapsulating agents are Underwriters Laboratories Inc. (UL) Classified for use on both Class A and Class B fires. They are considered wetting agents, and therefore their requirements can be found in NFPA 18.[3] They greatly enhance water's ability to absorb heat, thereby reducing the amount of time necessary to extinguish the fire.

APPLICATIONS

Applications for encapsulating agents are varied. Extinguishing capabilities include wood structures, wildland combustibles, paper mills, recycling centers, and rubber tire–manufacturing facilities. Rapid cooling of the fire and reduction of the temperature of surrounding surfaces helps prevent reignition of combustible vapors.

Encapsulating agents are also effective on Class B fuel fires/spills (including polar solvents—fuels soluble in water). Gasoline, diesel, crude oils, aviation, acetone, heptane, ethanol and ethanol-blended fuels, industrial alcohols and solvents, and refinery byproduct incidents

Micelle ■ A molecule that has both a water-insoluble (hydrophobic) component and a water-soluble (hydrophilic) component.

Hydrophobic ■ The tendency of a nonpolar substance to aggregate in an aqueous solution and exclude water molecules.

Hydrophilic ■ A polar substance that has a strong affinity to water.

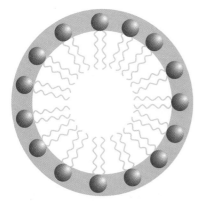

FIGURE 3.3 In this micelle, an outer ring of red heads (hydrophilic component) is attracted to water, whereas the internal, hydrophobic blue tails are attracted to oil

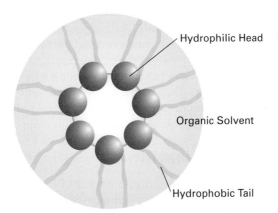

FIGURE 3.4 Micelles can also have a structure that is reversed. This occurs in a Class B hydrocarbon liquid fire scenario

have all been successfully suppressed and mitigated. The encapsulating solution encloses both the hydrocarbon liquid and vapors, rendering them nonflammable and nonignitable.

These agents can be premixed and stored inside the engine apparatus water tank and proportioned at the pump panel to supply hose lines for attack on structural fires. In addition, these agents can be metered out into a supply hose line to feed engine apparatus-mounted monitors or aerial ladder nozzles. For smaller-scale fires, encapsulating agents can be employed through use of a portable fire extinguisher.

Fixed extinguishing system applications entail firefighters using their engine apparatus and hose to supply heat-encapsulating agent to a sprinkler system via its fire department connection (FDC). Agent manufacturer concentration ratios must be adhered to commensurate with the fuel load and magnitude of the fire.

INTERNATIONAL PAPER PLANT: CORDELE, GA

In the mid to late 1990s, the International Paper Plant located in Cordele, Georgia, experienced a hydraulic pit press fire. Materials ignited included hydraulic oil, paper, and pulp. Initial attempts to extinguish the fire using water were ineffective. Firefighters then applied foam, which also proved futile. A mutual aid fire company arriving on the scene carried encapsulating agent. They applied the agent at a 3% solution. The fire was extinguished as the solution penetrated into the hydraulic-soaked pulp to douse the flames and prevent reignition.

When the hydraulic pit press was refurbished, the plant managers decided to install a fixed fire protection system using an encapsulating agent. The system consisted of a deluge system with a 150-gallon (568 L) bladder tank holding the agent feeding a 3 in. (76 mm) diameter riser. The system has performed successfully each time there has been an incident that required a discharge of the system.

Wet Chemical Extinguishing Agent

These agents are used in both pre-engineered, fixed systems and handheld portable fire extinguishers. Wet chemical restaurant suppression systems are the result of many years of experience in the design of firefighting systems. They are ideally suited to meet the fire protection demands associated with modern food processing and catering facilities. Cooking grease fires have long been studied for their reignition characteristics. The fire can reoccur if the fuel is not cooled below its autoignition temperature (see Figure 3.5).

PHYSICAL PROPERTIES

Fixed wet chemical extinguishing systems were developed in the early 1980s for the suppression of kitchen fires involving cooking oils, fats, lards, and greases (Class K combustibles). The wet chemical is composed of a premixed aqueous (40%–60% water by weight, depending upon the manufacturer) solution. The water is mixed with potassium carbonate, potassium acetate–based chemicals, or a combination of these forms of extinguishing agent. It provides fast flame knockdown by cooling burning materials below their ignition temperature, displacement of oxygen, and suppression of vapors. Fire damage is contained within the area of origin. The agent also cools hot metal surfaces of cooking equipment to help prevent reignition of lingering combustible vapors (see Table 3.2).[4]

Saponification ■ A process in which a chemical agent combines with cooking grease to form a soapy layer at the surface to seal off the fuel from oxygen. It is an endothermic (heat-absorbing) reaction.

Saponification

Animal fats contain high amounts of fatty acids compared to vegetable oils. Dry chemical extinguishing agent is alkaline in makeup. When an alkaline agent is discharged over fatty acids, it forms a **saponification** (soapy) blanket. This soap layer helps to seal off the burning material from ambient oxygen and smothers the fire.

Vegetable oils, with lower fatty acid and cholesterol content, are now replacing animal fats in the frying and processing of foods. This greatly reduces the production of

FIGURE 3.5 Commercial kitchen with pre-engineered wet chemical nozzle system. *Source:* Ronald R. Spadafora

saponification and the effectiveness of dry chemical, which is the major reason for the development of a wet chemical agent.

Vegetable oils also have a higher autoignition temperature (685°F [362°C]) than most animal fats and therefore must be heated to higher temperatures. Energy-efficient cooking appliances are highly insulated to maintain vegetable oils at these high temperatures. This equipment stays hot for longer periods of time, making it more difficult to secure against reignition of vapors. Fires involving these oils destroy soap blankets and are more difficult to extinguish. The need for the cooling effect of water is evident.

TABLE 3.2	Characteristics of Wet Chemical Extinguishing Agents
AGENT PROPERTY	**VALUE RANGE**
Storage life	12 years
Freeze point	10°F to −40°F (−12°C to −40°C)
Boiling point	215°F to 230°F (102°C to 110°C)
Specific gravity	1.2–1.4
pH	7.8–13

Source: "Characteristics of Wet Chemical Extinguishing Agents" from Operation of Fire Protection Systems. Copyright © 2008. Used by permission of National Fire Protection Association.

National Association of Fire Equipment Distributors

The National Association of Fire Equipment Distributors (NAFED) was established in 1963. Its objective is to improve technical competence in the fire protection industry. Tests conducted by this Chicago-based organization in 1978 and again 10 years later in 1988 attempted to evaluate pre-engineered, commercial cooking fire extinguishing equipment utilizing dry chemical agent. These tests revealed a steady decline in efficiency over the 10-year period between tests. This downturn was partly attributed to the change in the methods of preparing food and the replacement of traditional commercial cooking equipment (deep fat fryers, grills, griddles, and broilers) with modern, energy-efficient equipment.

Dry chemical suppression systems for commercial cooking establishments are no longer manufactured. The original manufacturers therefore do not provide replacement parts and components for required service maintenance and repairs. In addition, many state-licensed fire equipment distributors who maintain pre-engineered fire suppression systems refuse to work on existing dry chemical systems in commercial kitchens.[5]

Underwriters Laboratories Inc.

Pre-engineered dry chemical suppression systems for the protection of commercial cooking equipment, plenums, and ducts were developed in the 1960s. As commercial cooking operations, appliances, and supplies changed over the years, UL recognized the need for a new set of standards. During this time, UL developed a series of fire tests for these systems designed to duplicate the potential fire hazard. Previous tests only simulated fires in commercial cooking appliances; they did not use actual cooking equipment. The tests established a standard for the type of extinguishing agent, system components, and fire detection system.

When wet chemical extinguishing systems were first introduced, they provided solutions to the problem of effectively extinguishing commercial kitchen fires. On November 21, 1994, UL adopted a new standard: *Fire Testing of Fire Extinguishing Systems for Protection of Restaurant Cooking Areas* (known as UL-300). Manufacturers of fire suppression systems for this type of application who wanted to sell UL Listed* fire protection equipment after this date were required to resubmit their systems to UL for testing.

UL-300 STANDARD

As of November 21, 1994, all new, pre-engineered commercial cooking equipment was required to conform to the UL-300 standard. In addition, existing systems needed to be reinstalled or upgraded to conform if the system was no longer protecting against the type of hazard it was originally installed to protect against. This determination can be made by the authority having jurisdiction (AHJ) or in some cases by the insurance company for the commercial establishment. Existing systems from before November 21, 1994, may be acceptable if installed in the original location, if they remain in compliance with the original listing, and if they continue to offer fire protection for the original equipment without any changes.

The UL-300 standard takes into consideration cooking appliance design, cooking agent ignition characteristics, and "worst-case scenario" fire simulations. Tests performed under this standard are more sophisticated, because genuine cooking appliances are used. The test procedures create a higher heat release rate (HRR) fire involving fryers, griddles, ranges, char broilers, and woks. Only wet chemical systems have been able to meet the UL-300 standard. Today, the fire protection industry continues to educate local AHJs and encourages end users to upgrade their commercial cooking pre-engineered fire suppression systems to meet the current UL-300 standard.[7]

UL-300 was created and adopted in conjunction with changes made by the National Fire Protection Association (NFPA) in Standard 17 (*Standard for Dry Chemical Extinguishing Systems*) and NFPA Standard 17A (*Standard for Wet Chemical Extinguishing Systems*). Wet chemical systems must comply with NFPA Standard 96 (*Standard for Ventilation Control and Fire Protection of Commercial Cooking Operations*).[8]

---

*Manufacturers of products submitted for testing enter into an agreement with UL for the use of their Listing mark and are required to maintain a level of quality that complies with UL's requirements.[6]

Extinguishing Systems

Wet chemical systems extinguish fires using the local-application method. This is accomplished through the installation of fixed, pre-engineered fire suppression systems that deliver wet chemical agent directly onto the hazard. In addition, portable fire extinguishers are required to be used in conjunction with these systems to support the extinguishment objective. System hardware components are similar to dry chemical systems. Piping delivers the agent to nozzles arranged to atomize the solution and distribute it upon the equipment being protected. Pre-engineered systems are proprietary, defined by predetermined flow rates, pipe sizes/lengths, nozzle pressures, and amount of agent required.

LOCAL APPLICATION

Wet chemical pre-engineered extinguishing systems are highly effective on fires in commercial cooking appliances. They are also used to protect restaurant ventilating equipment (hoods and ductwork). Wet chemical pre-engineered systems require no special distribution components but include a hardware package consisting of control panels, agent container, expellant cylinder, piping, fixed temperature-sensing fusible link detection system, releasing device, manual actuator, nozzles, and gas/electric shutdown devices (see Figure 3.6).

Wet chemical extinguishing agent is stored inside a container adjacent to the protected equipment (see Figure 3.7). The suppression agent flows through system piping (black, chrome plated, or stainless steel), from which it is discharged into the plenum and duct areas and onto cooking appliances. Discharge nozzles, constructed of brass, stainless steel, or a combination of both, are narrow in diameter and must have metal or rubber blow-off caps to keep their tips free of grease buildup.

Wet chemical agent is sprayed at a low velocity to avoid violent reactions and the splattering of burning oil and grease that could cause flame spread and injury to personnel. Wet chemical agent will not produce toxic gases when used as an extinguishing agent.

FIGURE 3.6 Manual pull station for wet chemical range hood protection system. *Source:* Ronald R. Spadafora

FIGURE 3.7 Cabinets containing wet chemical agent and expellant nitrogen gas. *Source:* Ronald R. Spadafora

A typical fixed system discharges approximately three to four gallons (11–15 L) of agent in 30 seconds. This amount has proven to be successful because of the characteristics of wet chemicals. Due to the effectiveness of wet chemical agent, the application of an additional supply on the hazard is not required. There is also no need to manually shut off the system. Postfire cleanup is readily accomplished by flushing the area with water or steam.

Pre-engineered systems are automatically activated by a fixed temperature-sensing fusible link detection system located within the ductwork or cooking appliance hood. The fusible link provides the benefit of early detection and quick response to the fire. The detection system triggers a spring-loaded, mechanical/pneumatic type of release mechanism, which initiates the pressurizing of the wet chemical agent. The wet chemical agent storage tanks have a working pressure of 100 psi (689 kPa). Gas flow and electrical power to the kitchen equipment is automatically shut off upon system activation. This action stops these two sources from continuing to feed and reignite the fire (see Figure 3.8). Systems can also be remotely activated manually via a mechanical pull station. This device is designed to be installed along an egress route from the cooking area a minimum of 10 ft. (3 m) and a maximum of 20 ft. (6 m) from the kitchen exhaust system.

Fire protection discharge can be very costly to a restaurant's owner and employees. Interruption downtime of the business can result in layoffs, lost customers, and costly service bills. Cleanup time and expense is often directly proportional to the quantity of agent and length of time the agent is discharged. Wet chemical systems discharge a limited quantity of agent that will be less costly to recharge and require less downtime. This limited quantity of agent also will not cause overflow, reducing cleanup time. Surfaces coated with wet chemical agent should be cleaned as soon as possible with soap and water to avoid staining and corrosion of appliances.

FIGURE 3.8 Mechanical gas flow shut off. *Source:* Ronald R. Spadafora

> **NOTE**
>
> Wet chemical fire protection systems have applications outside of commercial cooking establishments. Wet chemical agents that contain multiple salts to keep from freezing have been developed for vehicle fire use. They do not rely on added ethylene glycol or propylene glycol as a freeze-point depressant. They are effective as a fixed system (engine compartment) or portable fire suppression agent in vehicles operating in subfreezing temperatures (see Figures 3.9 and 3.10).

FIGURE 3.9 Engine compartment wet chemical protection system. *Source:* Ronald R. Spadafora

FIGURE 3.10 Wet chemical containers on flatbed for tractor engine compartment protection system.
Source: Ronald R. Spadafora

Inspection, Testing, and Maintenance

Wet chemical extinguishing systems require periodic inspection, testing, and maintenance, according to NFPA 17A. The physical condition of all stored extinguishing agent containers and expellant gas cylinders must be checked along with their connections to piping. Verification of stored pressure requirements must also be performed. Proper operation of pre-engineered commercial kitchen extinguishing systems necessitates testing on a semiannual basis. Tested elements include detection devices, alarms, releasing mechanisms, manual pull stations, and mechanical fuel shutoffs. In addition, fusible links must be maintained free of grease to ensure proper operation. Grease buildup can cause these devices to fail, negating the operation of the extinguishing system. Fire codes require that certified fire protection technicians, approved by the local AHJ, conduct all mandated inspections and tests.

Firefighters and fire prevention inspectors should conduct regularly scheduled examinations of pre-engineered commercial cooking extinguishing systems to determine condition and operability. Fire department personnel performing a visual check of a wet chemical system should evaluate the following items:

- Wet chemical and expellant gas container pressure gauges and refill dates
- Wet chemical cylinder connections from the distribution piping to the nozzles
- Status/condition of fusible links and actuators
- Manual pull station located adjacent to the means of egress from the room
- Whether tamper indicators and seals are intact
- Presence/position of nozzles protecting cooking equipment and fire spread
- Presence and condition of nozzle blow-off caps
- Cleaning maintenance records for filters, hoods, and ductwork

THE NEED FOR FIRE PREVENTION INSPECTIONS

A restaurant's sprinkler system controlled a fire, which started when cooking oil heating on a gas stove ignited, until firefighters arrived to extinguish it. The restaurant, located in a large, single-story strip mall, included a dining room, a service counter, and the kitchen in the rear. A wet pipe sprinkler system provided coverage throughout the mall, and a wet chemical range hood extinguishing system provided coverage in the restaurant's kitchen.

Firefighters located the fire in the stove area. Fire investigators determined that the fire originated when an employee heating cooking oil left the kitchen to perform another chore. Upon returning, the employee saw that the oil was smoking heavily and then saw it burst into flames.

Investigators examining the kitchen range hood fire extinguishing system discovered that although the system activated it did not control or extinguish the fire, because the cylinders containing the wet chemical extinguishing agent had been disconnected from the distribution piping to the nozzles. The system was heavily damaged from the fire's heat.[9]

Portable Fire Extinguishers

Wet chemical portable fire appliances are also known as Class K fire extinguishers. Two-gallon (8 L) stainless steel cylinders have a UL 2A: K rating. They are designed to be used in conjunction with wet chemical, UL-300 compliant, commercial kitchen fire suppression systems. Wet chemical agent used in portable fire extinguishers has been tested by UL using fire performance standard *ANSI/UL 711: Rating and Fire Testing of Fire Extinguishers*.

Class K portable fire extinguishers have a unique wand applicator at the end of the discharge hose. This allows the user to apply the agent as a fine mist, which helps prevent grease splash and fire reflash while cooling the cooking appliance equipment. It also allows the user to apply the agent exactly where aimed from a safe distance. Discharge time (corresponding to amount of agent) ranges from one to two minutes. Temperature range for storage is from 40°F to 120°F (4°C to 49°C). In the event of a fire, the primary pre-engineered suppression system should be activated first. The Class K portable fire extinguisher is generally required to be positioned within a 30 ft. (9 m) travel distance of commercial cooking equipment.

CHAPTER REVIEW

Summary

Water is a highly efficient and relatively inexpensive fire suppression agent. It is readily available in adequate quantities under most circumstances. Wetting agents contain chemical additives that provide water with better penetration and spreading characteristics when used on ordinary combustibles and certain liquid fuel fires. Wet chemical agent is specifically formulated to replace dry chemicals for the extinguishment of fires caused by the ignition of greases and oils during food processing and cooking. Pre-engineered commercial kitchen suppression systems use this agent to protect range top burners, grills, griddles, broilers, deep fat fryers, ventilation hoods, and ductwork.

Review Questions

1. What causes the molecular components of water to attract each other?
2. What unusual property allows the solid form of water to float on its liquid form?
3. Why is the low viscosity of water considered a disadvantage when it is used to fight fires?
4. Describe the physical structure of the micelle when used on a Class B hydrocarbon liquid fire.
5. Why are wetting agents added to water?
6. What classification of fire is wet chemical agent primarily designed to extinguish?
7. In what ways does wet chemical agent extinguish fire?
8. Explain what is meant by the term "saponification."
9. What did National Association of Fire Equipment Distributors (NAFED) testing determine to be the reason for a steady decline in performance of pre-engineered, commercial cooking fire extinguishing equipment that uses a dry chemical agent?
10. Name the organization whose fire testing using actual commercial cooking appliances led to the use of wet chemical pre-engineered extinguishing systems.
11. Name the two methods of application used to deliver wet chemical agent onto burning Class K materials.
12. Why do discharge nozzles, installed with pre-engineered wet chemical extinguishing systems, spray the agent at a low velocity onto the equipment they protect?
13. What feature of a Class K portable fire extinguisher allows it to apply the agent as a fine mist?

Endnotes

1. National Fire Protection Association (NFPA), *NFPA 18: Standard on Wetting Agents* (Quincy, MA: NFPA, 2011).
2. Ibid.
3. Ibid.
4. James D. Lake, "Chemical Extinguishing Agents and Application Systems," in *Operation of Fire Protection Systems* (Quincy, MA: NFPA, 2003), 608.
5. Pat Jaugstetter, "5 Simple Reasons to Upgrade to the Current UL-300 Standard," *Firewatch* September (2006): 28–30.
6. Jay A. Guy, "Wet Chemical Pre-Engineered Restaurant Systems and Water Sprinkler Systems: A Comparison," www.halcyon.com/NAFED/HTML/Wetsprk.html.
7. Underwriters Laboratories Inc., *UL-300 Standard: Standard for Fire Testing of Fire Extinguishing Equipment for Protection of Commercial Cooking Equipment*, 3rd ed. (Northbrook, IL: Underwriters Laboratories Inc. [Global], 2005), ANSI approved July 12, 2010.
8. Joe Bernanek, "An Overview of Pre-Engineered Fire Suppression Systems and Industry Changes That Affect Your Property," *Firewatch* June (2007): 30–32.
9. John R. Hall, Jr., *US Experience with Non-Water-Based Automatic Fire Extinguishing Equipment* (Quincy, MA: NFPA, 2012).

CHAPTER 4

Sprinkler and Water Spray Fire Suppression Systems

Source: Ronald R. Spadafora

KEY TERMS

Loading, *p. 56*
Listed, *p. 57*
Clapper valve, *p. 58*
Differential dry pipe valve design, *p. 58*
Fire Department Connection, *p. 59*

Jockey pump, *p. 63*
OS&Y valve, *p. 65*
Section valve, *p. 66*
Wall post indicator valve, *p. 67*
Post indicator valve, *p. 67*

Fusible link, *p. 69*
Bimetallic element, *p. 69*
Frangible bulb, *p. 69*
Frangible pellet, *p. 69*

OBJECTIVES

After reading this chapter, the reader should be able to:

- Quote performance statistics of sprinkler systems
- Describe the different types of sprinkler systems
- Be familiar with system components
- Analyze the design of an in-rack sprinkler system
- Understand the advantages of an early suppression, fast response sprinkler system

Introduction

Sprinklers are among the most popular and valuable fire protection systems. A fire sprinkler system cools down a fire and prewets surrounding areas by distributing water on them. Droplets released by a fire sprinkler system are designed to be large enough to penetrate into the seat of the fire without evaporating. This design effectively limits the magnitude of the fire and keeps it from spreading. Most deaths related to fire are a result of inhaling the products of combustion (smoke and toxic gases). Water droplets from sprinkler systems help to reduce these byproducts and weigh down smoke particles to make it safer for occupants to breathe and escape.

51

Sprinkler Systems

Fire sprinklers are active fire protection systems; they consist of a water supply that provides adequate pressure and flow rate to a water distribution piping system onto which sprinkler heads are connected. They were first used only to protect factories and large commercial buildings, but today residential and small building systems are available at an affordable price. Sprinkler systems are intended to either control or suppress fire. Control mode sprinkler applications are designed to limit the heat release rate (HRR) of the fire. They prewet surrounding combustibles to prevent fire spread. The fire is not extinguished until the burning combustibles release all their energy or manual extinguishment is performed by firefighters.

Sprinkler systems installed today are designed using a density and area approach. The design density is a measurement of how much water per square foot of floor area should be applied to a given area. *NFPA 13: Standard for the Installation of Sprinkler Systems* requires the design professional to examine three factors: occupancy hazard classification, water density requirement to protect the occupancy's hazards, and adequate water supply to meet demand. When these factors are determined, the designer can then make decisions about parameters that include the type of sprinkler system to install, pipe layout and size, the type of sprinkler heads, and sprinkler head location and spacing. Changes to any one of these three factors due to major renovations or change in occupancy hazard classification may require a change in design parameters.

PERFORMANCE STATISTICS

The following statistics are based on reported 2007 to 2011 structural fires, an estimated 10% of which showed sprinklers present. These estimates are projections based upon the detailed information collected in Version 5.0 of the US Fire Administration (USFA) National Fire Incident Reporting System (NFIRS) and the annual National Fire Protection Association (NFPA) fire department experience survey. Confined fires are not included in the analysis of reliability or effectiveness of automatic extinguishing equipment, because fires reported as confined fires are usually reported without sprinkler performance details or as fires too small to activate operating equipment.

Sprinklers operated in 91% of all reported structure fires large enough to activate sprinklers. This statistic excludes buildings under construction and buildings without sprinklers in the fire area. When sprinklers operated, they were effective 96% of the time, resulting in a combined performance of operating effectively in 87% of all reported fires for which sprinklers were present in the fire area and the fire was large enough to activate them. When sprinklers fail to operate, the reason most often given (in 64% of failures) is shutoff of the system before fire began. This can occur in the course of routine inspection or maintenance.

When sprinklers operate but are ineffective, the reason usually has to do with an insufficiency of water applied to the fire, either because water does not reach the fire (44% of cases of ineffective performance) or because not enough water is released (30% of cases of ineffective performance).[1]

ONE MERIDIAN PLAZA FIRE

At approximately 2023 hours on February 23, 1991, a smoke detector was activated on the 22nd floor of the 38-story Meridian Bank Building, also known as One Meridian Plaza. The building was undergoing major construction/renovation work, including the installation of additional sprinkler system protection. The fire was reported to the Philadelphia Fire Department (PFD) at approximately 8:40 p.m. and burned upward for more than 19 hours, completely consuming eight floors of the building.

One Meridian Plaza was a steel-frame structure with poured concrete floors over metal decks. It had three-hour-fire rated columns, two-hour-fire rated beams and floor/ceiling systems, and one-hour-fire rated corridors and tenant separations. Shafts, including stairways, were of two-hour-fire rated construction. The roof consisted of a one-hour-fire rated assembly. The skyscraper had three underground levels, 36 above-ground floors, two mechanical equipment room floors (12 and 38), and two rooftop helipads. The building was rectangular in shape, approximately 243 ft. (74 m) in length by 92 ft. (28 m) in width (approximately 22,400 ft.^{2} (2081 m^2). Site work for construction began in 1968, and the building was completed and approved for occupancy in 1973. The fire began on the 22nd floor in linseed oil–soaked rags. It burned out of control and was very difficult for firefighters to place under control due to lack of electrical power as well as insufficient water pressure coming from the building's standpipe system.

Only the floors located below grade were protected by automatic sprinklers at the time of construction. At the request of selected tenants, sprinklers were installed on several floors during renovations, including all of the 30th, 31st, 34th, and 35th floors and parts of floors 11 and 15. Limited-service sprinklers, connected to the domestic water supply system, were installed in part of the 37th floor. The building owners had plans to install sprinklers on additional floors as they were renovated.

The PFD committed approximately 316 personnel operating 51 engine companies, 15 ladder companies, and 11 specialized units, including EMS units, to the 12-alarm fire. The incident was managed by 11 battalion chiefs and 15 additional chief officers under the overall command of the Fire Commissioner.

The fire caused three firefighter fatalities and injuries to 24 more. The three firefighters who died were attempting to ventilate the center stair tower. They radioed a request for help and provided their location, which was on an upper floor of the building. Extensive search and rescue efforts proved futile. Their bodies were later found on the 28th floor. The firefighters had exhausted all of their air supply. At the time of their deaths, the 28th floor was not burning but had an extremely heavy smoke condition. After the loss of three firefighters, hours of unsuccessful attack on the fire, with several floors simultaneously involved in fire, and a risk of structural collapse, the Incident Commander withdrew all personnel from the building. The fire ultimately spread up to the 30th floor.

The fire was stopped when it reached the 30th floor, which was protected by automatic sprinklers. Ten sprinkler heads activated. The vertical spread of the fire was stopped solely by the action of the automatic sprinkler system, which was being supplied by fire department apparatus pumpers. The 30th floor was not heavily damaged by fire, and most contents were salvageable. The fire was declared under control at 3:01 p.m., February 24th. At the time, it was the largest high-rise office building fire in modern American history[2] (see Figure 4.1).

FIGURE 4.1 Meridian Bank fire. *Source:* George Widman/Associated Press

Types of Fire Sprinkler Systems

Fire sprinkler systems protect a variety of buildings and property. Over the years many different types have been developed. This section will review wet, dry, deluge, and preaction sprinkler systems. Each system has its own unique set of characteristics that protect both people and property. Wet pipe systems are always filled with water. The water in the pipes is under a moderate amount of pressure. When a sprinkler head activates, the pressurized water in the pipes is immediately released, providing a faster reaction time than any other type of system. The wet pipe system is also the most reliable, boasting cost savings for easy installation and low maintenance. In unheated enclosures or areas where room temperatures approach freezing, antifreeze solution is added to protect water pipes from cracking.

In a dry pipe system, the pipes are filled with compressed air. When the sprinkler head activates, a valve releases the compressed air through the piping and out of the sprinkler head. Once air is released, the pressure in the pipe decreases, allowing water to fill the system. Dry pipe systems, however, have a slower reaction time but are commonly installed where pipes may freeze.

Deluge systems have all open sprinkler heads and therefore system piping is at atmospheric pressure. The heads do not have heat-sensing operative devices. Water is not present in the piping until the system operates—upon activation of a detection system. Deluge sprinkler systems provide a simultaneous application of water over the entire area being protected.

Preaction systems are hybrids of wet, dry, and deluge systems; their makeup depends upon the objectives the design engineer is looking to accomplish. Piping for preaction sprinkler systems may or may not be pressurized with air. This type of system has an additional fire-detection device that will recognize a fire before sprinklers activate. When this action occurs, the main water valve for the system opens, allowing water to flow through the piping up to the sprinkler head. Upon activation, however, it will then operate much like a wet pipe system. Preaction systems are generally installed in areas in which accidental wet pipe system activation could cause major damage to valuable storage commodities, books, manuscripts, archived materials, and works of art.

WET PIPE AUTOMATIC SPRINKLER SYSTEMS

The most commonly installed fire sprinklers are automatic wet pipe systems. This type of system is used where the temperature is maintained at minimum of 40°F (4°C) to prevent the system from freezing. The only operating components of these simple systems are the sprinkler heads and the alarm check valve. Sometimes a combination of a wet pipe and a dry pipe system may be used when part of the building cannot be heated.

The sprinkler head orifice is normally closed by a disc or cap. This cap is held in place by a heat-sensitive releasing element. A rise in temperature to a predetermined level causes the sprinkler head to operate. Water is then discharged. Sprinkler heads are installed at standard intervals on the piping. If more than one head opens, the water pattern of two adjoining heads overlaps.

Automatic sprinklers are very effective for life safety. The downward force of the water lowers the smoke level in the room. The sprinklers also serve to cool the smoke. This makes it possible for building occupants to survive in the area much longer than they could if the room were without sprinklers. When sprinklers are installed, there are rarely problems getting water to the base of the fire. Most standard sprinkler systems automatically sound an alarm when a sprinkler head discharges water. This alarm is usually an audible signal in the building (see Figure 4.2).

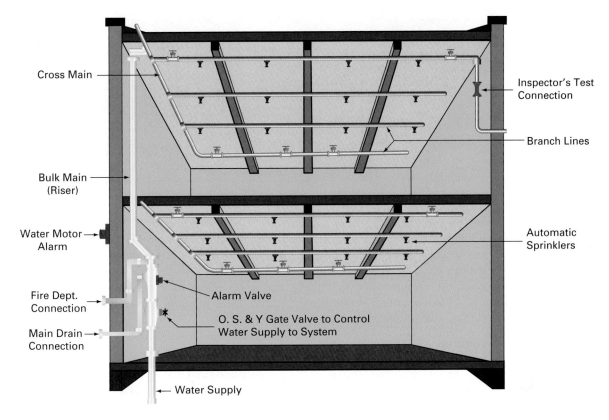

FIGURE 4.2 Wet pipe automatic sprinkler system

Cold-Weather Valves
Automatic wet sprinkler systems should not be completely shut off and drained to avoid freezing during cold weather. Depending upon the authority having jurisdiction (AHJ), however, parts of the sprinkler system may be shut down. These shutoff valves are called cold-weather valves.

Antifreeze Solutions
In locations at which temperatures can drop below 40°F (4°C), automatic wet sprinkler systems may require an antifreeze solution. The antifreeze solution is a mixture of chemicals designed to prevent water from freezing. Glycerin and propylene glycol are two approved antifreeze solutions. Antifreeze solutions are expensive and may be difficult to maintain. Antifreeze is therefore normally used for small, unheated areas.[3]

To ensure that antifreeze solution will control a fire and also provide the desired level of freeze protection, it is essential for the solution to be at the correct concentration and mixed properly. Maintenance of antifreeze systems must be performed regularly to ensure that proper concentrations throughout the system exist. Antifreeze solutions are heavier than water. Sampling from the top and bottom of the system helps to determine if an antifreeze solution has settled. Antifreeze can also separate from water due to poor mixing. When this occurs, the antifreeze solution will exhibit a higher concentration in the lower portion than in the upper portion of the system. If antifreeze concentration is too low near the water connection but acceptable near the top portion of the system, this can indicate that the system is becoming diluted near the water supply.

> **NOTE**
>
> Antifreeze solutions with concentrations of propylene glycol exceeding 40% and concentrations of glycerin exceeding 50% have the potential to ignite when discharged through automatic sprinklers. Consideration should be given to reducing the acceptable concentrations of these antifreeze solutions by using a safety factor.[4]

Sediment

When wet pipe sprinkler systems operate, the water flow in the piping is at a relatively low velocity compared to a dry pipe system. This allows for the accumulation of water sediment and foreign objects in the system piping. Preventive measures should be taken to avoid this accumulation by flushing the supply mains to these systems at least semiannually. Moreover, other water supply sources (gravity tanks, pressure tanks, suction tanks, etc.) should be cleaned periodically.[5]

Loading

Environmental conditions may exist that cause the build-up of foreign materials on the sprinkler heads. This situation can prevent sprinkler heads from functioning properly. This accumulation is commonly called **loading**. Foreign material insulates the sprinkler head and keeps it from operating at the design temperature. Furthermore, if the material is allowed to harden, it may prevent the sprinkler head from activating. Loaded sprinkler heads should be replaced with new sprinkler heads rather than attempting to clean them, because cleaning may damage the heads or cause them to leak.

Sprinkler heads designed to be exposed to corrosive conditions are often covered with a protective coat of wax or lead by the manufacturer. Care should be taken to make sure that the protective coating is not damaged when handling or installing the heads. Corrosive vapors can make automatic sprinkler heads inoperative or reduce the speed of operation. They can also damage the delicate components of the sprinkler heads. In most cases, this type of corrosive action takes place over a long time period. For this reason, the sprinkler heads must be carefully inspected for signs of corrosion.

Applications

Automatic wet pipe sprinkler systems are recognized by NFPA standards *13D: Standard for the Installation of Sprinkler Systems in One- and Two-Family Dwellings and Manufactured Homes* and *13R: Standard for the Installation of Sprinkler Systems in Low-Rise Residential Occupancies* as life safety fire protection systems.

RESIDENTIAL SPRINKLER SYSTEMS

These types of systems incorporated new fast-response technology in the late 1970s. Research demonstrated that in order to make small domestic water supplies effective for sprinkler protection and to control a residential fire before small rooms could fill up with toxic smoke a sprinkler needed to be considerably more sensitive to heat than the standard sprinklers. Residential sprinklers were simultaneously developed by sprinkler manufacturers while appropriate product testing criteria were being developed by Underwriters Laboratories Inc. In addition, installation recommendations were also developed by the NFPA Committee on Automatic Sprinklers. Residential sprinklers can now be installed in conformance with NFPA 13D and NFPA 13R. Residential sprinklers provide fast response to fire as well as a special water distribution pattern and cooling capabilities.[6]

Loading ■ The buildup of foreign materials and contaminants on a sprinkler head.

QUICK-RESPONSE SPRINKLERS

Quick-response sprinklers (QRS) are tested under the same product criteria as standard sprinklers but also exhibit the fast-response characteristics of **Listed** residential sprinklers. Some manufacturers developed these sprinklers by installing the residential sprinkler operating mechanism into a standard sprinkler frame. Others successfully submitted residential sprinklers to UL for testing under the criteria for standard sprinklers. In both cases, UL designated these special sprinklers as QRS. By establishing a separate Listing category, UL has identified QRS as different from both standard and residential sprinklers. Unlike residential sprinklers, however, QRS are not required to have special cooling and distribution abilities. All new health care and assisted-living facilities are required to install QRS heads.[7]

Listed ■ All sprinklers installed in conformance with National Fire Protection Association (NFPA) standards must be Listed. A Listing means that the manufacturer has successfully submitted the product to an independent laboratory for testing against an established standard of quality. Both Underwriters Laboratories Inc. and Factory Mutual Research have established special Listing categories for fire sprinklers.

Dry Pipe Automatic Sprinkler System

Dry pipe sprinkler systems are installed where a wet pipe system cannot be heated to prevent freezing. The pressure in the piping is controlled automatically by an air maintenance device. The system uses standard sprinkler heads. When a sprinkler head is activated by the heat from a fire, the pressure is reduced in the piping. The drop in air pressure causes the dry pipe valve to open. Water then flows into the piping and out of the activated sprinklers. This water flow initiates a local alarm to alert occupants in the building of the fire. (see Figure 4.3).

AIR SUPPLY

Dry pipe valves must have a reliable air supply that is available at all times. The compressed air supply for the dry pipe system is usually provided by an electrically driven air compressor, which is automatically controlled to turn on and off at the designed minimum

FIGURE 4.3 Dry pipe automatic sprinkler system main control valve. *Source:* Ronald R. Spadafora

and maximum pressures, respectively. The air compressor must be adequately sized to be able to bring the system air pressure up to the design level in 30 minutes or less.

DRY PIPE DIFFERENTIAL VALVE DESIGN

Dry pipe sprinkler systems are more complicated than wet pipe systems and require more inspection, maintenance, and testing. A typical dry pipe system model uses differential water and air pressure on a single **clapper valve**. The clapper valve prevents the pressurized water in the fire mains from entering the sprinkler system piping.

The air pressure in the system piping is lower than the water supply pressure. A **differential dry pipe design** is used to prevent the greater water supply pressure from forcing water into system piping. This is achieved via the surface area of the clapper valve exposed to the system side being much larger than the surface area of the clapper valve facing the water supply. The larger surface area spreads out the force, correlating to less pressure requirements. Most dry pipe valves are designed so that a moderate air pressure will hold back a much greater water pressure.

When an automatic sprinkler is exposed for a sufficient time to a temperature at or above its temperature rating, it activates and allows the air in the piping to vent. As the air pressure in the piping drops, the pressure differential across the dry pipe valve changes, allowing water to enter the piping system. Water flow from sprinklers needed to control the fire is delayed until the air is vented from the sprinklers.

Differential valves use a 5:1 or 6:1 (system:supply) surface area ratio. This translates into 5 or 6 psi (34 or 41 kPa) of water pressure being held back by 1 psi (7 kPa) of air pressure. A safety factor, usually about 20 to 25 psi (138 to 172 kPa), is maintained above the trip pressure to help prevent false activations of the system.[8]

Dual arrangements have two separate but connected air and water clapper valves. This type of dry pipe valve has an intermediate chamber situated between the air and water seats. It is open to atmospheric pressure. This chamber has a drain that leaks out water that may pass through the water seat. More modern dry pipe valves have priming water above the clapper assembly, with the sprinkler system air pressure above it. This water provides a positive set and seal of the air and water clapper valves.

A latching mechanism is always required for dry pipe valves. The latch prevents the clapper assembly from swinging back into the waterway and causing potential damage to the valve and piping. Dry pipe valves must be manually reset after activation by removing the valve housing cover, releasing the clapper assembly from the latch, and manually resetting the valve seats.

LOW-DIFFERENTIAL DRY PIPE VALVES

Low-differential dry pipe valves are designed so that the air or nitrogen pressure is only 15 to 20 psi (103 to 172 kPa) above the static water supply pressure. The tripping point of the valve is normally about 10% less than the water pressure. The major advantage of the low-differential dry pipe valve is the reduced initial velocity of the water flowing into the system compared to conventional valves. Another benefit of this type of valve is the speed of operation which allows water to travel quickly through piping to the fused sprinkler heads.

QUICK-OPENING DEVICES

In a dry pipe system, there is a delay between the opening of a sprinkler and the discharge of water which may allow the fire to spread and more sprinkler heads to fuse. This difficulty may be partly overcome by the installation of quick-opening devices (QODs). These components are used to reduce the time needed to open the clapper valve and allow water into the system. An accelerator is a QOD that automatically activates when a drop of 2 psi (14 kPa) in air pressure is detected in the system. It changes the water and air pressure balance in the system and allows the water to be forced through system piping rapidly.[9]

Clapper valve ■ A valve with a hinge on one side that allows water to flow in only one direction.

Differential dry pipe valve design ■ A valve that uses larger surface area under relatively low air pressure to hold back greater water pressure.

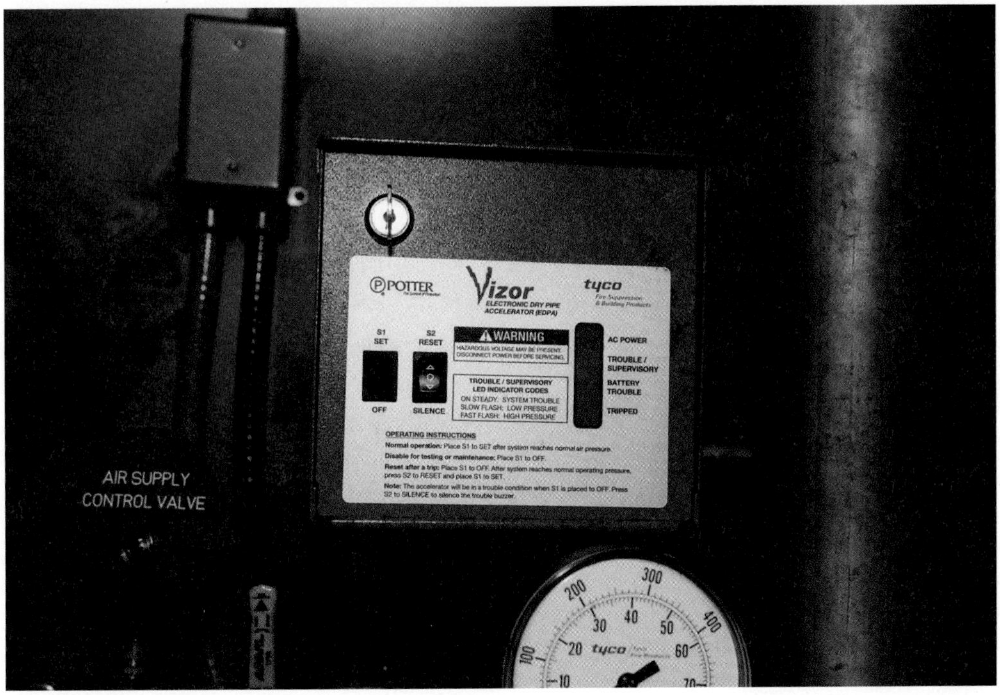

FIGURE 4.4 Electronic dry pipe sprinkler accelerator. *Source:* Ronald R. Spadafora

ACCELERATOR

Accelerators are normally designed with multiple pressure chambers. They trip the dry pipe valve more quickly by sensing the pressure drop as a result of the fusing of a sprinkler head. These devices introduce air pressure taken from immediately above the dry pipe clapper valve into their intermediate chamber, which is directly below the air clapper of the dry pipe valve. This additional pressure breaks the differential and opens the valve immediately. Following activation, the accelerator must be reset manually (see Figure 4.4).

APPLICATIONS

Dry pipe systems are the second most common sprinkler system type. They are most often used in unheated warehouses, loading dock areas, attic spaces, parking garages, and refrigerated coolers. Dry pipe system valve rooms, however, must be heated.

Nonautomatic Dry Pipe Systems

In nonautomatic dry pipe systems, all piping is normally dry, and water is supplied by firefighters by pumping water into the system through a **fire department connection** (FDC). Some nonautomatic dry pipe systems are supplied by the manual operation of a water-control valve. There are several nonautomatic dry pipe application systems:

- *Open sprinkler heads or fixed spray nozzles:* These are fixed local-application systems installed for the direct protection of electrical transformer vaults and similar types of hazardous areas.
- *Exterior exposure sprinklers:* These systems use open sprinkler heads to create an external water curtain on the walls of a building.
- *Foam:* A foam–water sprinkler or nozzle system is an application system that discharges a mixture of water and low-expansion foam concentrate, resulting in a foam spray. These systems are commonly used in occupancies associated with high-challenge fires, such as flammable liquid storage facilities and aircraft hangars.

Fire department connection ■ An inlet sprinkler system connection that allows fire department apparatus pumpers to supplement the system with water through hoses.

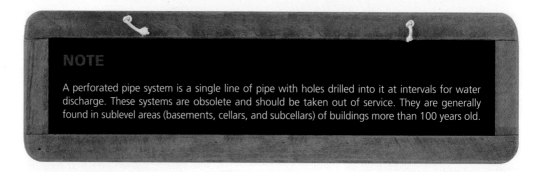

NOTE

A perforated pipe system is a single line of pipe with holes drilled into it at intervals for water discharge. These systems are obsolete and should be taken out of service. They are generally found in sublevel areas (basements, cellars, and subcellars) of buildings more than 100 years old.

Deluge Sprinkler System

A deluge system (see Figure 4.5) is applicable for extra-hazard occupancies in which rapid fire spread can be expected. Deluge systems are connected to a water supply through a deluge valve that is commonly released via a detector signaling a fire alarm control unit (FACU). When the deluge valve is activated, water enters the piping and discharges through all of the sprinkler heads in the system. The deluge valve has a mechanical latch that allows it to remain open once tripped and does not reset. Deluge systems are used in chemical storage/processing facilities and industrial plants, where plenty of water is required over a large area to cool combustibles. They may also be installed in personnel egress paths or building openings to slow travel of fire.[10]

Some deluge systems incorporate the use of a foam concentrate to mix with water and form a foam solution that, when discharged, can provide a protective blanket of finished firefighting foam to help control the development of a fire. Fixed foam deluge systems are used in the protection of aircraft hangers.

FIGURE 4.5 Deluge sprinkler system.
Source: Ronald R. Spadafora

FIGURE 4.6 Preaction sprinkler system.
Source: Ronald R. Spadafora

Preaction Sprinkler System

Preaction systems (see Figure 4.6) are designed to protect against accidental situations in which there is a danger of serious water damage to valuable property. This destruction is usually caused by damaged sprinkler heads or broken piping. In a standard or single-interlock preaction system, the water supply is contained by a preaction valve. This valve is connected to a supplemental detection system. Water will not enter into the piping system until the detection system is activated and will not be applied on the fire until sufficient heat causes the individual sprinkler head to fuse.

In a double-interlock system, pressurized air or nitrogen is added to the piping. The pressure on the preaction valve is constantly monitored. If the pressure changes due to a pipe leak, for example, an alarm will sound, but the system will not activate; the valve will remain closed. Once a detection system is activated, however, the valve is released, allowing water into the sprinkler piping. Water will not be discharged until sufficient heat causes individual sprinkler heads to fuse. Activation of either the fire detectors alone or sprinkler heads alone, without the concurrent operation of the other, will not allow water to enter the piping. Because water does not enter the piping until a sprinkler fuses, double-interlock systems are considered dry systems in terms of water delivery times.[11]

The preaction sprinkler system has several advantages over a dry pipe system. Its valve opens sooner, because the detector devices react to fire conditions faster than the sprinkler heads. Arming the system prior to sprinkler activation allows water to be positioned at the sprinkler heads and ready for quick discharge onto the fire when needed. This scenario allows the system to apply water on the hazard more quickly, leading to less property damage.

Detectors and devices used to activate preaction valves include smoke detectors, gas-detecting systems, hydraulic, electric, and manual release, and automatic signals from other fire protection and safety systems.

Inspection, Testing, and Maintenance

Automatic fire sprinkler systems are required to be inspected, tested, and maintained in accordance with *NFPA 25: Standard for the Inspection, Testing, and Maintenance of Water-Based Fire Protection Systems*. The requirements contained in NFPA 25 are based on the type of sprinkler system installed.

In order to follow the requirements of NFPA 25, it is important to understand the meaning of the terms *inspection*, *testing*, and *maintenance* derived from the standard.

INSPECTION

Inspections are conducted by visually examining the system. Inspections check for physical damage and determine operating condition.

TESTING

Testing consists of periodic procedures to determine the status of a system. Testing includes physical checks involving water flow devices, fire pumps, and alarms. These tests are in addition to original acceptance tests at time intervals specified in NFPA 25.

MAINTENANCE

Maintenance is work performed by a licensed sprinkler contractor or equivalent certified person to keep equipment operable or to make repairs.

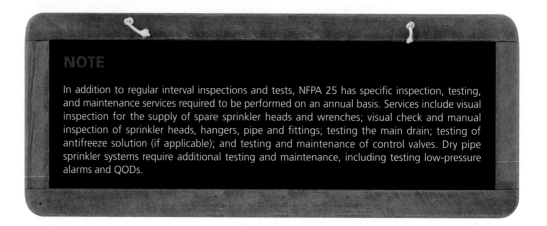

NOTE

In addition to regular interval inspections and tests, NFPA 25 has specific inspection, testing, and maintenance services required to be performed on an annual basis. Services include visual inspection for the supply of spare sprinkler heads and wrenches; visual check and manual inspection of sprinkler heads, hangers, pipe and fittings; testing the main drain; testing of antifreeze solution (if applicable); and testing and maintenance of control valves. Dry pipe sprinkler systems require additional testing and maintenance, including testing low-pressure alarms and QODs.

System Components

In the sections ahead, we will cover the various types of components that make up fire sprinkler systems.

FIRE PUMPS

A fire pump is dedicated to fire protection. It is a part of a fire sprinkler system's water supply and can be powered by electric, diesel, or steam. The pump intake is connected to either the public underground water supply piping or a static water source (gravity or suction tank). It provides enhanced water flow and pressure to the sprinkler system risers. The centrifugal fire pump is the standard pump currently used in fire protection systems. It is preferred because it is reliable and compact, requires low maintenance, and can be powered by a variety of motor drivers (see Figure 4.7).

FIGURE 4.7 Foreground, centrifugal fire pump with electric motor driver.
Source: Ronald R. Spadafora

The water available to the centrifugal pump must always be under pressure, because it cannot lift water to provide a supply source. A water tank can be used if the container supplies the pump via gravity. An example of this type of water source is called a suction tank. As the water passes through the pump, it reaches a rotating impeller. This impeller is designed to grab the water on the inlet side of the pump and then discharge the water under increased pressure into the fire protection system.

The fire pump is commonly found inside a room that is fire resistant or constructed of noncombustible material. The pump room should be located as close as possible to the fire protection system and kept clean and accessible at all times. No storage is permitted in this room. The fire pump, motor driver, and control equipment must also be protected against possible interruption of service. The temperature inside the pump room is required to be maintained above 40°F (4°C) at all times to prevent freezing of the water. The pump room should only be used for fire protection functions and not for general plant operations.

Automatic pumps have their motors activated by controllers when there is a drop in water pressure or water flow in the fire protection system. The controllers are adjusted so that they are not too sensitive to pressure and flow changes. If an electric motor driver is used, a standby power generator is sometimes required. If an engine motor driver is used, it will require fuel storage tanks. Manually operated fire pumps are often found in industrial and manufacturing occupancies that have employees certified to use these systems on the premises at all times.

Often found on sprinkler systems, a **jockey pump** is small, yet it is an important part of the fire pump's control system. It is connected to a fire sprinkler system in order to maintain a certain pressure. It is sized to ensure that if a sprinkler head activates there will be a system pressure drop sensed by the fire pump's automatic controller, which will start the fire pump. Jockey pumps are generally multistage centrifugal pumps.

Jockey pump ■ An auxiliary pump used to maintain system pressure without starting the main pump.

FIRE PUMP INSPECTION, TESTING, AND MAINTENANCE

Fire pumps deteriorate over time and are also vulnerable to corrosion, tampering, and accidental impact damage. Because they are used infrequently, fire pumps must be systematically inspected to determine their condition, operability, and need for routine maintenance. Visual inspection of fire pump components (motor, shaft couplings, and valves) is critical. Defective components can hinder the pump from operating or cause the pump to overheat. Another essential task of fire pump inspection is ensuring that the fire pump's couplings are aligned properly. Couplings that are misaligned are far more likely to fail and could cause disruption of service.

Fire pumps must undergo regular preventative maintenance to ensure that they maintain their effectiveness. One of the most common maintenance measures is proper bearing lubrication. Well-maintained fire pumps are reliable and play an important role in ensuring sprinkler system efficiency. Regular inspection and testing are generally performed according to the requirements found in NFPA 25.

ALTERNATE POWER

Fire pumps powered by a reliable electrical supply are generally only required to have a single source. Often, however, the code official will decide if an alternate power supply is required, as per *NFPA 20: Standard for the Installation of Stationary Pumps for Fire Protection*. At least one alternate source of power (normally independent of the primary power supply) will be required when the primary power source cannot be considered reliable and the height of the structure that the fire pump serves is beyond the pumping capacity of fire department apparatus. An alternate source of power is not required, however, if a backup motor–driven fire pump is installed in accordance with NFPA 20.

FIRE DEPARTMENT CONNECTION (FDC)

An FDC is normally located on the exterior of the building; through it, fire apparatus can pump supplemental water into a fire protection system (sprinkler, standpipe, or test connection) or another system furnishing water for fire extinguishment to supplement existing water supplies. The *International Fire Code* (IFC) stipulates that FDCs be installed on the street side of the building and be fully visible and recognizable from the street or point of fire department vehicle access or as otherwise *approved* by the fire chief.[12]

Fire department connections are required by the IFC to be marked with minimum signage (made of metal with raised letters at least one-inch high). The code also requires the use of specific language, such as "automatic sprinklers." Supplemental information is also needed when an FDC does not serve the entire building or is interconnected with another building. Generally, the FDC manufacturers rate each 2.5 in. (65 mm) hose connection as delivering 250 gpm (946 L/min.) to the system at 100 psi (689 kPa)[13] (see Figure 4.8).

VALVES

Sprinkler systems that are supplied from a public water main have a control valve located between the building and the water main in a box that is recessed into the sidewalk. It allows for the entire system to be shut down. AHJs can facilitate firefighters locating this control valve by requiring a sign to be placed on buildings that indicates the number of feet directly perpendicular from the sign to the valve. The sign might read, "Shutoff for Sprinkler System Located 9 Feet Opposite This Sign," or something similar. A special key is required to operate this valve.

FIGURE 4.8 Automatic sprinkler system fire department connection.
Source: Ronald R. Spadafora

GATE AND CONTROL VALVES

Sprinkler systems are excellent for controlling fires. However, they can cause water damage if they are not shut down soon after the fire has been controlled. No control valve on the system on the fireground, however, should be closed except on the order of the Incident Commander. In large facilities, it may be difficult for the fire department to locate the control valve to shut down the system. Preplanning this information will be extremely helpful, as will inputting the control valve's whereabouts into communication databases for responding units. If the fire is "under control" and interior conditions permit, building maintenance personnel may be designated to close the control valve.

Gate valves may also be provided in sprinkler water distribution systems. They allow the sprinkler system to be shut off for repairs or maintenance. Such valves are normally of a nonrising stem type. They are operated using a special key wrench. A valve box is located over the valve to keep it free from dirt. The valve box also provides a convenient access point for the valve wrench to reach the valve nut.

The main control valve for the building sprinkler system is commonly located inside the structure in a secure room at the level at which the water supply enters the building. The principle type of control valve is known as the **outside screw and yolk (OS&Y) valve**. When the wheel handle of this valve is turned, it directly raises and lowers the gate of the valve by interacting with its stem. The stem of the valve itself raises and lowers visibly outside the body of the valve, and the wheel handle remains in a stationary position (see Figure 4.9). As the operator rotates the wheel handle counterclockwise, the stem lifts out of the wheel handle, opening the gate valve. A clockwise rotation of the wheel handle moves the stem back into the gate and closes the valve. When the stem is no longer protruding beyond the wheel handle, the gate valve is closed.

Outside screw and yoke valve ■ An OS&Y (outside screw and yoke) valve functions by opening and closing via a gate, which lowers both in and out of the valve.

FIGURE 4.9 Main OS&Y valve. *Source:* Ronald R. Spadafora

Section valve ■ Used to divide a sprinkler system into zones or areas.

OS&Y **section valves** are used to divide a sprinkler system into zones or areas. They are normally open and monitored. The section valve is also used to manually close the water supply, for isolated maintenance, for example. Firefighters should always check that section valves are in the full open position when sprinkler water flow and pressure is inadequate during sprinkler operation (see Figure 4.10).

FIGURE 4.10 OS&Y section valve. *Source:* Ronald R. Spadafora

FIGURE 4.11 Post indicator valve.
Source: Ronald R. Spadafora

An additional control valve for the building sprinkler system may be mounted to the outside wall. This device, known as a **wall post indicator valve** (WPIV), provides a wheel handle for opening and closing the valve. The wheel handle is normally secured in the open position (when the sprinkler system is active) by a chain and lock. An indicator window on the valve's body denotes whether the valve is open or closed. Another control valve, attached to an upright post, is known as a **post indicator valve** (PIV). The building or section of the building controlled by the valve is usually marked on the post. The position of this valve (open or closed) is shown through an indicator window in the post. On some posts, a padlock must first be opened to release the operating wrench or wheel handle (see Figure 4.11).

Wall post indicator valve ■ The control valve for a building sprinkler system that may be mounted to the outside wall.

Post indicator valve ■ A control valve attached to an upright post.

VALVE SECURITY

Valve security is typically provided by locking all sprinkler control valves in the open position with hardened steel chains and padlocks. WPIVs and PIVs should have the movement of their operating wheel handles or wrenches severely restricted. Access to keys should only be available to building maintenance personnel and authorized individuals trained in the operation of the sprinkler system. When deliberately cut or broken by hand tools to shut down a valve, however, chains and padlocks will not provide supervisory notification to monitoring personnel.

PIPING ARRANGEMENT

Typical piping arrangement for an automatic sprinkler system is categorized as follows:

- *Branch lines:* Horizontal piping supplied by the riser from the cross main. Sprinkler heads are installed directly onto the branch lines.
- *Cross main:* The first main supplied from the sprinkler riser. Cross mains directly service a number of branch lines.
- *Feed main:* A transfer main that is supplied by the riser to supply a separate area not supplied from the cross main.

Chapter 4 Sprinkler and Water Spray Fire Suppression Systems

- **Riser:** A vertical supply pipe that initially supplies the entire sprinkler piping network. At the sprinkler riser is generally where a transition from underground piping to interior piping occurs. Additionally, the sprinkler clapper valve, one-way check valve, FDC, alarm valve, main drain, and other components are attached to the riser.

SPRINKLER HEADS

The sprinkler head distributes water over a defined fire hazard area. Each sprinkler, other than the deluge type, operates by actuation of its own temperature linkage. The typical sprinkler consists of a frame, thermally operated linkage, cap, orifice, and deflector. Sprinkler heads are screwed into the branch line piping at standard intervals. Sprinkler heads are factory tested to withstand high pressures without damage or leakage. If properly installed, there is little danger of the sprinkler not operating unless it is damaged.

The shape of the deflector determines the spray pattern of the water discharged from the sprinkler head. The force of the water against the standard deflector creates a heavy spray that is directed outward and downward. This pattern is the best way to extinguish and control a fire as well as prevent its spread. In general, this is an umbrella-shaped spray pattern. At a distance of 4 ft. (1.2 m) below the deflector, the spray covers a circular area with a diameter of approximately 16 ft. (5 m) when the sprinkler is discharging 15 gpm (57 L/min.). Sprinkler discharge patterns must not be obstructed by building structural elements, furnishings, or storage.

There are over 50,000 different variations of sprinkler heads. Sprinklers manufactured after January 1, 2000, are required to have a Sprinkler Identification Number (SIN). Sprinkler heads manufactured prior to that date shall be replaced as required with sprinkler heads of similar characteristics such as orifice size, temperature rating, and deflector orientation (see Figure 4.12).

COMPONENTS

- **Frame:** The frame provides the main structural component that holds the sprinkler together. The frame keeps the thermal linkage and cap in place. It also supports the deflector during discharge. Frame styles include standard and low profile, flush, and concealed mount. Some frames are designed for extended spray coverage, beyond the range of normal sprinklers. Special coatings are available for areas subject to high corrosive effect. Selection of a specific frame style is dependent on the size and type of area to be covered, anticipated hazard, visual impact features, and atmospheric conditions.

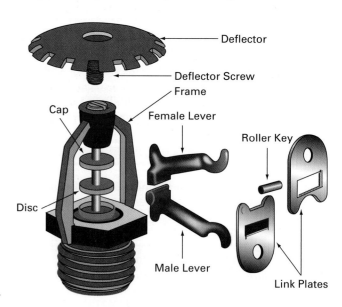

FIGURE 4.12 Standard sprinkler head components

- *Thermal linkage:* The linkage is the component that controls water release. Under normal conditions, the linkage holds the cap in place and prevents water flow. When the link is exposed to heat, however, it weakens and releases the cap covering the orifice. Common linkage types include soldered metal levers as well as frangible glass bulbs and solder pellets. When the operating temperature is reached, there will still be a time lag (approximately 30 seconds to four minutes) prior to activation of the sprinkler head. This is due to the time required for linkage fatigue. Conventional responding sprinklers typically operate close to the three- to four-minute time lag, whereas QRS operate more rapidly. Selection of a sprinkler response characteristic is dependent upon the existing risk, acceptable loss level, and desired response action.

RELEASE MECHANISMS FOR SPRINKLER HEADS

The **fusible link** sprinkler has a two-piece metal component held together by a solder with a predetermined melting point. The link is attached to two levers that are holding down the cap atop the sprinkler orifice. When the solder melts, the levers pull the link apart and are released from the sprinkler. This sequence of events allows water pressure in the piping to push the cap from the orifice to initiate discharge.

Sprinkler heads that can automatically cycle on and off have a **bimetallic element** in the form of a disk made of two distinct metals as the heat-sensitive element. When the sprinkler is off, the disk maintains pressure on a piston assembly. During a fire, however, when the sprinkler's design temperature rating is reached, the disk flexes and opens. This action releases pressure on the piston assembly and triggers water to discharge from the sprinkler. When the temperature of the bimetallic element is reduced, its shape returns to normal. This action forces the piston into the closed position and stops water discharge.

The **frangible bulb** sprinkler has a small glass bulb (filled with liquid and air) positioned between the orifice cap and the sprinkler frame. Heat from a fire causes the liquid to expand against the air, causing the glass bulb to break and allowing for water discharge.

A **frangible pellet** sprinkler has a metal rod between the orifice cap and sprinkler frame. The rod is held in place by a pellet of solder under compression. When the solder melts, the rod is released from the orifice cap. The cap is then displaced by the water pressure in the sprinkler piping network.

Fusible link ■ A heat-actuated release mechanism for a sprinkler head consisting of a two-piece metal component held together by a solder with a pre-determined melting point.

Bimetallic element ■ A device formed of two metals that are bonded together, each of which has a different coefficient of thermal expansion.

Frangible bulb ■ A breakable glass bulb filled with glycerin or alcohol, used to hold the cap over the orifice of a sprinkler head in place. When heated, fluid expands, causing glass to break and thereby activating the sprinkler head.

Frangible pellet ■ A breakable release mechanism for a sprinkler head with a metal rod between the orifice cap and sprinkler frame. The rod is held in place by a pellet of solder under compression.

- *Cap:* The cap positioned over the sprinkler head orifice provides a watertight seal. It is held in place by the thermal linkage. Operation (failure) of the linkage causes the cap to move from its position and permits water flow. Caps are constructed solely of metal or a metal with a Teflon disk.
- *Orifice:* The machined opening at the base of the sprinkler frame is the orifice. Water flows from the piping and out of this opening to impact the deflector. Most orifice openings are .5 in. (12.7 mm) diameter, with smaller bores available for residential applications and larger openings for higher hazards.
- *Deflector:* The deflector can be found mounted on the frame opposite the orifice. The purpose of this component is to break up the water stream discharging from the orifice into a more efficient extinguishing pattern. Deflector types determine how the sprinkler is mounted. Common sprinkler mountings include upright (water flows upward toward the mounted deflector above the pipe), pendent (water flows downward toward the mounted deflector below the pipe), and sidewall (water is discharged in a lateral position from a wall). The sprinkler must be mounted as designed to ensure proper action. Selection of a particular style is often dependent upon physical building constraints.[14]

(See Figures 4.13 and 4.14.)

FIGURE 4.13 Upright sprinkler heads project upward into a space or opening. They are used to provide better coverage between a building's structural elements or obstructions. They provide a circle-shaped spray pattern.

SIDEWALL SPRINKLERS

Sidewall sprinkler heads are used when sprinklers cannot be installed in the ceiling. These heads stand out from a wall and have special deflectors designed to discharge most of their water away from the nearby wall. The spray discharge pattern is semicircular, with a small portion of the water directed at the wall behind the sprinkler itself. Unless specifically listed for use in ordinary-hazard occupancies, sidewalls are restricted to use in light-hazard occupancies only.

There are different types of sidewall sprinkler heads. The two major types are vertical (pendent and upright) and horizontal. Pendent and upright sidewalls look like standard sprinkler heads. The difference, however, is that their deflectors are bent to direct the discharge spray in their unique pattern. These sidewall heads cannot be installed in the dead air space* located at the corners of rooms, where the wall and ceiling meet. Vertical sidewall deflectors also must be placed no more than 6 in. 152 mm) below a ceiling.

Horizontal sidewall sprinkler heads are more common. These sprinkler heads are designed to be installed along a wall or the side of a beam. Horizontal sidewalls are generally

FIGURE 4.14 Pendant sprinkler head. Water flows downward out the orifice toward the mounted deflector.
Source: Ronald R. Spadafora

---

*Dead air space is said to be 4 in. (102 mm) down from the ceiling and 4 in. (102 mm) out from the wall (this is the same dead space encountered for smoke detectors).

FIGURE 4.15 Horizontal sidewall sprinkler heads stand out from a wall. They provide a half-circle spray pattern. Note the smaller secondary deflector that sprays water back toward the wall in order to protect it. Sidewall sprinkler heads are used when sprinklers cannot be located in the ceiling.

used in place of pendent or upright sprinklers because of aesthetics, building construction, or installation considerations.[15]

Horizontal sidewalls are given more leeway when it comes to installation positioning. Their deflectors may be closer than 4 in. (102 mm) from the wall on which they are located, and they may be placed as far as 12 in. (305 mm) below a noncombustible ceiling when they are Listed for such a purpose[16] (see Figure 4.15).

In-Rack Sprinkler System

The in-rack storage of commodities is dangerous from a fire protection standpoint. Rack storage provides protection from ceiling-mounted sprinkler heads. It also creates air pockets that allow the fire to continuously breathe, thereby fanning the flames, so to speak. Warehousing practices since the mid-1940s have changed significantly. Warehouses have become larger, increasing their rack storage heights and combustible fire load. Ceiling sprinkler systems proved inadequate in their attempt to reach the origin of the fire. This was especially true when the fire started on one of the lower tiers.

Specifically designed for the protection of racked storage areas, in-rack sprinkler systems are manufactured to release water into a targeted area to control fire and minimize damage and downtime. Although in-rack sprinkler systems cannot prevent fires, they can contain a fire when it starts in a specific area and control it within a short period of time. The sprinklers are installed along a network of piping secured within the rack storage structure for adequate protection throughout the entire storage area.

These types of fire suppression systems are used extensively in rack structures that are automated and use mechanized lifting equipment controlled by a computer for loading and unloading stock. Detectors are placed within close proximity to stored materials and equipment for the quickest possible response in the event of a fire.[17] In-rack sprinkler heads are typically equipped with large water shields over their deflectors.[18]

In-rack sprinkler systems, however, do have their disadvantages. They require a large capital outlay during the initial installation. They also limit the flexibility of the storage arrangement and are inconvenient when racks must be moved or rearranged. In addition, in-rack sprinklers are more vulnerable to being damaged as products are put into or removed from the racks.[19]

Early-Suppression, Fast-Response Sprinklers

In the 1980s, early-suppression, fast-response (ESFR) sprinkler systems were developed as an alternative to in-rack systems. They are designed to release two to three times the amount of water of conventional sprinkler heads. Conventional heads discharge water at a rate of about 20 to 40 gpm (76–151 L/min.). ESFR heads discharge water at 100 gpm (379 L/min.). Their droplets are also larger, which gives them greater momentum than droplets emitted from conventional heads. Therefore, more water reaches the fire, allowing the fire to be controlled or extinguished.

In general, ESFR systems can be used in warehouses with storage that does not exceed 40 ft. (12 m) in overall height and those with a ceiling height of less than 45 ft. (14 m). Designs that include a combination of both ESFR systems and in-rack sprinklers, however, may allow for storage above these heights. ESFR systems are designed to protect a wide array of commodities. This provides more flexibility in warehouse operations when compared to traditional types of sprinkler systems.

CHAPTER REVIEW

Summary

Fire engineering and safety experts generally agree that automatic sprinklers represent the single most significant fire suppression system of a fire protection program. Properly designed, installed, and maintained, systems can overcome deficiencies in risk management, building construction, and emergency response. They may also enhance the flexibility of building design and use by increasing overall safety.

Review Questions

1. What stopped the vertical spread of fire at the Meridian Bank Building (One Meridian Plaza) in February 1991?
2. Environmental conditions may exist that cause foreign materials to build up on the sprinkler heads. This buildup can prevent sprinkler heads from functioning properly. What is this buildup commonly called?
3. Explain how an automatic dry pipe sprinkler system accelerator (quick-opening device) operates.
4. Why are deluge sprinkler systems installed to protect high-hazard areas, such as chemical storage facilities?
5. Why does a preaction sprinkler system activate sooner than a dry pipe system?
6. Why must the water available to a centrifugal fire pump be always under pressure?
7. Regarding the OS&Y valve, what occurs when the stem of the valve rises?
8. How can a firefighter determine the position (open or closed) of a WPIV?
9. Why is in-rack storage of commodities dangerous from a fire protection standpoint?
10. How do early-suppression, fast-response (ESFR) sprinkler systems differ from traditional sprinkler systems?

Endnotes

1. John R. Hall, Jr., *US Experience with Sprinklers* (Quincy, MA: NFPA, 2013).
2. J. Gordon Routley, Charles Jennings, and Mark Chubb, *Highrise Office Building Fire One Meridian Plaza Philadelphia, Pennsylvania* (USFA-TR-049/February 1991), US Fire Administration/Technical Report Series (Washington, DC: USFA, 1991).
3. Church Mutual Insurance Company, "Sprinkler System Type Impacts Maintenance and Precautions," *Risk Reporter for Religious Institutions* 10, no. 4 (2011).
4. National Fire Protection Association (NFPA), *NFPA 13: Standard for the Installation of Sprinkler Systems* (Quincy, MA: NFPA, 2013).
5. John L. Bryan, *Automatic Sprinkler and Standpipe Systems*, 2nd ed. (Quincy, MA: NFPA, 1990).
6. ABCO Peerless Sprinkler Corporation, "Frequently Asked Questions," www.abcopeerless.com/learning/FAQs/FAQs.html.
7. Walter S. Beattie, "Dry Pipe Sprinkler Systems—Inspection, Testing, & Maintenance," http://waltbeattie.com/2012/04/14/dry-pipe-sprinkler-systems-inspection-testing-maintenance.
8. Fire Department, City of New York, *Study Material for the Examination for the Certificate of Fitness for S-12 Citywide Sprinkler System* (New York, NY: New York City Fire Department, 2012).
9. Vanguard Fire and Security Systems, Inc., "Foam Sprinkler Systems," last modified January 26, 2009, see the Vanguard Fire & Security Systems, Inc. website.
10. Fire-Tech Engineers India, "Pre Action Systems," www.indiamart.com/firetech-engineers/products.html.
11. International Code Council (ICC), *International Fire Code* (Washington, DC: International Code Council, 2012).
12. David T. Phelan, "Fire Department Connections," *Fire Engineering* 164, no. 3 (2011).

13. Phoenix Fire Protection, "Fire Sprinkler Head Components," www.phoenixfp.com/fire-sprinkler-questions/67-fire-sprinkler-head-components.
14. Fire Department, City of New York, *Study Material for the Examination for the Certificate of Fitness for S-11 Residential Sprinkler System* (New York, NY: New York City Fire Department, 2010).
15. Martin Cruz, "Sidewall Sprinkler Head Specifications," see the eHow website.
16. Glenn P. Corbett, "Sidewalls: The 'Superhuman' Sprinkler Head?" *Fire Engineering* 149, no. 6 (1996), 36.
17. National Fire Protection Association (NFPA), *NFPA 231C: Standard for Rack Storage of Materials* (Quincy, MA: NFPA, 1998).
18. Joseph Hankins, "Choosing the Right Automatic Sprinkler," *Disaster Recovery Journal* 5, no. 2 (1992), 56–60.
19. Arthur E. Cote and Percy Bugbee, *Principles of Fire Protection* (Quincy, MA: NFPA, 1988).

Additional References

Doug Chartier, "Controlling Microbiologically Influenced Corrosion in Fire Sprinkler Systems," *Fire Engineering* 157, no. 11 (2004), 116–117.

David R. Blossom, "Fire Department Operations at Sprinklered Properties," *Fire Engineering* 157, no. 10 (2004), 82–92.

Federal Emergency Management Agency (FEMA), "US Fire Administration's National Fire Incident Reporting System (NFIRS 5.0) Fire Data Analysis Guidelines and Issues," www.usfa.fema.gov/downloads/pdf/publications/nfirs_data_analysis_guidelines_issues.pdf.

Donald Garner and Gary Keith, "What Firefighters Should Know About Automatic Sprinkler Choices," *Fire Engineering* 147, no. 6 (1994), 54, 58, 60, 62–63.

Jeffrey A. Harwell, "Fire in Sprinklered Texas Warehouse with High-Piled Storage," *Fire Engineering* 165, no. 4 (2012), 55.

H.C. Kung, "A Historical Perspective on the Evolution of Storage Sprinkler Design," *Fire Protection Engineering* 1st quarter (2011).

Dean J. Maggos, "Understanding Residential Sprinkler Systems and Basic Salvage," *Fire Engineering* 159, no. 1 (2006), 207–208.

National Fire Protection Association (NFPA), *NFPA 13D: Standard for the Installation of Sprinkler Systems in One- and Two-Family Dwellings and Manufactured Homes* (Quincy, MA: NFPA, 2013).

National Fire Protection Association (NFPA), *NFPA 13R: Standard for the Installation of Sprinkler Systems in Low-Rise Residential Occupancies* (Quincy, MA: NFPA, 2013).

National Fire Protection Association (NFPA), *NFPA 20: Standard for the Installation of Stationary Pumps for Fire Protection* (Quincy, MA: NFPA, 2013).

National Fire Protection Association (NFPA), *NFPA 25: Standard for the Inspection, Testing, and Maintenance of Water-Based Fire Protection Systems* (Quincy, MA: NFPA, 2013).

John Norman, "Automatic Sprinkler Systems: The 'Silent Sentinels,'" *Firehouse* 15, no. 10 (1990), 30, 32, 34, 36, 38, 40.

Norman J. Thompson, *Fire Behavior and Sprinklers* (Quincy, MA: NFPA, 1964).

CHAPTER 5

Standpipe Fire Suppression Systems

Source: Ronald R. Spadafora

KEY TERMS

Standpipe system, *p. 75*
Fire brigade, *p. 78*
Pressure-reducing device, *p. 78*
Pressure-reducing valve, *p. 78*
Pressure-restricting device, *p. 78*
Pressure-control valve, *p. 78*
Residual pressure, *p. 80*
Check valve, *p. 86*
Gate valve, *p. 86*

OBJECTIVES

After reading this chapter, the reader should be able to:

- Describe the different types of standpipe systems
- List the classifications of standpipe systems
- Understand the lessons learned from the One Meridian Plaza fire
- Explain the water supply sources for standpipe systems
- Assess inspection, testing, and maintenance of standpipe systems

Introduction

A **standpipe system** is a fire safety system that is designed to provide firefighters or building personnel with rapid access to water via connection point outlets for fire hoses in the event of a fire. Standpipes may be installed as standalone systems, or they may receive their water from the same source piping as sprinkler systems. Standpipes are normally installed in tall or large area buildings. They can, however, be put in other types of structures as well, depending upon hazards. Standpipes are classified according to design use.

Provisions for standpipe installation requirements can generally be found in *Model Building and Fire Codes* as well as *NFPA 14: Standard for the Installation of*

Standpipe system ■ An arrangement of piping, valves, hose connections, and related equipment installed in a structure, with hose connections located in such a manner that water can be discharged in streams through attached hose lines and nozzles. Piping installed in a building serves to transfer water from a supply source to hose connections at one or more locations.

FIGURE 5.1 Typical standpipe system

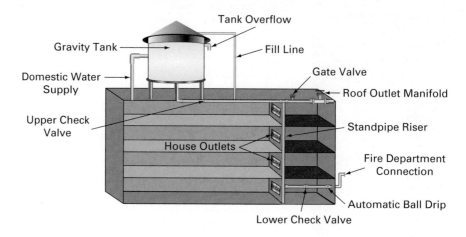

Standpipe and Hose Systems. Standards pertaining to the performance of care, use, servicing, and testing of standpipe systems and their components should be in accordance with *NFPA 25: Standard for the Inspection, Testing, and Maintenance of Water-Based Fire Protection Systems.*

Overview

The City of New York is the foremost "vertical" city in the United States and has more than one thousand high-rise buildings. Fires in these types of buildings are challenging, and firefighting success depends upon both the condition of the standpipe system and the experience and skill of the operators. Standpipe systems are also installed in locations with no access for fire department vehicles (such as parking garages and pier areas) or where excessive distance precludes the stretching of hose lines directly from engine apparatus.

Standpipe systems are an important part of fire protection in a building. They provide water that firefighters can manually discharge through hoses onto a fire. The piping system runs vertically and horizontally throughout the building (see Figure 5.1).

Types of Standpipe Systems

A standpipe system is intended to maintain sufficient water pressure inside a structure. It also is designed to provide enough water outlets for hose lines that hose stretches are not overly long and labor intensive. A standpipe system reduces liability in the event of a fire. Insurance companies usually offer a discount to owners of buildings with this kind of fire protection. NFPA 14 identifies various types of standpipes. A concise review of each type follows.

COMBINED (DUAL)

Combined standpipe and sprinkler systems carry water in a single riser for both fire protection systems. Each connection from a standpipe that is part of a combined system to a sprinkler system will have an individual control valve of the same size as the connection.

AUTOMATIC WET

A standpipe system that always contains water and that is attached to a water supply capable of supplying the system flow and pressure demand is known as an automatic wet standpipe. It requires no action other than opening a hose valve to provide water at hose connections. In some cases, a fire pump may be used to increase the water pressure. The wet standpipe system is the most commonly used in heated buildings in which there is no danger of the water in the piping freezing. Any part of the standpipe system that is exposed to freezing temperatures should be insulated.

MANUAL WET

A standpipe system that contains water at all times but relies on firefighters to supply the fire department connection (FDC) to meet system demand is known as a manual wet standpipe. Its water supply is not capable of meeting flow and pressure requirements. The purpose of the water supply is only to maintain water within the system and thereby reduce the time it takes to get water to the hose outlets

AUTOMATIC DRY

A standpipe system permanently attached to a water supply capable of supplying system demand is known as an automatic dry standpipe. This system is usually installed in a building that is not heated. It contains air or nitrogen under pressure in its piping. A dry pipe valve is installed to prevent water from entering the standpipe system and is designed to open when there is a drop in system air pressure. This action occurs when a hose outlet is opened. A control valve is installed at the automatic water supply connection. This valve should be kept open at all times. Special care must be taken when using an automatic dry standpipe system. The firefighter on the nozzle must never point it at the fire until all of the air has been bled from the system. Otherwise, pressurized air would be discharged onto the fire. This would cause the fire to burn more intensely.

MANUAL DRY

A standpipe system having no permanently attached water supply is known as a manual dry standpipe. It relies exclusively on the fire department apparatus to supply the FDC in order to provide the system with its required water flow and pressure demand. Pipes with no air or water in them feed the system. This system is commonly used in buildings that are not heated (see Figure 5.2).

FIGURE 5.2 Dry standpipe connection for building under construction.
Source: Ronald R. Spadafora

Chapter 5 Standpipe Fire Suppression Systems

Fire brigade ▪ Trained personnel, especially formed temporarily or called upon to assist a fire department in an emergency.

Pressure-reducing device ▪ Commonly referred to as a PRD, this device is installed at or near the standpipe hose outlet connection to limit static (nonflowing) and residual (flowing) water pressure. Older installations may have pressure-reducing devices with threads (mechanical-type) attached directly to the outlet. Firefighters should uncouple these devices prior to attaching their hose to the standpipe outlet.

Pressure-reducing valve ▪ Commonly referred to as a PRV, this device reduces water pressure under both static and residual conditions. It uses a spring-loaded valve assembly to modulate the position of the valve disc in the waterway.

Pressure-restricting device ▪ Limits water discharge through standpipe outlets by reducing pressure under flowing conditions. This is done via reducing the cross-sectional area of the hose outlet. Firefighters should use a screwdriver or similar tool to pry this device out from just inside the hose outlet prior to connecting their hose.

Pressure-control valve ▪ A pilot-operated device that uses water pressure within the system to modulate the position of a spring-loaded diaphragm within the valve to reduce downstream pressure under flowing and nonflowing conditions.

SEMIAUTOMATIC DRY

A standpipe system permanently attached to a water supply that is capable of supplying the system demand is known as a semiautomatic dry standpipe. This type of standpipe system is also known as a preaction system. Air is stored inside the pipes, which can be either pressurized or nonpressurized. Once an actuation device (manual pull station or a detector) is activated, water enters the system via the opening of a deluge valve. A manual valve is also installed at each hose connection outlet. Semiautomatic systems are designed for situations in which there is danger of serious water damage should the system piping become broken—for example, by moving equipment.[1]

Classes of Standpipe Systems

Standpipe systems are classified depending on who is expected to use the system. There are three classes of standpipe found within buildings: Class I, Class II, and Class III.

CLASS I

A Class I standpipe system provides a 2.5 in. (64 mm) hose connection for use primarily by trained personnel, **fire brigade** members, or by the local fire department. This class of system may or may not have a hose attached to its outlets (typically located in a stairwell, standing alone; see Figure 5.3), because the users will normally carry folded lengths of hose to the standpipe outlets from which firefighting operations will begin. Outlets can, however, be nonexposed (found inside cabinets). Due to the likelihood of excessive pressure on the lower floors being generated by fire department engines through the FDC as well as building fire pumps, these systems may be required to have a **pressure-reducing device** (see Figure 5.4), **pressure-reducing valve**, **pressure-restricting device**, or a **pressure-control valve** attached to help regulate the discharged pressure from the system. The fire hoses used from these outlets are 2.5 in. (64 mm) in diameter.

FIGURE 5.3 Class I standpipe outlet located inside a staircase of a high-rise building. *Source:* Ronald R. Spadafora

FIGURE 5.4 Mechanical pressure-reducing device threaded onto a standpipe outlet. *Source:* Ronald R. Spadafora

Engine company firefighters should, if possible, use only their department-issued hose for standpipe operations. Occupant-use hose is commonly small in diameter and therefore attached to the standpipe outlet by means of a reducer fitting. Additionally, in-house hose is often not regularly maintained and may fail under firefighting operational pressures. Ladder company firefighters may, under unusual circumstances, use this hose when necessary to protect life and/or facilitate fire suppression. When engine company firefighters encounter a reducer fitting, however, it must be removed to permit attachment of fire department 2.5 in. (64 mm) hose to the standpipe outlet.

 EIGHTH-ALARM FIRE

Chief Spadafora recalls arriving as a Fire Department of New York (FDNY) Battalion Chief on March 11, 1999, to a major structural fire at Broadway and 109th Street in Manhattan. The fire was in an occupied, fireproof, high-rise, multiple dwelling that was under large-scale renovation. The structure was 11 stories in height, and the fire originated on the first floor in a commercial occupancy (a restaurant). Fire spread onto scaffolding located at the rear of the building. The scaffolding was covered with combustible netting and ran the entire height of the building, spanning an open shaft and three apartments per floor. Upon arriving on the upper floors of the building, there was a dire need for water. The numerous floors and apartments involved in the fire (three apartments on every floor), initially precluded the luxury of having engine company firefighters working off of standpipes with fire department hose. Chief Spadafora ordered ladder company firefighters to operate in-house standpipe hose to keep the fire inside the apartments until engine company arrival. In-house hose lines also allowed ladder company firefighters to perform preliminary occupant searches inside apartments. The use of in-house hose at this fire contributed to a successful extinguishment operation.[2]

Residual pressure ■ A portion of the total available pressure that is not used to overcome friction or gravity while forcing water through pipe, fittings, fire hose, and adapters.

> **NOTE**
>
> Pre-1993 standpipe systems were designed for a maximum of 65 psi (448 kPa) at the uppermost or most hydraulically remote floor outlet. This took into account that two lengths (100 ft. or 30 m) of 2.5 in. (64 mm) diameter hose line with a 1 1/8 in. (29 mm) smooth-bore tip could flow 250 gpm (946 L/min).[3] Since then, standpipe system design has changed. In more recent editions of NFPA 14, the design pressure is based on a **residual pressure** of 100 psi (689 kPa). Fire protection engineers must be aware of local fire department needs and capabilities when approving standpipe systems. The local authority having jurisdiction (AHJ) is permitted to adjust pressure design based upon first-responder requirements. The design pressure should be based on the amount of hose and nozzle type the first arriving company will be deploying at a fire.
>
> Nozzle pressure is adjusted by use of a wheel handle at the hose outlet valve. This requires coordination between the engine company officers and firefighters both at the nozzle—advancing on the fire—and at the hose outlet valve. Radio communications are essential. As a rule of thumb, three lengths (150 ft. or 46 m) of 2.5 in. (64 mm) hose requires 70 psi (483 kPa) and four lengths (200 ft. or 61 m) of 2.5 in. (64 mm) hose requires 80 psi (552 kPa) at the outlet with water flowing.[4]

CLASS II

A Class II standpipe system (see Figure 5.5) is designed for use by the occupants of a building. A length of hose and the nozzle are connected to the standpipe outlets, making this system readily available for use. The Class II system allows occupants to safely escape the area of concern by using the hose to provide a safe passageway, protecting

FIGURE 5.5 Class II standpipe outlet located on a floor of commercial occupancy. *Source:* Ronald R. Spadafora

the exit route. The system is not designed for fighting fires in the traditional firefighter sense. The hose size is generally 1.5 in. (38 mm) in diameter. This smaller-diameter hose is less difficult to maneuver and control when in operation than 2.5 in. (64 mm) hose. Smaller hose, however, correlates to greater friction loss and the possibility of insufficient nozzle pressure and water flow onto the fire. Firefighters can, however, hook-up to this type of standpipe to extinguish a fire. This involves disconnecting the occupant hose, removing the reducer fitting, and exposing the standpipe's 2.5 in. (64 mm) outlet.

Most Class II systems are designed to provide a minimum pressure of 65 psi (448 kPa) and are limited to a maximum 100 psi (689 kPa) of residual pressure at the hose valve outlet. It is essential that firefighters disengage any PRVs that may be installed. Many Class II standpipe systems do not have a fire department connection in order to supplement the water supply at the hose outlet.[5]

PRESSURE REDUCING VALVES

The examination of the One Meridian Plaza fire provides valuable insight regarding the importance of understanding the workings of PRVs. Significant information from this fire relates to the vulnerability of fire protection systems that are installed to support fire-suppression efforts. A major problem in the operation of the standpipe system had significant ramifications, leading to life loss and large-scale fire damage.

Improperly installed standpipe PRVs provided inadequate pressure for fire department hose streams using 1.75 in. (44 mm) hose and automatic fog nozzles. PRVs were installed to limit standpipe outlet discharge pressures to safe levels. The PRVs, however, were set too low to produce effective hose streams. Tools and expertise needed to assist the Philadelphia fire department in their attempt to adjust the valve settings did not become available until too late.

Initial hose-line firefighters carrying out orders from the Tactical Command Post established on the 21st floor of the building were unsuccessful in extinguishing fire on the floors above. Firefighters were not able to enter onto the 22nd floor due to the intense heat and low water pressure. The PRVs at the standpipe outlets provided less than 60 psi (414 kPa) discharge pressure. This low pressure for the automatic fog nozzle being used was insufficient to develop effective fire streams. The PRVs at One Meridian Plaza were field adjustable through the use of a special tool. It was not until several hours into the fire, however, that a technician knowledgeable in the adjustment technique arrived at the fire scene to regulate the pressure on several of the PRVs in the stairways.

CLASS III

This system may be used by either professional firefighters or by occupants of the building. The hose may be adjusted to either 1.5 in. (38 mm) or 2.5 in. (64 mm) in diameter. This is done by attaching the proper fittings to the hose line. A Class III standpipe is a hybrid version of Classes I and II. This type of standpipe system is very common in buildings in which occupant load is consistent.[6]

Firefighting Water Supply Operations

Development of adequate hose streams using standpipe systems starts with the engine company chauffeur (apparatus pump operator) supplying the standpipe FDC with water at 100 psi (689 kPa) plus an additional 5 psi (34 kPa) per floor above grade to the outlet being used inside the structure. The 100 psi (689 kPa) takes into account the friction loss in two lengths of 3.5 in. (89 mm) diameter supply hose into the FDC as well as the friction loss due to path of water flow up the standpipe riser through piping and fittings. If more lengths of either supply or attack hose are required, the pump operator must increase the pump discharge pressure accordingly.

FIRST INTERSTATE BANK FIRE

On May 4, 1988, a fire destroyed five floors (floors 12 through 16) of the 62-story First Interstate Bank building in downtown Los Angeles. Each of the four stairwells of the building contained a combined (dual) standpipe with a pressure-reducing valve (PRV) at each landing. The four standpipes were supplied by two fire pumps, one diesel and one electric. The building's two fire pumps were shut down by a sprinkler contractor, however, to facilitate connecting a new sprinkler system. Additionally, the combined standpipe system was drained. Low water pressure hampered initial interior attack firefighting operations. It was estimated by fire department officials that more than 2,500 gpm (9,464 L/min) was delivered by multiple hose lines that were supplied by engine company apparatus pumps via the building's fire department connections and that it took nearly 400 firefighters to control the fire. A lesson learned from this fire was that fire department engine apparatus must pump at sufficient pressure to enter the building's combined standpipe system. In the case of the First Interstate Bank fire, the system pressure was set at 585 psi (4,033 kPa).[7]

NOTE

Fire department engine apparatus pumpers must operate at pressures that provide effective hose streams. There are two classifications, conventional and high-pressure, as discussed ahead.

Conventional pumpers: Two-stage 1,000 gpm (3,785 L/min.) or 2,000 gpm (7,571 L/min.) pumpers.

High-pressure pumpers: A pumper with a third-stage capability. The third stage can supply 500 gpm (1,893 L/min.) at 700 psi (4,826 kPa).

Apparatus with two-stage pumps when operated in series position (first-stage impeller discharges into the second-stage impeller) are used to develop higher pressures at lower engine revolutions per minute (rpm). Fire departments in cities with very tall buildings, such as New York, have high-pressure pumpers stationed in high-rise districts. Similarly, the Chicago fire department has pumpers with high-pressure capability. Standard operating procedures (SOPs) recommending pump pressures in excess of 500 psi (3,447 kPa) into FDCs have been established.

Water Supply

There are multiple sources of water that are potentially available to firefighters during a fire. These water supply sources are discussed ahead.

PUBLIC WATERWORKS CONNECTION

The street main supplies water using the water pressure in the public water works system. Sometimes a street main may not be connected to the system if it is located too far away from the building.

FIRE PUMPS

The fire pump is usually connected to the public water main for most of its water supply. Fire pumps are designed to take the water from a supply source and then discharge the water into the standpipe system under pressure. They can be used as primary or auxiliary

FIGURE 5.6 Brick and steel trestle supporting a wooden gravity tank. *Source:* Ronald R. Spadafora

sources of water supply for standpipe systems. They may be automatic or manually activated. Fire pumps can also be designed to draw water from a suction tank and deliver it into the system when needed. They may also be used in combination with gravity tanks that supply a standpipe system. Moreover, an FDC can be connected to a fire pump.

When manually activated pumps are installed, they are used in combination with a gravity tank and/or a pressure tank. These tanks are designed to operate when there is a pressure drop within the fire protection system. The operation of the gravity or pressure tank and the initiation of its supervisory signal should trigger the fire pumps being activated by personnel on the premises. Alternate sources of power requirements are determined as per *NFPA 20: Standard for the Installation of Stationary Pumps for Fire Protection*.

GRAVITY TANKS

Gravity tanks are used for water storage and supply. They are made of wood, steel, or concrete. Gravity tanks are used as a primary or secondary water supply source for standpipe systems. A gravity tank uses the force of gravity to deliver water to the standpipe system. Tanks may be located on the tops of buildings or raised on tall supporting towers, situated at least 25 ft. (8 m) above the highest standpipe hose outlet that it supplies (see Figure 5.6). To ensure adequate hose line pressure at the top floors of buildings, gravity tanks are sometimes used in combination with pressure tanks (see Figure 5.7).

The water pressure in a gravity tank system depends on the elevation of the tank. This is a major advantage over other kinds of systems. For every 1 ft. (0.3 m) the tank is above the discharge outlet, an additional 0.433 psi (3 kPa) of water pressure is achieved. A full tank of water is needed to ensure that the standpipe system works properly during a fire. Gravity tanks range in capacity from 5,000 gallons (18,927 L) to more than 100,000 gallons (378,541 L). Keeping the tank full also prevents wooden tanks from shrinking and steel tanks from rusting.

It is best if gravity tanks are used only for fire protection and for no other purpose. Tanks used for domestic purposes and fire protection need to be refilled more often. These

FIGURE 5.7 Combination gravity tank and pressure tank installation

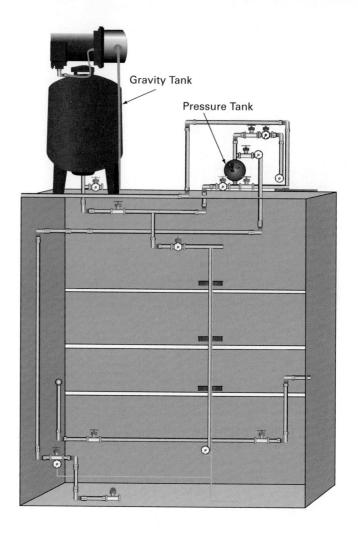

dual purpose tanks become settling basins for sediment mixed in with the water. This sediment can then be drawn into the piping, resulting in the standpipe system becoming clogged. This can lead to reduced flow within the standpipe system.

Gravity tanks may be exposed to very low temperatures. During winter months in cold climates, all parts of the gravity tank must be insulated or heated to keep the water from freezing. Several methods are used to heat the tank and the pipe that supplies the water. Hot water may be circulated by gravity, or steam can be discharged directly into the tank. In addition, steam coils can be installed inside gravity tanks in order to protect them. The tank can be severely damaged if the water inside the tank freezes. The temperature of the water should always be at least 40°F (4°C). Ice should not be allowed to build up on the gravity tank. The extra weight of the ice can weaken its supports and cause the tank to collapse without warning. Moreover, falling icicles may cause damage or injury. It is essential that the tank is properly heated, insulated, and carefully maintained.[8]

PRESSURE TANKS

Pressure tanks (see Figure 5.8) are enclosed water vessels of limited size. Air pressure in the tank permits the discharge of water into the standpipe system. A pressure tank may be a primary or secondary water supply source for a standpipe system. A pressure tank is normally installed in an enclosed structure. The temperature in the enclosure is kept at 40°F (4°C) or above. The heated structure may be located on any floor in the building or

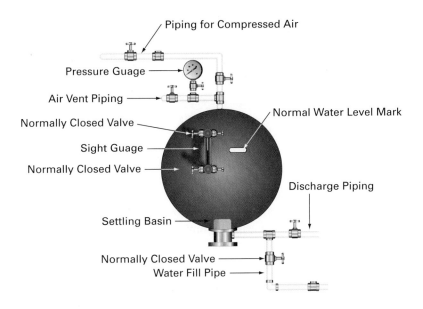

FIGURE 5.8 A standard pressure tank

on the roof in a tank house (see Figure 5.9). Pressure tanks are usually kept approximately two-thirds full of water and one-third full of pressurized air. The acceptable air pressure inside the tank may vary from 15 psi to 80 psi (103 to 552 kPa). An air compressor with automatic controls is required for maintaining the air pressure. Pressure tanks will be provided with approved, closed-circuit high- and low-water and high- and low-air alarms. The maximum capacity of pressure tanks is typically 9,000 gallons (34,069 L). Some standpipe systems require more water. If necessary, several pressure tanks can be used in combination to supply the system. The inside of pressure tanks must be inspected every few years. They need to be maintained free of rust and foreign materials.[9]

FIGURE 5.9 Pressure tanks inside a tank house.
Source: Ronald R. Spadafora

Fire Department Connection

Similar to a sprinkler system, a fire department connection is installed on a standpipe system. The connection is used by the fire service to pump water into the system. FDCs must always be accessible, and each connection should be equipped with a **check valve** to keep water from flowing back down and into the supply hose. Some FDCs have double check valve assemblies (two check valves with tight seals in a series between two control valves) located inside the building. The chances of both check valves leaking at the same time is remote; therefore, a double-check-valve system provides better protection against water backflow, which can distort the color, taste, and smell of the public water supply. Double check valves must also be accessible for easy examination and maintenance.

For automatic standpipe systems, the FDC allows the apparatus pump operator to provide supplemental water into the standpipe system. For manual dry standpipe systems, however, the FDC is the primary means of supplying water into the system by the fire service. Hose outlets at each floor can also be used by firefighters to supply standpipes under circumstances in which the FDC is clogged and nonfunctional.

> **Check valve** ■ A one-way valve that allows fluid to flow through it in only one direction.

Valves

Different types of valves operate differently and serve different functions. These valve types are discussed ahead.

HOSE OUTLET VALVES

At selected locations in the building, the outlet piping is connected to a hose. These connections are controlled by hose valves. The hose valve must be manually opened by the firefighter. No water is allowed into the hose until the valve is opened.

CHECK VALVES

A check valve is used to prevent backflow for safety reasons and to ensure that the water flows in the system in only one direction.

SWING CHECK VALVES

These valves also prevent backflow. They allow full, unobstructed flow and automatically close as pressure decreases. They fully close when water flow stops. A swing check valve is normally used in systems with a gate valve, because of the low pressure drop across the valve.

ALARM CHECK VALVES

This type of check valve lifts from its seat when water flows into a standpipe system. Devices attached to these valves are designed to initiate an alarm by a drop in water pressure in the standpipe system. The alarm signals the occupants of the building that the standpipe system has been activated.

GATE VALVES (NONRISING STEM)

Gate valves of the nonindicating type (nonrising stem) are provided in water distribution systems to allow for segments of the standpipe system to be shut off for repairs or maintenance without reducing protection over a wide area. Such valves typically require a special key wrench to operate. A valve box is located over the valve to keep dirt from it and to provide a convenient access point for the valve wrench to reach the valve nut.

> **Gate valve** ■ A valve that opens by lifting a round or rectangular wedge (gate) out of the path of the fluid.

OUTSIDE SCREW AND YOKE GATE VALVES

Similar to sprinkler systems, Outside Screw and Yoke (OS&Y) gate valves may be installed at several locations along the system. The OS&Y valves can be used to shut down the entire system (main valve) or just a part of the standpipe system (section valves).

DRAIN VALVE

Each standpipe system may also be fitted with a drain valve. The drain valve is located at the lowest point on the standpipe system. The drain valve is opened when the standpipe system has been used, tested, or repaired.

Hose Stations

Hose stations are located at each story of the structure. They are typically required by code to be of sufficient number that all portions of the building can be reached by hose streams from nozzles attached to approximately two lengths (100 ft. [30 m]) of hose. Hose outlets are protected by metal caps. At each hose outlet supplied by a standpipe system, hose is usually stored on a rack firmly supported and placed between 5 to 6 ft. (1.5–1.8 m) above the floor or landing. The racks should be located where the sun or excessive heat will not damage the hose.

A nozzle designed to discharge water is commonly attached to the lead end of the hose. In general, nozzles on 2.5 in. (64 mm) hose will have a smooth-bore, 1 in. (25 mm) or 1 1/8 in. (29 mm) tip. If this type of installation may result in damaged equipment, the hose and nozzle may be stored inside a fire cabinet readily located near the hose outlet valve. Fire cabinet doors may have a transparent viewing panel to allow the hose and equipment to be clearly visible. A permanently attached sign with distinctive, colored letters reading "FIRE HOSE" is required to indicate the equipment contained therein.

Standpipe Kit

A standpipe kit is a bag or carrying case designed to hold the tools and adapters needed to make a standpipe hookup an efficient operation for firefighters (see Figure 5.10). Although lengths of hose do not necessarily fit into the kit, they can be considered part of it. Four lengths of 2.5 in. (64 mm) hose, bundled to allow a free hand for navigating staircases, is essential to ensure that firefighters have enough hose to reach the seat of the fire.

Alarm Devices and Supervisory Switches

Standpipe systems are designed with features that provide important information concerning system activation and control valve closure. Water flow alarm devices send signals upon water movement through piping to a fire alarm control unit (FACU) for alarm appliance activation. Tamper switches emit a supervisory signal upon control valve closure to notify building personnel of shutdown.

Water Flow Alarms

Water flow alarms give warning to occupants and building personnel of the actual flow of water in piping. Water-operated alarm devices must be located near the alarm valve, dry pipe valve, or other water control valves in order to avoid long runs of connecting pipe.

Supervisory (Tamper) Switches

Tamper switches provide valve security by electronically supervising control valves through the fire alarm system. They initiate a signal if the control valve is closed. Tamper

FIGURE 5.10 The contents of a standpipe kit include hose spanners, in-line pressure gauge, hand control wheel for outlet valve, vice grips, adjustable pipe wrench, fittings, control nozzle with 1 1/8 in. smooth-bore tip, subway emergency exit key, and a wooden chock.
Source: Ronald R. Spadafora

switches do not, however, prevent the closing of the valve. In addition, they can become defective and inoperative with no visible indication of a malfunction.

Inspection, Testing, and Maintenance

INSPECTION

Standpipe systems must be inspected regularly to ensure serviceability. The inspector should check visually that the system is free from corrosion and that there is no obvious physical damage to the system. Special effort must be made to detect evidence of tampering with the system and missing components. A visual inspection of all parts of the fire pump and the controlling equipment should also be conducted. This inspection includes the condition and reliability of the power supply. If any problems are discovered with the equipment, immediate action must be taken to correct them. All valves and connections to water supply sources should be inspected and checked to ensure that they are in the correct position.

Hose outlets are examined to see if pressure-restricting devices (PRD) are present. In addition, hose should be visibly inspected to ensure it is in good condition, and that all attached equipment is operable. When examining hose, cuts, abrasions, and dried-out gaskets could cause it to be placed out of service. Fire cabinets must be unobstructed.

TESTING

FDCs for a standpipe system are typically required to be hydrostatically tested by code at standard time intervals. The standpipe system itself is also subject to flow and pressure tests to demonstrate its functionality for fire department use. All major defects in the system discovered during testing should be immediately reported to the building owner. High-priority defects include an empty water supply tank, a break or a leak in the system's piping, an inoperative water supply valve, a defective FDC, or any repair necessitating a partial or total shutdown of the system. Fire pumps must be fully tested to make sure that the pump, motor driver, controller, power supply, and all other components are working properly.

FIGURE 5.11 A standpipe manifold located at the top floor of a fireproof building, just inside the stairwell. *Source:* Ronald R. Spadafora

At the top of the highest riser a three-way manifold equipped with three 2.5 in. (64 mm) hose valves with hose valve caps should be provided. This roof manifold is used for extinguishing fires on the roof or in adjacent buildings. It is also used when testing the water flow in the standpipe (see Figure 5.11).

MAINTENANCE

NFPA 25 is the standard used for inspection, testing, and maintenance of standpipe systems. Compliance with the standard helps to ensure serviceability and effectiveness during fire incidents. Testing requirements include hose outlets, pumps, piping, and valves.

One or more competent, certified employees are required to perform the maintenance tasks. A detailed record should be kept of each assignment for examination by any representative of the local fire department. Fire pumps are maintained in order to ensure the reliable operation of the pump in the case of an emergency. The pump should be activated regularly according to the manufacturer's specifications to verify that it is working properly. Care should be taken to make sure that the pump does not overheat while conducting the test. A centrifugal pump relies on the water supply for cooling and lubrication. Therefore, it should never be operated if the pump is not being supplied with water. The operating stems of (OS&Y) valves should be lubricated as needed. The valve must be completely closed and reopened, not only to check its operation but also to distribute the lubricant throughout the valve.

Gravity tanks and pressure tanks also require regular maintenance. Tank volume must not fall below its designed water level. Maintenance duties may include painting a steel tank on a regular basis to prevent rusting. Prior to painting, all loose, old paint, rust, scale, and other surface contamination should be removed and patch repair made where needed. Wooden gravity tanks require hoop and grillage upkeep. In addition, silt must be removed from gravity tanks during interior maintenance operations to avoid accumulation to the level of the tank outlet.

CHAPTER REVIEW

Summary

A standpipe is a rigid water pipe system—built into a structure in a vertical position—to which fire hoses can be connected. Standpipe systems allow water to be applied manually on a fire from nearly any location. They are commonly found in tall and large floor area buildings. Some older buildings have a designated riser to supply their standpipe system, whereas many newer buildings have a combined system with a single riser supplying both the standpipe and the sprinkler system. Standpipes can be automatic, with their own water supply, or manual dry, depending totally on the local fire service to flow water through the system. Standpipe classifications include systems designed for firefighters and trained personnel as well as standpipes for use by untrained occupants.

Review Questions

1. What type of standpipe system always contains water and is attached to a water supply capable of supplying the system demand?
2. How is an automatic dry standpipe system designed to open?
3. What classification of standpipe provides a 2.5 in. (64 mm) hose connection for use primarily by trained personnel, fire brigade members, or the local fire department?
4. How does a pressure-restricting device limit water discharge through standpipe outlets?
5. What was the lesson learned regarding combined (dual) standpipe systems at the First Interstate Bank fire?
6. Why is hose 1.5 in. (38 mm) in diameter generally attached to standpipe outlets of a Class II standpipe system?
7. What was the major problem in the operation of the standpipe system at the historic One Meridian Plaza fire?
8. What do gravity tanks rely on to move their water?
9. Why is it best that a gravity tank be used for no other purpose but fire protection?
10. In general, what is the maximum water capacity of pressure tanks?
11. What is the primary means of supplying water to a manual dry standpipe system?
12. Gate valves installed in standpipe systems provide what major benefit?
13. Fire cabinets contain what kind of firefighting equipment?

Endnotes

1. Minnesota State Fire Marshal, "Quick Response—Standpipes," https://dps.mn.gov/divisions/sfm/programs-services/Documents/Quick%20Response%20Newleter/QR1106Standpipes.pdf.
2. Pat McKnight, "11 Floors of Fire," *WNYF* 59, no. 2 (1999).
3. Dan Speigel, "Standpipe Operations: The Basics," *Fire Engineering* 159, no. 10 (2006).
4. John P. Grasso, "The In-Line Pressure Gauge," *Fire Engineering* 149, no. 2 (1996).
5. LAFIRE.com, "Los Angeles Fire Department Historical Archive," www.lafire.com/famous_fires/1988-0504_1stInterstateFire/ExSummary/LAFD-ExecutiveSummary.htm.
6. Mark Van Der Feyst, "Standpipe Systems," *Firefighting—In Canada*, www.firefightingincanada.com/content/view/13831/213/.
7. Glenn P. Corbett, "Standpipe Systems," *Fire Engineering* 146, no. 3 (1993).
8. Ronald R. Spadafora, "What Firefighters Should Know About Gravity Tanks," *WNYF* 72, no. 2 (2012): 18–21.
9. Ronald R. Spadafora, "What Firefighters Should Know About Pressure Tanks," *WNYF* 72, no. 3 (2012): 18–19

Additional References

Fire Department of the City of New York (FDNY), *Study Material for the Certificate of Fitness for Consolidated Exam for Standpipe Systems With Pressure Tanks, Sprinkler Systems With Pressure Tanks, And Sprinkler Systems With Pressure Tanks Complex-Wide (F-96)* (New York, NY: New York City Fire Department, 1999).

Fire Department of the City of New York (FDNY), *Study Material for the Examination for the Certificate of Fitness for Standpipe System, (S-13) City Wide Standpipe System (Excludes Personal Supervision of Multizone Systems) and (S-14) Standpipe for Multi-Zone System (Personal Supervision of Multizone Systems F.C.905.1.1)* (New York, NY: New York City Fire Department, 2013).

Dave Mettauer, "Standpipe Systems Explained," *Firewatch* December (2007): 8–9.

National Fire Protection Association (NFPA), *NFPA 14: Standard for the Installation of Standpipe and Hose Systems* (Quincy, MA: NFPA, 2013).

National Fire Protection Association (NFPA), *NFPA 25: Standard for the Inspection, Testing, and Maintenance of Water-Based Fire Protection Systems* (Quincy, MA: NFPA, 2011).

CHAPTER 6

Foam and Foam Fire Suppression Systems

Source: Ronald R. Spadafora

KEY TERMS

Foam, *p. 92*
Boilover, *p. 93*
Foam concentrate, *p. 95*
Foam solution, *p. 95*
Heat resistance, *p. 95*
Burnback resistance, *p. 95*
Fuel resistance, *p. 95*

Surfactant, *p. 96*
Knockdown, *p. 96*
Alcohol resistance, *p. 96*
Solvent, *p. 97*
Polymer, *p. 97*
Foam chamber, *p. 98*
Foam monitor, *p. 98*

Foam maker, *p. 99*
Wildland/urban interface, *p. 102*
Overhaul, *p. 103*
Thermocouples, *p. 104*
Venturi effect, *p. 107*

OBJECTIVES

After reading this chapter, the reader should be able to:

- List the physical characteristics of foam
- Explain the general principals of foam extinguishing systems
- Analyze the differences between chemical and mechanical foam; protein and synthetic foam; low- and high-expansion foam; and Class A and Class B foam
- Describe compressed air foam system (CAFS) technology
- Review the general principles of foam extinguishing systems
- Understand fixed foam systems and their components
- List elements of portable foam systems and firefighting

Foam ▪ A firefighting foam is a stable mass of small, air-filled bubbles with a lower density than oil, gasoline, or water. Foam is made up of three components (water, foam concentrate, and air). When mixed in the correct proportions, these three ingredients form a homogeneous foam blanket.

Introduction

Foam has been used as a fire extinguishing medium for flammable and combustible liquids since the early 1900s. A properly formed and applied foam blanket can extinguish a flammable or combustible liquid fire by the combined mechanisms of cooling, separating

92

the flame/ignition source from the product surface, suppressing vapors, and smothering. It can also be used proactively in fuel-spill situations to float over the hazard and secure the scene, thereby facilitating mitigation. Unlike foam, water is heavier than many hydrocarbon fuels. When applied directly on a typical hydrocarbon fuel fire or spill, water will sink to the bottom and have minimal beneficial effect on extinguishment or vapor suppression. In addition, if the liquid fuel heats up to above 212°F (100°C) the water may boil below the fuel surface. This reaction is known as **boilover**. It can lead to fuel exploding out of its contained area and spreading the fire. Firefighting foam is an important fire extinguishing agent for all potential ordinary combustible and flammable/combustible liquid hazards.

Boilover ■ The expulsion of hot liquid from a storage tank as a result of water boiling below the fuel surface.

Overview

Firefighting foam consists of small, air-filled bubbles with a lower density than water, oil, or gasoline. It consists of three components: water, foam concentrate, and air. Foam concentrate is proportioned into water to create a foam solution. This solution is then commonly mixed with air (aspirated) to produce firefighting foam, which flows readily over fuel surfaces.

Typical use locations, including aircraft hangars and loading racks, require an automatic detection and control system. In these areas, heat and flame detectors are installed to provide activation information to a fire alarm control unit (FACU).

Physical Characteristics

To be effective, foam must contain the right mix of physical characteristics. Important properties for successful extinguishment include the following:

- *Free flow:* Ability of the foam blanket to spread across a fuel surface and around obstacles to achieve complete coverage.
- *Consistency:* Good cohesion properties are needed to maintain the blanket effect but not be so thick as to hinder its ability to flow.
- *Expansion:* The expansion of the solution is critical to the effectiveness of foam. The foam's ability to retain its water content relies on the integrity of the blanket bubbles. A foam blanket that is not properly expanded will drain rapidly, providing negligible protection against the heat of the fire.
- *Heat resistance:* Demonstrates strong opposition to the destructive effects of heat radiated from a fire as well as hot objects.
- *Fuel resistance:* Minimal fuel pickup to withstand saturation and burning.
- *Vapor suppression:* Foam must be capable of containing flammable vapors generated by the fire in order to break the fuel-oxygen-heat components of the fire triangle and to minimize the risk of reignition.
- *Alcohol resistance:* Alcohol fuels have a high affinity to water. Foam blankets that are not alcohol-resistant will be destroyed, because their makeup is more than 90% water.

General Principals of Foam Extinguishing Systems

Foam is used for firefighting purposes by forming a blanket on the surface of burning liquids or spills. This covering prevents flammable vapors from leaving the surface and inhibits oxygen reaching the fuel. The fire will cease to exist when the fuel and oxygen are separated by a properly applied foam blanket. Additionally, the water content in the foam has a cooling effect not only on the flaming fuel but on surrounding materials. This aids in the prevention of reignition. Either fresh water or seawater may be used. Water containing contaminants, such as detergents, oil residues, or corrosion inhibitors, may have a negative effect on foam quality. Foams are generally more stable when generated with lower water temperatures. The typical concentrate works best when water temperature range is between 35°F and 80°F (1.7°C and 27°C; see Figure 6.1 and Table 6.1).

FIGURE 6.1 Foam extinguishes fire in several ways: Cooling (water content); separating flame/ignition source from product (fuel) surface; suppressing vapors; and smothering (excluding oxygen in the air).

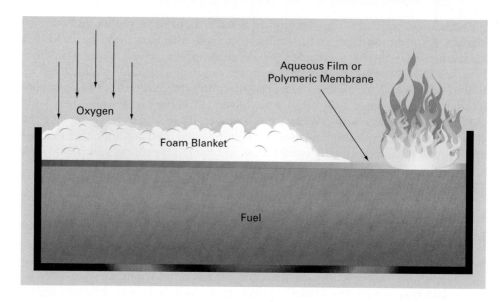

TABLE 6.1	How Foams Extinguish Fire
Cooling (due to water content)	
Smothering (by preventing air and flammable vapors from combining)	
Separating (by intervening between the fuel and the fire)	
Suppressing (by preventing the release of flammable vapors)	
Insulating (from radiant and convection heat by means of dead air space in bubbles)	

Foam should be light enough to float on flammable liquids yet heavy enough to resist winds. If flammable liquids have spilled, fires can be prevented by covering the spill promptly with a foam blanket. Additional foam may be necessary from time to time to maintain the blanket for extended periods until the spill has been mitigated. Foam should ideally flow freely and quickly cover burning fuel. It should also have good cohesive properties in order to stay together to form a vapor-sealing foam blanket. Successful foam blankets retain enough water to provide a sustained seal. Rapid loss of water causes the foam to dry out and break down from high heat. Clean air should be used to aerate the foam solution.

Foam should not be used on electrical fires. If, however, electrical current can be shut down, then a foam blanket can be used to extinguish the resulting fire, which is typically Class A or Class B. Foam also is not recommended for use on materials that may be stored as liquids but are normally gases under ambient temperatures. Examples include propane and vinyl chloride. Foam should also not be employed on Class D materials that react with water, such as magnesium, titanium, potassium, lithium, calcium, zirconium, sodium, and zinc.

Water streams physically disrupt a foam blanket, and they should only be used for cooling adjacent surroundings. Plunging a foam stream directly into the fire can splash the fuel, causing the fire to spread and combustible vapors to escape. Foam nozzle application should be a gentle (spray-type) stream to reduce the mixing of foam and fuel. Straight foam streams directed into the middle of a fuel spill should be avoided.

Chemical and Mechanical Foam

There are two basic types of foam: chemical and mechanical. Chemical foam is considered obsolete today because of the many containers of powder required to extinguish even small fires. Chemical foam is a mixture of two powders, sodium bicarbonate (an alkali) and aluminum sulfate (an acid), water, and a stabilizer produced in a foam generator. The stabilizer was added to give the foam a tenacious quality and to increase its shelf life. Premixed foam powders were stored in cans and introduced into the water stream during hose-line firefighting operations. A device called a foam hopper was used for this application. Handheld, portable foam fire extinguishers used the same two chemicals in solution. The extinguisher was activated by inverting it, allowing the chemicals to mix and react. The foam created was a stable solution of small bubbles containing carbon dioxide (CO_2).[1]

In the 1940s, another type of foam was developed. A liquid protein–based **foam concentrate** (soy protein) was mixed with water. This type of foam is called mechanical foam. It is also produced by mixing a foam concentrate with water at the appropriate concentration. This **foam solution** is then aerated and agitated to form a bubble structure.

Protein and Synthetic Foam

Protein foams contain natural proteins as the foaming agents. Unlike synthetic (man-made) foams, protein foams are biodegradable. They flow and spread slower but provide a foam blanket that has more **heat resistance** and is more durable. Protein foams include regular protein foam, fluoroprotein foam, film-forming fluoroprotein foam, and alcohol-resistant film-forming fluoroprotein foam.

Regular protein (P) foam has been used since World War I. It is available in either a 3% or a 6% concentrate. It is derived from naturally occurring sources of protein, such as hoof and horn meal or feather meal. The protein meal is hydrolyzed in the presence of lime and converted to a mixture of amino acids. Additional components are added, such as foam stabilizers, corrosion inhibitors, antimicrobial agents, and freezing-point depressants. It will produce highly stabilized mechanical foam. Because of this stability, however, it is slow moving when used to cover the surface of a flammable liquid. Regular P foam must always be used with an air-aspirating type of discharge device. When antifreeze is added, foam can be produced in subfreezing temperatures down to –10°F (–23°C).

FIREFIGHTING FOAM: CONCENTRATE RATIOS

Foam concentrates are designed to be mixed or proportioned with water at specific ratios. Six-percent (6%) foam concentrates are mixed with water at a ratio of 94 parts water to six parts foam concentrate.

EXAMPLE:
To make 100 gallons (379 L) of foam solution, you would need to mix 6 gallons (23 L) of foam concentrate with 94 gallons (356 L) of water.
Using this example with 3% foam concentrate, what is the ratio of water to foam concentrate?

Answer: Mix 3 gallons (11L) of foam concentrate with 97 gallons (367 L) of water.

Foams derived from protein concentrates have several advantages. In general, they have excellent **burnback resistance** capabilities. In addition, they have good heat stability and resist reignition very well. They are used successfully in the extinguishment of Class B (hydrocarbon) fires and can be found in fire protection systems protecting flammable and combustible liquids. Drawbacks of protein foam include lack of fluidity and limited **fuel resistance**. These negative characteristics correlate to reduced effectiveness and a rapid breakdown of the foam blanket.[2]

Fluoroprotein (FP) foam was developed in the 1960s by National Foam Inc. It is also available in either a 3% or 6% concentrate. This product is manufactured using the same

---

Foam concentrate ■ A liquid concentrate supplied from the manufacturer that, when mixed with water in the correct proportion, forms a foam solution. Two of the most common concentrations are 3% and 6% foams. These values are the percentages of the concentrate to be used in making the foam solution. For example, if 3% concentrate is used, three parts concentrate must be mixed with 97 parts water to make 100 parts foam solution. If 6% concentrate is used, six parts concentrate must be mixed with 94 parts water.

Foam solution ■ A solution of water and foam concentrate after they have been mixed together in the correct proportions.

Heat resistance ■ The ability of a foam to resist the destructive effects of heat radiated from any remaining fire from the liquid's flammable vapor and any hot metal wreckage or other objects in the area.

Burnback resistance ■ The ability of the finished foam to resist direct flame impingement such as would occur with partially extinguished petroleum fire.

Fuel resistance ■ The ability of a foam to resist fuel pickup so that the foam does not become saturated and burn.

Surfactant ■ A performance additive that can greatly reduce the surface tension of water when used in very low concentrations.

Knockdown ■ The time required for a foam blanket to spread across a fuel surface or around obstacles in order to achieve complete extinguishment.

Alcohol resistance ■ The ability of the foam blanket to defend against an alcohol's affinity for water. A foam blanket is more than 90% water.

method as regular P foam but with the addition of fluorocarbon **surfactants**. The addition of these surfactants in the concentrate improves the performance over P foam. FP foam is more resistant to fuel contamination/pickup due to its oil-rejecting properties; therefore, it can be discharged directly onto the fuel surface, and the foam blanket will not become saturated by fuel vapor.

Moreover, FP foam can be delivered via a subsurface method of forcing expanded foam into the base of a storage tank containing a hydrocarbon fuel. The expanded foam enters via the bottom of the tank and then floats up through the fuel to the surface, where it covers the liquid with a foam blanket.* In general, it is better than regular P foam, because its longer blanket life provides better safety when entry is required by firefighters for rescue. FP foam has fast **knockdown** characteristics and is compatible with dry chemical extinguishing agents that typically destroy regular P foam. Air-aspirating discharge devices are necessary for its use, creating a fluid foam blanket. When the water is mixed with antifreeze, it produces foam in subfreezing temperatures.

Film-forming fluoroprotein (FFFP) foam is a derivative of synthetic, aqueous film-forming foam (AFFF) and FP. These types of concentrates are based on FP formulations to which an increased quantity of fluorocarbon surfactants has been added. FFFP foam concentrates were developed with the goal of quick fire knockdown similar to AFFF for certain applications. They are able to apply a very thin film atop a fuel spill to prevent the release of combustible vapors. FFFP is not as effective as AFFF on aircraft crash and highway fuel-spill fires, however. FFFP foam also does not have the burnback resistance of FP foam when used on fuel-in-depth fires.

FFFP, when properly applied, however, offers finished foam that is long lasting, has good fuel-shedding ability, and has good resistance to heat. It is also compatible with dry chemical extinguishing agent in the event that a coordinated dual attack is required. FFFP foam can be generated with either air-aspirating or non-air-aspirating nozzles. When used through a non-air-aspirated nozzle, FFFP foams do not provide expansion ratios as good as those of AFFF when used through the same type of nozzle. They are available in either a 3% or 6% concentrate.

Alcohol-resistant film-forming fluoroprotein (AR-FFFP) foam builds on the positive attributes of regular P foam and FP foam. This protein foam has synthetic additives as part of the concentrate. It can be used on fires and spills involving hydrocarbon fuels. AR-FFFP is designed with **alcohol resistance** in mind. This concentrate is used at incidents involving polar solvents (liquids with a strong affinity for water). AR-FPPP is produced from a combination of synthetic stabilizers, foaming agents, fluorochemicals, and alcohol-resistant membrane-forming additives.

Polar solvents (acetic acid) and water-miscible fuels, such as alcohols (methanol and ethanol), ethers, and acetones, are destructive to non-alcohol-resistant foams. AR-FFFP (3%) acts as a conventional AFFF on hydrocarbon fuels, forming an aqueous film on their surface. On alcohol-type fuels, the polymeric membrane of AR-FFFP (6%) separates the foam from the fuel and prevents the destruction of the foam blanket. Newer formulations of AR-FFFP are designed to be used at 3% on both types of fuels. Overall, AR-AFFF is a versatile foam offering good burnback resistance, fire knockdown capabilities, and high fuel tolerance on both hydrocarbon and alcohol fuel fires.

Synthetic foams are based on surfactants and provide better flow and faster knockdown of flames compared to protein foams, but they have limited postfire security. Two of the most widely used synthetic foams are AFFF and alcohol-resistant aqueous film-forming foams (AR-AFFF).

---

*A High Back-Pressure Foam Maker (HBFM) is used for this "in-line" foam delivery method. HBFMs are typically designed for subsurface foam injection into hydrocarbon fuels stored in vertical cone roof storage tanks. Flow rates are from 300 gpm (1,136 L/min) to 550 gpm (2,082 L/min) at 150 psi (1034 kPa) inlet pressure.

AFFF was developed in the mid-1960s by the US Navy. It is water-based and frequently contains hydrocarbon-based surfactants, fluorochemical surfactants, and **solvents**. AFFF is available in 1%, 3%, or 6% concentrates. This synthetic foam has a low viscosity and therefore has the ability to spread easily and rapidly over the surface of most hydrocarbon-based liquid fuels. A water film forms beneath the foam, which cools the liquid fuel and prohibits the formation of flammable vapors. This provides dramatic fire knockdown, an important factor in fighting aircraft fires. AFFF is self-healing and will re-cover open areas in the foam blanket caused by agitation of the flammable liquid surface. It is also used on flammable liquid spills to prevent ignition.

AFFF's low viscosity makes it more effective on hydrocarbon liquids with high surface tension (diesel and kerosene) than on hexane and high-octane gasoline with lower surface tension. Water drains rapidly from the foam bubbles, leading to rapid fire extinguishment. Long-term sealing of combustible vapors and burnback resistance are sacrificed, however.

AFFF requires a very low-energy input to produce high-quality foam. Therefore, it can be applied through a wide variety of foam-delivery systems. This versatility makes AFFF an ideal choice for airports, refineries, manufacturing plants, and municipal fire departments.

Alcohol-resistant aqueous film-forming foams (AR-AFFF) are based on AFFF chemistry to which a **polymer** has been added. They are resistant to the water-absorbing action of polar solvents because they form a protective film or layer between the burning surface and the foam (see Figure 6.2).

Unlike AFFF, which can be sprayed directly onto the fire, AR-AFFF must be bounced off of a surface and allowed to flow down and over the liquid to form its membrane. When used on hydrocarbon fuels, AR-AFFF produces the same rugged aqueous film as a standard AFFF agent. It also provides fast flame knockdown and good burnback resistance when used on hydrocarbons and polar solvent fuels. Overall, AR-AFFF is the most versatile type of foam available today and is available in 1%, 3%, or 6% concentrates.

> **Solvent** ▪ A substance that is capable of dissolving or dispersing one or more other substances.

> **Polymer** ▪ A large, long-chained molecule consisting of many similar units (monomers) bonded together.

FIGURE 6.2 Fireboat monitor with AR-AFFF-generating capability.
Source: Ronald R. Spadafora

TABLE 6.2 Properties of Firefighting Foam

PROPERTY	P	FP	FFFP	AFFF	AR-AFFF
Knockdown	Fair	Good	Good	Excellent	Excellent
Heat Resistance	Excellent	Excellent	Fair	Good	Good
Fuel Resistance (Hydrocarbons)	Fair	Excellent	Good	Moderate	Good
Vapor Suppression	Excellent	Excellent	Good	Good	Good
Alcohol Resistance	None	None	None	None	Excellent

Source: "Properties and Comparisons of Fire Fighting Foam Types" from *A Firefighter's Guide to Foam.* Used by permission of National Foam, Inc.

The properties of firefighting foam are dependent on the type of foam used. Table 6.2 is a composite of the foams discussed and the characteristics they possess.[3]

Low-Expansion Foam

Low-expansion foams are an aggregate of gas-filled bubbles made by mixing air into a water solution containing a foam concentrate. These agents are of lower density than flammable and combustible liquids, which allows the foam to float over a burning liquid surface and form a continuous blanket that seals volatile combustible vapors from access to air. Low-expansion foam agents have expansion ratios up to 20:1.

Low-expansion foam systems are used when a blanket of foam is needed to float on the horizontal surface of a flammable or combustible liquid. They have limited vertical or three-dimensional surface protection applications and are ideal when coating the fuel and displacing oxygen in the air are the preferred methods of extinguishment. Their blankets have high water content, with fine bubbles of relatively high weight and small volume. Low-expansion foam provides good flow ability, a vapor-tight seal to prevent reignition, and is resistant to flames and heat. It can be generated with either fresh- or seawater.

FIRE AND SPILL APPLICATIONS

Low-expansion foam is produced in foam playpipes, or stationary/mobile foam installations. They are used for extinguishing fires in both solid materials and liquids due to exceptionally good flow characteristics. The foam distributes itself over the entire surface of the fire within a very short time.

Used for mitigation, low-expansion foam suppresses the emission of flammable vapors. The flammable material remains covered by a gas-tight, insulating-and-cooling foam layer for a long period. Firefighting operations using low-expansion foam can be accomplished successfully from a distance due to the foam's wide projection range. Low-expansion foam can be used in both aspirated and nonaspirated form as an aqueous solution. Examples of this feature can be found inside aircraft hangars and at fuel storage depots. These foam concentrates are also suitable for use in sprinkler and deluge water spray systems.[4]

Design requirements provided in *NFPA 11: Standard for Low-, Medium-, and High-Expansion Foam* for low-expansion foam are concerned with the bulk storage of fuels and chemicals and the vessels they are contained in. Large storage tanks with surface areas greater than 400 ft.^{2} (.09 m^2), such as fixed cone roof tanks and floating roof tanks, are addressed in NFPA 11.

Protection options for bulk storage tanks may include discharge application devices, such as a **foam chamber**, **foam monitor**, or fixed nozzles. The design densities provided in NFPA 11 are for hydrocarbon or nonmiscible (nonmixable) fuels and are specific to the

Foam chamber ■ A foam discharge device normally used on fixed (welded) cone roof fuel storage tanks. Mounted to the exterior of the tank, with a deflector located on the interior of the tank. The expanded foam discharges against the deflector, which directs the foam solution to the interior tank wall, where it falls to the liquid surface and spreads across it.

Foam monitor ■ An application device that delivers a large foam stream and that is mounted on a stationary support that either is elevated or is at grade.

discharge device used. Miscible liquids require a specific Listing of the foam concentrate for the fuel or chemical and discharge device.

Diked areas (linear obstructions) built around bulk storage tanks are also protected by low-expansion foam systems. They are built to inhibit the free flow of fuels or chemicals from a tank that has ruptured or is leaking. Dikes provide a wall of protection from the fuel flowing outside a manageable fire area. A spreading fuel fire in a diked area can create a more complex fire hazard as there may be more than one tank located inside the dike. **Foam makers** and hose lines can be used to protect these areas. If foam makers are installed, the distance they can be placed apart at the dike wall is determined by their flow rate.

High-Expansion Foam

High-expansion foams control fires by cooling, smothering, and reducing the oxygen content by steam dilution. They are formulated to retard drainage and provide stability. They have an extremely high foam volume and low water weight. The foam blanket created contains large amounts of air and is therefore considered "dry" foam. Its primary fire extinguishing property is separating fuel from the flames and its insulating characteristics. In addition, a relatively high deterioration rate allows minute drops of water in the bubbles to be released and evaporate instantly when applied to a fire. This occurrence generates steam with an expansion ratio (water to vapor) of approximately 1:1,700, displacing oxygen in the ambient air and causing cooling.[5]

High-expansion foams have an expansion ratio of over 200:1. They are suitable for enclosed spaces in which quick filling of a large area is required. They will flow down openings and fill compartments, spaces, and crevices, replacing air in these areas. High-expansion foam concentrate, when used in combination with a high-expansion foam generator, creates a superior foam blanket with tough adhesiveness and good stability (see Figure 6.3). Most generator systems produce expansion ratios between 400:1 and 1000:1. The foam generators apply foam to large areas in total flooding and three-dimensional applications, such as

Foam maker ■ A device designed as an air-aspirating discharge foam applicator. It is used principally for the protection of dikes, open floating (pontoon) roof storage tanks, fixed roof storage tanks, as well as wherever low-velocity foam streams are desired. Foam makers generally are installed in the line of a fixed foam fire protection system, mounted either horizontally or vertically.

FIGURE 6.3 Portable high-expansion foam generator.
Source: Ronald R. Spadafora

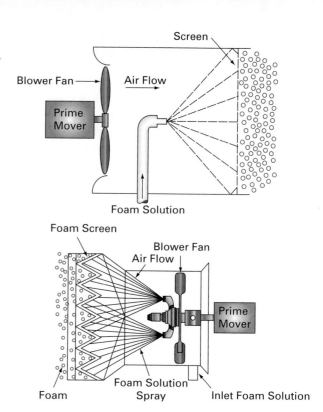

FIGURE 6.4 Portable high-expansion foam-generator components

warehouses and ship cargo holds. Foam generator components include a motor, blower fan, foam solution inlet pipe, and foam screen (see Figure 6.4). High-expansion foam nozzles are also available. High-expansion foam can be created with either fresh- or seawater. In addition to being a superior foaming agent, high-expansion foam also has a wetting ability to increase the penetrating effect of water on deep-seated, ordinary (Class A) material fires. High-expansion foam is available in a 2% concentration.

FIRE AND SPILL APPLICATIONS

High-expansion foam may be used for both total area flooding and local direct application. In addition, besides being effective on hazard types involving Class A and B fires, this type of foam has been used successfully in controlling liquefied natural gas (LNG) fires by blocking heat from the flames to the LNG and thereby reducing the vaporization rate. High-expansion foam is also effective in reducing vapor concentrations from unignited LNG and other hazardous, low-boiling-point gaseous products, such as ammonia spills. Additional common applications suited for high-expansion foam include:

- Aircraft hangars
- Hazardous waste storage
- Paper product and tire warehouses
- Flammable liquid storage
- Mining
- Ship holds and engine rooms
- Power stations
- Gas turbine generators
- Cable tunnels
- Engine test cells

- Transformer rooms
- Basements, cellars, subcellars, and enclosed spaces
- Communications switching stations

FIREFIGHTER SAFETY CONCERNS

Fixed (permanently installed) high-expansion foam generators found in industrial, manufacturing and storage facilities will fill the entire volume of a protected area in a matter of minutes (see Figure 6.5). Firefighters entering the area after a high-expansion foam system discharge must use caution. The burning fuel is blanketed with foam but may still be burning. Entering an area too early after discharge can have adverse effects, such as providing oxygen to the fire and reigniting the material. Another safety factor that must be considered is the lack of visibility when traveling through the foam bubbles. A firefighter entering the foam is essentially blind to voids along the ground as well as mounted or suspended objects. Firefighters should use search ropes to stay oriented and hose streams to break down the foam blanket while travelling through the area.

NFPA 11 instructs anyone entering an area in which high-expansion foam has been discharged to use a fog spray from a hose line to cut a path through the foam.

Positive pressure, self-contained breathing apparatus (SCBA) always should be worn before entering an enclosed space that has been filled with foam, regardless of the stage of the fire. Asphyxiation deaths and injuries to firefighters occur most often at the beginning and end of a fire operation. Deadly carbon monoxide and other toxic gases can become trapped inside the foam blanket. Also, the oxygen in the enclosed space may have been replaced by carbon dioxide generated from the combustion process. Moreover, oxygen may have been replaced or reduced to dangerous levels by steam generated when the water of the foam blanket evaporates.

FIGURE 6.5 Three large high-expansion foam generators at roof level protecting a roll paper warehouse building.
Source: Ronald R. Spadafora

Class A Foam

Class A foams were developed in the mid-1980s for fighting wildfires. The lower surface tension of the water assists in the wetting and saturation of Class A fuels. This aids fire suppression and can prevent reignition. Favorable experiences led to its acceptance for use on structural fires, rubbish, and inside dumpsters. Class A foam is not designed to be used on Class B fires. During **wildland/urban interface** fires, Class A foam is used as a protective barrier to cover a structure that is in the path of a fast-moving wildland fire. The foam prevents the structure from reaching its ignition temperature. It is also capable of preventing airborne embers from igniting the structure. This technique is also used to create a firebreak by blanketing trees, brush, and foliage. The foam raises the moisture level, and the rapid movement of the fire is slowed or halted.

> **Wildland/urban interface** ■ The zone of transition between unoccupied land and human development.

Class A foam, when mixed in correct proportions with water, changes two properties of the water. As a wetting agent or surfactant, it will decrease its surface tension and allow for better penetration into burning material. It will also create a blanket of foam, allowing it to cling to vertical as well as horizontal surfaces. These two characteristics permit water to absorb more heat.

Class A foam is typically used at very low concentrations. Proportioning percentages range from 0.1% to 1% by volume of water. Unlike Class B foams, proportioning accuracy and application rates are not as critical to the performance of the foam. Class A foam, however, will not work better at rates higher than 1% concentration. At higher proportional rates, the surface tension of the foam solution actually begins to increase. For fires in structures during the Incipient phase, a concentration of 0.50% is most effective. Fires in the Growth phase of burning may require firefighters to use a 1% concentration.

GENERATING PROCESS

Class A foam is mechanical foam. The foam concentrate is usually purchased in 5 or 55 gallon (19 L or 208 L) containers or drums. When the concentrate is diluted with water, it forms a foam solution that when mechanically agitated with air creates a firefighting foam blanket over the surface of the fire. Foam-generating systems include eductors, proportioners, and foam bubble–generating devices (foam nozzles, air-aspirating nozzles, and compressed-air systems).

FIREFIGHTING OPERATIONS

The majority of working fires that municipal fire departments respond to are Class A fueled. Using Class A foam correctly can provide a tool to increase the effectiveness of the application of water on fires. This can result in increased firefighter safety, better operational efficiency, and reduced property damage. Less fire and water damage relates to a reduction in environmental and financial impact on the community. Rural departments can also benefit from using Class A foam, because many nonmunicipal fire departments do not have a hydrant water distribution system in their fire district. This technology can also enhance the effectiveness of water shuttle operations.

Compressed-Air Foam System

A compressed-air foam system (CAFS) is defined as a standard water-pumping system that has an entry point from which compressed air can be added to a foam solution to generate firefighting foam of variable consistencies. The air compressor also provides energy, which helps propel the compressed air foam farther than aspirated or standard foam nozzles and other foam-delivery equipment. CAFS components include a water source, centrifugal pump, foam concentrate tanks, an air compressor (or pressurized air cylinders), a direct-injection foam-proportioning system on the discharge side of the pump, a mixing chamber, and control systems to ensure the correct mix of water, concentrate, and air. A CAFS is designed to extinguish Class A fires.

GENERATING PROCESS

CAFS injects air into the foam solution at the water pump. The air and foam solution mix as they move through the hose line. An air-aspirating nozzle at the end of the hose line is not required. For initial attack on a structural fire, a CAFS is normally operated at 0.3%. At this percentage, a wet (melted ice cream)-textured foam is created, allowing the agent to readily flow and spread easily over burning materials. A 0.2% setting produces a fluid foam consistency used for **overhaul** operations. When protecting exposed structures, a 1% concentration is very effective. This produces a dry, slow-draining foam blanket of shaving cream consistency. Using CAFS makes firefighter hose operation easier, because the hose is much lighter: approximately two-thirds of what is inside it is air. In addition, the apparatus fire pump operator's job is easier, because pump pressure is commonly uniform for multiple hose lines regardless of the length of the layout. A pressure governor preset at 90 psi (621 kPa) is all that is needed. High-quality streams can be created with either a smooth-bore or fog nozzle.[6]

Overhaul ▪ Decay phase firefighting operation to prevent fire extension and rekindles.

FIREFIGHTING OPERATIONS

Structural firefighting with a CAFS can assist firefighters in many ways. A reduction in fire-suppression times is a major benefit. Lighter hose lines allow firefighters to stretch, place, and operate their attack lines rapidly and with less stress and strain. The enhanced reach of the hose stream results in engine firefighters staying farther away from the seat of the fire and thereby enhancing their safety. A CAFS also decreases the surface tension of water, allowing it to penetrate into smoldering materials more effectively.

Firefighters working with a CAFS should also be aware of its limitations. A CAFS requires the same water flow rate as plain water. It does not reduce flow-rate requirements. Moreover, the enhanced efficiency of the water stream can produce large amounts of steam. In an enclosure, this can result in an untenable environment. In addition, a high-energy CAFS hose line has a strong nozzle reaction. Pistol-grip hose nozzles should be used and eye protection worn by all members on the hose line.

USE OF A CAFS IN FIXED SYSTEMS

The use of a CAFS in fixed systems is not generally recommended. Sprinkler systems are not hydraulically built to discharge compressed-air foam. The piping and sprinkler heads are only designed to flow a prescribed density of water per unit area. In general, when foam is delivered into a fire department connection (FDC) there is always the potential of the agent flowing back past the check valve into the potable water supply of the community. This can create contamination and other environmental issues (see Figure 6.6).

Standpipe systems that depend solely on the water supply from the local fire department or fire brigade can be used if they are not combined standpipe/sprinkler systems and not equipped with pressure-reducing valves (PRV). Rack hose inside standpipe cabinets must not be used. Firefighters should employ hose and delivery appliances specially made for a CAFS. Once the firefighting operation is concluded, the standpipe should be thoroughly flushed with plain water to remove residual foam ingredients. Building management should also be informed of firefighting activities; further maintenance and flushing may be required.[7]

FIGURE 6.6 Fixed compressed-air foam system (CAFS)

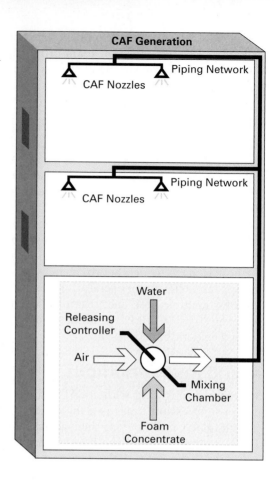

 CAFS CONTROLLED BURN TESTS

In 2001, three controlled burn tests were conducted in Palmdale, CA, by the Los Angeles County Fire Department (LACFD) in an attempt to examine and compare three extinguishing agents: water, Class A foam, and a CAFS with Class A foam concentrate. Test structures consisted of three one-story, wood-framed, single-family vacant structures. Each of the test structures had an identical floor plan consisting of six rooms. Interior walls were lath-and-plaster construction; the exterior stucco walls had been removed prior to the tests. All window glass was removed and replaced with plywood and all structures had composition shingle roofing. Each structure was furnished with identical new furniture, including beds, dressers, dining room tables and chairs, bookcases, upholstered couches, and various plastic items. **Thermocouples** were placed in the interior of each structure to detect temperatures at various locations.

All three firefighting operations were conducted using the same LACFD engine apparatus equipped with a 1,500 gpm (5,678 L/min.) single-stage centrifugal pump with 200 ft. (61 m) of attack hose line attached to it. A combination nozzle was used in the water and Class A foam tests, and a 1 in. (25 mm) smooth-bore nozzle was used in the CAFS test. Firefighters employed during the tests were trained in the use of foam for interior attacks, and the same attack team was used in each test. Fires were started by igniting furnishings at several locations. When the average interior room temperature reached between 550°F and 850°F (288°C and 454°C), firefighters outside the building started pulling the plywood panels away from the windows to simulate glass heat failure.

Firefighting operations for all three test fires started from a position in front of the structure and directed a stream through an open window or door. The firefighters then moved across the front of the structure or around to one side to direct a stream through another opening. The flow rates for water and the Class A foam fire tests were equal, at 90 gpm (341 L/min.); the flow rate for the CAFS fire test

Thermocouples ■ Temperature-measuring sensing devices.

was also 90 gpm (341 L/min.), with 30 ft.^{3} (0.8 m^3) of air. Foam concentrations were set at 0.5% for the Class A foam test and 0.2% for the CAFS test. The results from the tests are as follows:

Test Results

AGENT	QUANTITY	FIRE KNOCKDOWN TIME
Water	75 gallons (284 L/min.)	50 seconds
Class A foam	44 gallons (167L/min.)	25 seconds
CAFS	16 gallons (61 L/min.)	11 seconds

A CAFS was superior to either water or Class A foam. In addition, it was determined that the heat-absorbing properties of foam reduced the average interior temperatures significantly faster than water. It took 1 minute and 28 seconds and 1 minute and 25 seconds for the average interior temperature to drop from 600°F to 200°F (316°C to 93°C) using a CAFS and Class A foam, respectively, compared to 6 minutes and 3 seconds with water.[8]

Class B Foam

Class B foams are designed for Class B-type fires. This foam floats atop liquid fuel. Class B foams are also efficient in suppressing the vapors from unignited spills of these liquids.

There are many types of Class B foam. Each is made for a specific use. Many are used in fixed installations at which a known fuel hazard is present. Different fuels will affect the foam blanket produced by Class B foams in different ways. The same foam used to extinguish one flammable liquid might be totally ineffective against another due to its chemical composition. Often, an on-site industrial fire brigade can identify its risks more easily than a municipal department. Therefore, they can match the foam type to the risks known should a fire occur. Municipal fire departments, however, store and use Class B foam concentrate that can be effective on multiple flammable liquid types that are commonly encountered. This is because a municipal department may not know what type of fuel hazard it may be facing until it arrives on scene. They normally carry AR-FFFP and AR-AFFF.

GENERATING PROCESS

Class B foam concentrate must be mixed with a percentage of water that will produce enough solution to extinguish a combustible/flammable liquid fire or cover the spill. For many types of foam, the eductor rate for hydrocarbons is 3% and 6% for polar solvent liquids. Some Class B foams use 3% eductor rates for both hydrocarbons and polar solvents. Nozzles should be preset to 95 gpm (360 L/min.) and attached to no more than 350 ft. (107 m) of hose line. Class B foam should be supplied with 200 psi (621 kPa) inlet pressure.

FIREFIGHTING OPERATIONS

The decision to apply Class B foam on a liquid fire or fuel spill should be thoroughly thought out. The Incident Commander (IC) should determine if he or she has enough foam and foam concentrate on hand. Initiating foam operations on a spill fire that is too big to cover completely with the foam available, for example, may cause the fire to burn-back and consume the foam blanket that was applied. Waiting until more foam arrives before starting foam operations may be the best decision. Other considerations to think about are whether the foam type available is compatible with the fuel involved. Is extinguishing a spill fire the right tactic to use? Should firefighters let the fire burn and protect nearby property in order to exhaust the fuel?

> **NOTE**
>
> Application rates are dependent upon many factors, including the foam concentrate used, the type of fuel and whether or not it is on fire, and whether the fuel is contained.
>
> Minimum application rates for Class B foams are recommended in NFPA 11. Increasing these minimum recommended foam application rates will generally reduce the time required for extinguishment. In addition, application rates less than the minimum recommended will commonly extend the time required for extinguishment, or, if too low, the fire may not be successfully controlled or extinguished.
>
> Firefighters using Class B foam on liquid fires and spills must take numerous operational and safety precautions to ensure a successful and safe effort. First responders should ensure that the Incident Command Post (ICP) and Staging Area for resources is upwind and uphill from the incident. Identification of the product involved is essential to making the right decision as to what type of foam, if any, should be applied. All first responders must be using and wearing correctly prescribed personal protection equipment (PPE). Areas within the path of flowing fuel and vapors must be evacuated. Use air monitors to help determine the effectiveness of the foam blanket in inhibiting vaporization. If it can be accomplished safely, attempt to confine the hazard to as small an area as possible. Ensure that exposed structures and facilities are properly protected. Potential ignition sources must be eliminated.[9]

Fixed Foam Systems

Fixed Class B foam extinguishing systems are a complete installation piped from a central foam station. The foam is discharged through stationary applicator devices on the hazard being protected (see Figure 6.7). Major components of fixed foam systems follow.

FIGURE 6.7 Class B foam standpipe hose station.
Source: Ronald R. Spadafora

FOAM CONCENTRATE STORAGE TANKS

Foam concentrate is stored in tanks ready to supply the proportioning system. Except for a pressure-vacuum vent, the tank is kept closed to the atmosphere. A partially empty tank allows a large liquid concentrate surface to interact with air, allowing excessive evaporation and condensation. This situation degrades the foam concentrate and permits corrosion of the tank shell. The concentrate tank is, therefore, generally kept filled with liquid to ensure prolonged storage life of the concentrate.

BLADDER TANKS

Bladder tank systems use a pressure-rated container with a nylon-reinforced bladder to store the foam concentrate. System water pressure is used to squeeze the bladder, providing foam concentrate at the same pressure to the proportioner. With bladder tank systems, no external power source is required, and little maintenance is needed (see Figure 6.8).

PROPORTIONING DEVICES AND EDUCTORS

Proportioning is the introduction of the proper percentage of foam concentrate into a flowing stream of water to produce a foam solution. Most firefighting foam-proportioner devices are intended to mix foam concentrate with fresh- or saltwater in percentages from 1% to 6%. Foam concentrate proportioners are an integral component of any type (fixed, apparatus, or in-line) of foam extinguishing system.

Proportioners use the **venturi effect** to pull foam concentrate into the water stream. Water flows past a tapered venturi opening, creating a vacuum or negative pressure that draws the foam concentrate through a metering valve. This valve controls the amount of concentrate allowed to flow into the water stream. A ball check valve prevents water from flowing back into the pickup tube and the concentrate container (see Figure 6.9).

In-line proportioners with pickup tubes, also termed eductors, present a simple method for firefighters to proportion foam concentrate and water at the proper percentage at fire operations. The eductor can be attached to the pump panel discharge of the fire apparatus

Venturi effect ■ The reduction in fluid pressure that results when a fluid flows through a constricted section of pipe. Water is introduced, under pressure, at the inlet of the eductor. The eductor reduces the orifice available for the water to pass, so it must speed up to get through. This creates a pressure drop that, in turn, puts suction on the pickup tube. As the foam concentrate is pulled up the tube, it passes through a metering valve that allows the correct percentage to be introduced into the water stream. In most cases, the metering valve can be adjusted to select a 1%, 3%, or 6% foam solution.

FIGURE 6.8 Close-up inside a foam bladder tank.
Source: Ronald R. Spadafora

FIGURE 6.9 Tapered venture proportioner piping above a 2,500 gallon (9,464 L) low-expansion foam concentrate tank.
Source: Ronald R. Spadafora

or connected in a hose lay. The pickup tube is placed into a foam concentrate container. When using an in-line eductor, it is very important to follow the manufacturer's instructions concerning inlet pressure, maximum lengths of hose, and appropriate nozzle. Eductors must be thoroughly cleaned using water after being used. If they are not flushed out, they can clog due to hardened foam deposits and concentrate residue. This can negate the functionality of the eductor.

PUMPS

Pump skids (or balanced-pressure pump-proportioning systems) are an alternative to bladder tank systems. They automatically proportion foam concentrate over a wide range of flows and pressures without manual adjustment. They are commonly used to protect large, flammable liquid storage facilities. Pump skids are self-contained foam systems used in combination with atmospheric foam storage tanks. Foam concentrate is pumped from the storage tanks to a proportioner, and an automatic pressure-balancing valve regulates the foam concentrate pressure to match the water pressure.

FOAM CHAMBERS AND FOAM MAKERS

Foam chambers and foam makers are air-aspirating devices designed to protect flammable liquid storage tanks. Foam chambers apply low-expansion foam directly onto the fuel surface inside storage tanks with minimum foam submergence and fuel agitation. The effectiveness of the foam blanket is, therefore, enhanced. They are compatible for use with all types of foam concentrates. Foam makers can be installed at both the top (surface application systems) and bottom (sub-surface application systems) of fuel storage tanks. Other types of foam makers are installed in foam-distribution piping, from which expanded foam is directed to specific hazards, such as dike areas.

SPRINKLER FOAM-WATER HEADS AND NOZZLES

Sprinkler foam-water heads are available from various manufacturers in both non-air-aspirated and air-aspirated versions. Non-air-aspirated sprinklers using AFFF agents are more economical and can be used in either open-head (deluge) or closed-head (wet) sprinkler systems. Air-aspirated sprinkler heads are required for protein and fluoroprotein foams in deluge systems. Foam-water nozzles are more expensive than foam-water sprinkler heads. They are air-aspirated foam-water devices that provide a better aerated foam discharge compared to foam-water sprinkler heads.

FOAM GENERATORS

Foam high-expansion generators operate by coating their screens with high-expansion foam solution while air is blown through the screens to produce expanded foam. Because of its high expansion ratio, little water is required to generate large quantities of foam, thereby reducing the potential for hazardous runoff or water damage.

FOAM MONITORS

Foam monitors or turrets are permanently installed foam-discharge devices capable of being aimed that project large amounts of foam substantial distances. They normally are mounted on a rotating base that allows the projection of foam in a 360-degree circle around the monitor platform. They can be of the oscillating type, automatically moving and applying their foam stream in an arc-like pattern, using water pressure as the energy source. Remote (electrically)-controlled monitors are also manufactured. Other monitors are operated manually using a tiller bar to control direction and elevation. Various air-aspirating and non-air-aspirating nozzles are also available for use with monitors. Foam solution is distributed from a proportioning device to the monitors through a system of pipes and valves (see Figure 6.10).

FIGURE 6.10 Foam monitor with tiller bar for manual control of stream at electrical substation.
Source: Ronald R. Spadafora

FIRE DETECTION AND CONTROL EQUIPMENT

In many applications, including aircraft hangars and loading racks, the fire protection requirements include an automatic detection and control system. In these locations, heat or flame detectors are installed to provide input to a fire alarm control unit (FACU).[10]

Fixed Foam System Applications

Fixed foam systems protect hazards where flammable liquids are present. These hazards are common to many industries, including petrochemical, chemical, oil and gas, aviation, marine/offshore, manufacturing, electrical utilities, military, and transportation.

SUBSURFACE INJECTION FOR PERMANENT ROOF LIQUID STORAGE TANKS

Subsurface injection fixed foam fire extinguishing systems can be used for permanent roof combustible and flammable liquid storage tank protection. Finished foam is applied at the base of the tank well below the surface of the liquid. Low-expansion foam is used for this type of application. New tank design involves dedicated foam piping installed directly into the bottom of the tank, with nozzles properly spaced. Designers installing a fixed foam extinguishing system into existing tanks may tap directly into the tank's product line in order to circumvent having to empty the tank and drill additional openings.

SURFACE APPLICATION FOR PERMANENT ROOF LIQUID STORAGE TANKS

Permanent roof combustible/flammable liquid storage tanks may also be protected with fixed surface-application foam fire extinguishing systems. These types of systems have discharge outlets/applicators permanently mounted above the fuel surface (see Figure 6.11). They are designed to flow a thin blanket of low-expansion foam over the surface of the stored fuel.

FIGURE 6.11 Foam makers for fixed roof fuel storage tank. *Source:* Ronald R. Spadafora

FIGURE 6.12 Foam piping running up to a foam chamber atop the roof of a fuel storage tank.
Source: Ronald R. Spadafora

SEAL PROTECTION FOR FLOATING ROOF LIQUID STORAGE TANKS

A floating roof sits atop the combustible/flammable fuel surface inside of the storage tank. This buoyant structure rises and falls as liquid is added or removed from the tank and is designed to inhibit fuel vapors from gathering above the surface of the liquid and below the roof. The void space between the edges of the floating roof and the perimeter of the tank, however, is vulnerable to the collection of fuel vapors. This is the area or seal in which fire can occur and the focus of the fixed low-expansion foam fire extinguishing system.

Above-seal foam protection is provided through foam chambers and foam makers. These foam components can be wall mounted or roof mounted depending upon design (see Figure 6.12). A circular steel foam dam along the perimeter of the tank holds the foam blanket in place. This type of system is recommended to protect combustible seals. When the underside of the seal requires foam protection, system piping will penetrate the seal. This type of design may not need a foam dam if the seal is capable of holding the foam that has been discharged into it.

AIRCRAFT HANGARS

Fixed foam fire protection systems for aircraft hangar applications are designed to provide control and extinguishment of Class A or Class B fires. Most fire protection systems for aircraft handling facilities are designed in accordance with *NFPA 409: Standard on Aircraft Hangars*, *NFPA 11*, *NFPA 14: Standard for the Installation of Standpipes and Hose Systems*, and *NFPA 16: Standard for the Installation of Foam-Water Sprinkler and Foam-Water Spray Systems*.

There are four types of foam fire protection systems used for the protection of an aircraft hangar. These systems may be used separately, or they may be combined.

- *Ceiling design overhead foam-water sprinkler system:* These types of systems can be preaction, wet, or deluge. If used with protein or fluoroprotein low-expansion foam concentrate, the system has foam-water nozzles with an air inlet that allows air to be introduced into the foam solution before it contacts the stream deflector.

When used with AFFF low-expansion foam concentrate, however, foam-water sprinkler heads are used. These non-air-aspirating foam-discharge devices do not have an air inlet between the orifice and the deflector. AFFF requires less energy to aerate, but the foam produced is less aerated compared to protein or fluoroprotein concentrate with foam-water system nozzles.

- *Under-wing design monitor system:* A water oscillating monitor (WOM) automatically moves from side to side when discharging low-expansion foam directly onto the hangar floor. They are preset to oscillate over a given angle of elevation, range, and stream pattern to provide a specified flow rate over a designated area under the aircraft. The WOM has a nozzle that is mounted on a manifold several feet above the hangar floor. The stream is kept low enough so as not to hit the wing. The design objective is to provide full coverage beneath the wings to the center of the fuselage. At least two monitors are needed, located perpendicular to the fuselage of the aircraft, to provide unobstructed foam protection (see Figure 6.13).

  Stationary monitor systems are used where aircraft or maintenance equipment in the hangar could interfere with the normal operation of the WOM-type monitor. This type of system has preset nozzles mounted on a manifold or individual units located 3 ft. (1 m) above the hangar floor. Stream pattern and range is designed to flow under the wing of any aircraft.

- *Foam hose-line system:* Low-expansion foam hose-line systems are supplementary extinguishing systems designed to suppress incipient fires within aircraft hangars. Provisions are made for operating a minimum of two foam hand lines. Each system must flow a minimum 60 gpm (227 L/min.) with sufficient nozzle pressure for a minimum discharge duration of 20 minutes each. Hose is installed on approved racks or reels, and the hose is fitted with a foam nozzle or a combination-type nozzle designed to produce foam or water spray. These hand hose systems can be supplied from a connection to the sprinkler system.

- *High-expansion foam system:* High-expansion foam systems are installed in aircraft hangars in lieu of low-expansion systems to provide total-area flooding, three-dimensional fire protection. They are also used when environmental disposal

FIGURE 6.13 Water oscillating monitor (WOM) inside an aircraft hangar.
Used with permission of Tyco Fire Protection Products

FIGURE 6.14 Aircraft hangar with ceiling-mounted, high-expansion foam generators. Used with permission of Tyco Fire Protection Products

concerns are a major consideration, because the expansion ratio of high-expansion foam relative to low-expansion foam translates to considerably less concentrate per unit volume.

These systems are designed to cover the protected area to a depth of 3 ft. (0.9 m) with a foam blanket within one minute. High-expansion foam systems must have sufficient foam concentrate and water to operate at the design application rate for a minimum of 12 minutes. The foam blanket does not have to fill the entire volume or space being protected. For example, when a high-expansion foam system is installed inside a hangar, the foam height has to be only 10% higher than the hazard height. High-expansion foam generators do not directly project a foam stream. Rather, the distance from the generator is relatively short. The generators should be mounted at the ceiling or on exterior walls of the hangar where outside air only is used to generate the foam. In some cases, however, if high-expansion foam generators are mounted on exterior walls, it may not be possible to achieve the desired coverage. It is therefore recommended, that high-expansion foam generators be installed so that the flow of foam on the floor is delivered to areas beneath aircraft wings and fuselage sections (see Figure 6.14).

Truck Loading Racks

A truck loading rack is used to dispense combustible and flammable liquids. Loading racks can handle from two to more than a dozen trucks at one time. Most loading racks in the United States are equipped with fixed foam systems, activated either by automatic sensors or manually by means of an emergency switch. Low-expansion foam protection is used, and these systems are designed in accordance with NFPA 11 and NFPA 16.

The loading or unloading of tank trucks with combustible/flammable liquids can be a hazardous operation. The primary method to protect loading rack areas is to use foam-water sprinkler heads or nozzles mounted overhead at the roof or canopy level of the loading/unloading area. The system is commonly a deluge type. The discharge duration for any foam-water system covering a loading/unloading rack storage area is a minimum of 10 minutes. Supplementary, low-level, directional foam-water spray nozzles may also be included in the design. They are installed at the point of product transfer and beneath the vehicles to provide rapid fire extinguishment and dispersal of fuel. These foam-water spray nozzles are highly recommended if tank trucks are bottom loaded. Moreover, many tank trucks are built with aluminum storage tanks. A spill fire originating underneath the aluminum tank shell can cause it to weaken and fail.

CHAPTER REVIEW

Summary

All foam systems, regardless of size, consist of a foam concentrate supply, proportioning device, water supply, and foam discharge applicators. All the components must function properly to ensure successful system performance. Firefighting foam applications include both fixed foam extinguishing systems and portable systems.

Review Questions

1. Name the three components that make up finished firefighting foam.
2. What is the expulsion of hot liquid from a storage tank as a result of water boiling below the fuel surface called?
3. What favorable characteristic is obtained from fluoroprotein foam's oil-rejecting properties?
4. What prevents an AR-FFFP foam blanket from being destroyed by polar solvents and water miscible fuels?
5. What is the maximum expansion ratio of low-expansion foam agents?
6. High-expansion foams are used for what types of applications?
7. How is Class A foam used during wildland/urban interface fire applications?
8. How does a CAFS make firefighter hose operations easier?
9. What area of a floating roof liquid storage tank is vulnerable to the collection of fuel vapors?
10. What is the design objective of a water oscillating monitor (WOM) when used for aircraft protection?
11. Name the reasons that high-expansion foam systems are installed in aircraft hangars in lieu of low-expansion systems.
12. What equipment and tools should firefighters wear and use upon entering an area in which high-expansion foam has been discharged?

Endnotes

1. National Foam, Inc., *A Firefighter's Guide to Foam* (Exton, PA: National Foam, Inc., 1990).
2. Fire Suppression Systems Association, "Foam Concentrates for Fixed Systems," www.bfpe.com/pdf/foam_concentrates.pdf.
3. D.N. Meldrum, National Foam System, Inc., *Fighting Fire with Foam: Basics of Effective Systems* (Lionville, PA: Society of Fire Protection Engineers, 1979).
4. R. Craig Schroll, *Industrial Fire Protection Handbook*, 2nd ed. (Boca Raton, FL: CRC Press LLC, 2002).
5. Ibid.
6. Dominic Colletti, *The Compressed Air Foam Systems Handbook* (Royersford, PA: Lyon's Publishing, 2006).
7. Janet Wilmoth "Bubbles Still Beat Water," *Fire Chief* 51, no. 5 (2007): FF3.
8. David F. Peterson, "Class B Foam: The Neglected Tool," *Fire Engineering* 154, no. 10 (2001): 161.
9. A. Koetter Fire Protection LLC, "Ansul Foam Fire Protection Systems," www.koetterfire.com/engineered-fire-systems/ansul-fire-foam-system.php.
10. American Bureau of Shipping (ABS), *Guidance Notes on Fire-Fighting Systems,* www.eagle.org/eagleExternalPortalWEB/ShowProperty/BEA%20Repository/Rules&Guides/Current/141_FireFightingSystems/Pub141_FireFighting.

Additional References

Robert Burke, *Fire Protection Systems and Response* (Boca Raton, FL: CRC Press, 2008).

Chemguard, "Fixed or Semi-Fixed Fire Protection Systems for Storage Tanks," www.chemguard.com/pdf/design-manuals/D10D03192.pdf.

Dominic J. Colletti, "Using Class A Foam for Structure Firefighting, Part 2: Foam Chemistry," *Firehouse* 23, no. 7 (1998): 66–68, 70.

Dominic J. Colletti, "Using Class A Foam for Structure Firefighting, Part 3: Foam Generation Hardware," *Firehouse* 23, no. 8 (1998): 52–55.

Dominic J. Colletti, "Using Class A Foam for Structure Firefighting, Part 4: Compressed Air Foam Systems" *Firehouse* 23, no. 9 (1998): 122–125.

Glenn Corbett, ed., *Fire Engineering's Handbook for Firefighter I and II* (Tulsa, OK: PennWell Corporation—Fire Engineering, 2009).

Arthur E. Cote, *Operation of Fire Protection Systems—A Special Edition of the Fire Protection Handbook* (Quincy, MA: NFPA, 2003).

Jeff Cotner, "Our Department Has Class A Foam; Now What?" *Fire Engineering* 160, no. 7 (2007): 121–122.

A. Maurice Jones, Jr., *Fire Protection Systems* (Clifton Park, NY: Delmar Cengage Learning, 2009).

National Fire Protection Association (NFPA), *NFPA 11: Standard for Low-, Medium-, and High-Expansion Foam* (Quincy, MA: NFPA, 2010).

National Fire Protection Association (NFPA), *NFPA 13: Standard for the Installation of Sprinkler Systems* (Quincy, MA: NFPA, 2013).

National Fire Protection Association (NFPA), *NFPA 16: Standard for the Installation of Foam-Water Sprinkler and Foam-Water Spray Systems* (Quincy, MA: NFPA, 2011).

National Fire Protection Association (NFPA), *NFPA 409: Standard on Aircraft Hangars* (Quincy, MA: NFPA, 2011).

Craig H. Shelley, Anthony R. Cole, and Timothy E. Markley, *Industrial Firefighting for Municipal Firefighters* (Tulsa, OK: PennWell Corporation—Fire Engineering, 2007).

CHAPTER 7

Carbon Dioxide

Source: Ronald R. Spadafora

KEY TERMS

Global warming potential, *p. 119*
Greenhouse gas, *p. 119*
Ozone-depletion potential, *p. 119*
Atmospheric lifetime, *p. 119*
Phase diagram, *p. 119*
Triple point, *p. 120*

Critical temperature, *p. 120*
Critical point, *p. 120*
Sublimation, *p. 120*
Asphyxiation, *p. 121*
Cryogenic, *p. 121*
Dewar, *p. 121*

Drainboard, *p. 122*
Dip tank, *p. 123*
Surface fire, *p. 125*
Deep-seated fire, *p. 125*
Baghouse, *p. 125*

OBJECTIVES

After reading this chapter, the reader should be able to:

- List the physical and chemical properties of carbon dioxide (CO_2)
- Explain carbon dioxide fire extinguishment applications
- Compare total-flooding and local-application systems
- Evaluate the risks of using carbon dioxide
- Describe the two major types of carbon dioxide storage systems
- Describe firefighting strategy and tactics

Introduction

Carbon dioxide (CO_2) is used throughout the world for non-water-based fire extinguishing systems in the fire protection industry. It suppresses fire without leaving behind residue that can damage sensitive equipment. This result can lead to commercial and industrial occupancies getting back in business in a timely fashion. Carbon dioxide is a plentiful, noncorrosive gas. It is commonly compressed and cooled to its liquid form for storage and transportation. Upon release, it discharges as a gas under its own pressure, giving the appearance of steam. Its low temperature causes water in the air to crystallize, creating "snow" particles that aid in heat absorption. The discharge of carbon dioxide on a fire is designed to displace atmospheric oxygen, which supports combustion. The lack of free oxygen necessitates that building occupants be trained to immediately evacuate the area in which carbon dioxide systems will activate.

Overview

The benefits of carbon dioxide gas for fire extinguishment have been known for a long time. In 1914, the Bell Telephone Company of Pennsylvania installed portable CO_2 extinguishers for use on energized electrical wiring and equipment. By the 1920s, automatic fire extinguishing systems using carbon dioxide were available. In 1928, the National Fire Protection Association (NFPA) began work on formalized requirements for carbon dioxide extinguishing systems. By 1960, halon extinguishing systems were introduced and began making significant inroads within the market, but carbon dioxide continues to be used as an extinguishing agent in many applications, including gas and flammable liquid fires as well as fires involving electrically energized equipment (because it is electrically nonconductive). It also has limited use in fires involving Class A materials. Carbon dioxide is not effective in extinguishing fires in chemicals containing their own oxygen (cellulose nitrate). It is also not used for compounds (metal hydrides) or reactive metals (magnesium, sodium, and potassium), because they decompose carbon dioxide.[1]

Properties of Carbon Dioxide

Carbon dioxide is a colorless, tasteless, and odorless gas that is soluble in water. It is an attractive fire suppressant for several reasons. As a gas, it provides its own pressurization for discharge from its storage container. Because it is a gas under ambient conditions, it also provides three-dimensional fire extinguishment. Carbon dioxide can penetrate into the seam openings of burning equipment, enhancing its extinguishment efficiency. It is 1.5 times as heavy as air; when discharged into the atmosphere it will collect at low elevations (see Figure 7.1).

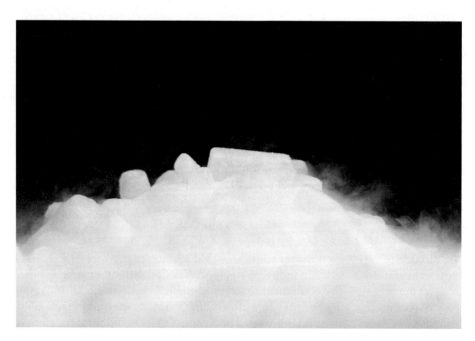

FIGURE 7.1 Carbon dioxide vapor is heavier than air and will accumulate in low areas. *Source:* Nikkytok/Shutterstock

| TABLE 7.1 | Physical Properties of Carbon Dioxide |

Chemical name: Carbon dioxide
Synonym: Carbon anhydride, carbonic acid gas, carbonic anhydride, dry ice
CAS registry number: 124-38-9
UN1013 (gas); UN2187 (liquid refrigerated); UN1845 (solid)

	US UNITS	SI UNITS
Chemical formula	CO_2	CO_2
Molecular weight	44.01	44.01
Vapor pressure* at 2°F (−16.7°C)	302 psig	2082 kPa
Specific gravity at 70°F (21.1°C) and 1 atm	1.522	1.522
Critical temperature	87.9°F	31.1°C
Critical pressure	1070.6 psia	7381.8 kPa abs
Triple point	−69.9°F at 75.1 psia	−56.6°C at 518 kPa, abs

*All psig values are referenced to 14.696 psia (101.325 kPa, abs).

PHYSICAL PROPERTIES

A look at Table 7.1[2] provides valuable information about the physical properties of carbon dioxide. This data will allow first responders to better evaluate its beneficial fire extinguishing characteristics as well as the hazards presented when discharged.

ENVIRONMENTAL PROPERTIES

Carbon dioxide has a **global warming potential** (GWP) of 1 over all time periods. GWP is a measure of how much heat a **greenhouse gas** traps in the atmosphere relative to the same mass of carbon dioxide and is evaluated for a specific timescale, commonly 20, 100, or 500 years. Carbon dioxide has zero **ozone-depletion potential** (ODP). It also has a variable **atmospheric lifetime** (ATL) that cannot be specified precisely. It is estimated in the order of five to 200 years.

Phases

With the exception of its reaction with water to form carbonic acid, carbon dioxide is generally an inactive compound. Figure 7.2 is a **phase diagram** that shows the relationship between temperature and pressure and the vapor, liquid, and solid phases of carbon dioxide. The solid phase can exist only at temperatures less than approximately −70°F (−57°C) in the pressure region above the solid curve known as the saturation curve. This demonstrates the relationship between temperature and pressure inside a container. The **triple point** of carbon dioxide is the temperature and pressure condition at which all three phases of carbon dioxide can exist in equilibrium. This triple point corresponds to a temperature of approximately −70°F (−57°C) and a pressure of about 75 pounds per square inch absolute (psia) or 10.9 kPa abs.* The **critical temperature** of carbon dioxide is 87.8°F (31°C). Above this temperature, carbon dioxide cannot be liquefied regardless of the pressure applied and is considered at its **critical point** (see Figure 7.2).

---

*Pounds per square inch absolute (psia) is used to denote that the pressure is relative to a vacuum rather than the ambient atmospheric pressure. At sea level, atmospheric pressure is approximately 14.7 psi. Therefore, a psia reading would include atmospheric pressure. Conversely, a pounds per square inch gauge (psig) reading indicates that the pressure is relative to atmospheric pressure. For example, a football pumped up to 13 psi above atmospheric pressure will have a pressure of 13 + 14.7 = 27.7 psia or 13 psig.

Global warming potential ■ A relative measure of how much heat a gas traps in the atmosphere compared to the amount of heat trapped by a similar mass of carbon dioxide.

Greenhouse gas ■ Gas that allows direct sunlight (ultraviolet) to reach the Earth's surface unimpeded yet absorb heat that is reradiated back into the atmosphere. This process allows less heat to escape. Many greenhouse gases occur naturally in the atmosphere, such as carbon dioxide, methane, and nitrous oxide, whereas others are synthetic (man-made). Synthetic greenhouse gases (GHGs) include the chlorofluorocarbons (CFCs) and the halocarbons (a group of gases containing fluorine, chlorine, and bromine).

Ozone-depletion potential ■ The ozone-depletion potential (ODP) of a chemical compound is the relative amount of degradation to the ozone layer it can cause; trichlorofluoromethane (R-11 or CFC-11) is fixed at an ODP of 1.0.

Atmospheric lifetime ■ Measurement of the time required to restore equilibrium following a sudden increase or decrease in its concentration in the atmosphere. Average lifetimes can vary from about a week (sulfate aerosols) to more than a century (carbon dioxide).

Phase diagram ■ A chart showing thermodynamic conditions of a specified substance at varying pressures and temperatures.

Triple point ■ The temperature and pressure at which a substance can exist in equilibrium in the liquid, solid, and gaseous states.

Critical temperature ■ The temperature above which a gas cannot be liquefied regardless of the pressure applied.

Critical point ■ At the critical point, there is no change of state when pressure is increased or if heat is added. The liquid and gas phases can't be distinguished.

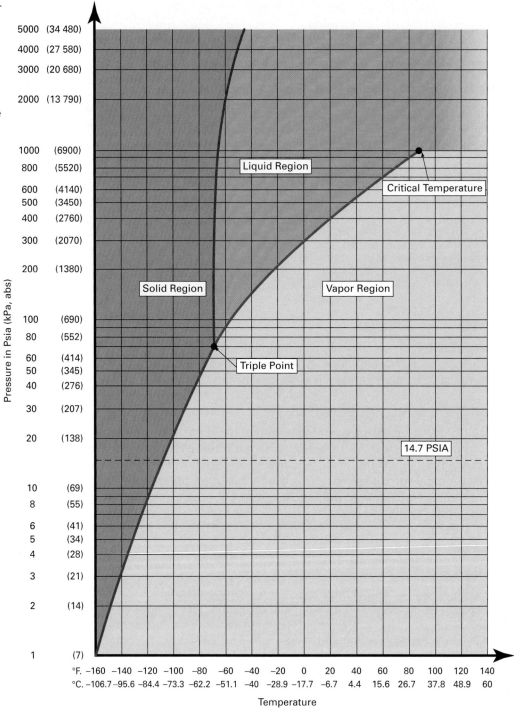

FIGURE 7.2 Carbon dioxide phase diagram

SOLID

Sublimation ■ The conversion of a solid directly into the gaseous state.

In its solid state, carbon dioxide is also known as dry ice. The solid region exists at temperatures less than −70°F (−57°C) and at pressures above the curved, dark line in Figure 7.2. Dry ice undergoes **sublimation** (direct phase change) to a gaseous state at −109.3°F (−78.5°C); this gaseous form is heavier than air and can accumulate in low-lying areas.

120 **Chapter 7** Carbon Dioxide

DANGERS OF DRY ICE

A 59-year-old, previously healthy man was checking on the status of a walk-in freezer in which ice cream was routinely stored at −10°F (−23°C). The previous day, the door on the freezer had been replaced due to a poor seal. To prevent the ice cream from melting during the repair, blocks of dry ice wrapped loosely in plastic were placed in coolers with lids that did not latch. The dry ice was not removed after the new door was installed.

Twenty minutes after entering the cooler, the victim was found collapsed inside the closed freezer with evidence of vomit in the airway. The victim was pulled partially out of the freezer and bystander cardiopulmonary resuscitation (CPR) was instituted. Emergency medical personnel arrived within six minutes. Advanced life support efforts were initiated. Upon arrival at the hospital, the victim's temperature was approximately 94°F (34°C), and further resuscitative efforts were terminated. The emergency physician became concerned after learning that the freezer had been repaired the day before and notified the Medical Examiner's (ME's) office. The ME visited the location of the freezer and determined that the new door was working properly. Death by **asphyxiation** due to carbon dioxide exposure was suspected, and the San Diego Hazardous Materials Incident Response Team (HIRT) was asked to investigate.

Upon opening the door of the freezer, HIRT members observed dry ice sublimating under the lids of coolers and spilling onto the floor. They deployed equipment to monitor the atmosphere in the walk-in freezer. Initial measurements with gas detectors showed a 13% oxygen concentration. The reading changed with the opening and closing of the door over the next 30 minutes, but the freezer always remained oxygen deficient. Based upon these readings, the HIRT estimated that the carbon dioxide levels inside the freezer could be as high as 40%. This case illustrates the lethal consequences of improper storage of dry ice.[3]

Asphyxiation ▪ A condition in which an extreme decrease in the concentration of oxygen in the body accompanied by an increase in the concentration of carbon dioxide leads to loss of consciousness or death.

LIQUID

Liquid carbon dioxide is also known as a **cryogenic** liquid. Cryogenic liquids are gases at normal temperatures and pressures but are kept in the liquefied phase by using very low temperatures and moderate pressures. Cryogenic liquids have boiling points below −238°F (−150°C). Although carbon dioxide has a slightly higher boiling point, it is sometimes included in this category. Carbon dioxide must be cooled below room temperature before an increase in pressure can liquefy it.

Different cryogens become liquids under different conditions of temperature and pressure, but all have two properties in common: They are extremely cold, and small amounts of liquid can expand into very large volumes of gas. The vapors released from cryogenic liquids remain very cold. They often condense the moisture in air, creating a highly visible fog. In poorly insulated containers, some cryogenic liquids actually condense the surrounding air, forming a liquid–air mixture.

Cryogenic ▪ Producing or related to low temperatures.

COMPRESSED GAS AND LIQUEFIED GAS PORTABLE CONTAINERS, CYLINDERS, AND VESSELS

A common use of carbon dioxide is to carbonate soft drinks. Businesses that serve fountain sodas and soft drinks use this gas. These occupancies include places of public assembly, bars, restaurants, sports arenas, fast-food chains, banquet halls, movie theaters, playhouses, cafeterias, and nightclubs. Traditionally, carbon dioxide being used in this fashion has not posed a hazard to first responders. Historically, high-pressure carbon dioxide gas cylinders have been common for this purpose, and therefore a relatively small amount of the gas has been stored.

Today, an insulated vessel called a **Dewar** is often purchased to store the carbon dioxide in its liquid form. Dewars are like thermos bottles. They help to minimize the loss of product that occurs when the liquid warms up and reverts back to a gas. Due to this constant change of phase from liquid to gas, however, long-term storage is not possible. A Dewar slowly releases gas if it is unused, completely emptying itself in approximately two weeks. The major upside to a Dewar is that one gallon (3.8 L) of liquid carbon dioxide equals 74 ft.³ or 2 m³ of gas! This, however, is a great danger to firefighters responding and operating at a carbon dioxide leak emergency, because large amounts of released CO_2 will quickly displace oxygen.

Dewar ▪ A container with double walls separated by a vacuum that is used to maintain substances at high or low temperatures. Named for its inventor, Sir James Dewar (1842–1923), a Scottish chemist and physicist.

MCDONALD'S RESTAURANT TRAGEDY

Carbon dioxide leaking into the walls of a Pooler, GA, McDonald's caused an 80-year-old woman to die from asphyxiation on September 7, 2011. It was the third time in six years that carbon dioxide caused a medical emergency at a US McDonald's. The woman died at a local hospital the day after carbon dioxide, used by the restaurant to carbonate their sodas, leaked into the women's bathroom. Another woman was also found unconscious in the restaurant bathroom. The carbon dioxide gas line ran from storage tanks located in a rear room of the McDonald's up and over the adjacent women's bathroom at ceiling level for approximately 25 ft. (7.6 m) before reaching the drink machines. A line used to funnel excess carbon dioxide out of the restaurant was disconnected. The gas flowed into the wall next to the women's bathroom instead of going outside to open air. The storage tanks were being refilled at the time of the incident. Nine people, including three firefighters, a McDonald's employee, and a family of three who tried to help the victims, were taken to the hospital with dizziness and trouble breathing.[4]

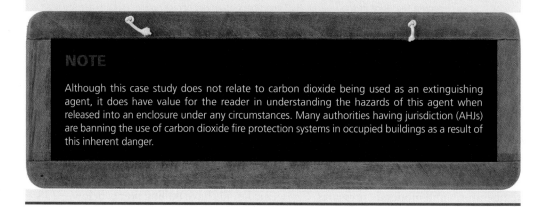

NOTE

Although this case study does not relate to carbon dioxide being used as an extinguishing agent, it does have value for the reader in understanding the hazards of this agent when released into an enclosure under any circumstances. Many authorities having jurisdiction (AHJs) are banning the use of carbon dioxide fire protection systems in occupied buildings as a result of this inherent danger.

VAPOR

Carbon dioxide is in its vapor phase when temperatures exceed $-120°F$ ($-84°C$) and pressures are less than indicated by the long, curved, dark line in Figure 7.2. It will not support combustion nor will it react with most materials.

Fire Extinguishment Applications

Carbon dioxide extinguishing systems have many applications. They are installed to protect computer rooms (subfloor), particle board chippers, equipment dust collectors, printing presses, dip tanks and **drainboards**, cable trays, electrical rooms, motor control centers, switch gear locations, paint spray booths, hooded industrial fryers, high-voltage transformers, nuclear power facilities, waste storage facilities, aircraft cargo areas, and vehicle parking areas. In addition, they are used internationally for marine applications in the protection of engine rooms, paint lockers, vehicle transport areas on cargo vessels, and flammable liquid storage areas. The steel and aluminum industries also depend upon carbon dioxide for fire protection. Many carbon dioxide systems in the metal-processing industry are rapid-discharge, direct local-application systems. This type of application is obtained by locating carbon dioxide storage containers close to the outlet nozzles. Carbon dioxide is also used to create an inerting atmosphere for handling flammable materials and for welding operations.

Drainboard ■ An inclined surface designed to allow flammable liquid residue from a dipped object to drain back into the dip tank.

Local-Application Extinguishing Systems

A major way in which carbon dioxide is used for fire extinguishment is known as local application. This type of system is generally appropriate for the extinguishment of surface fires in flammable liquids and gases. If an area is not able to contain carbon dioxide, local application may be the proper system to install. Concerns for occupant safety will also come into play in the decision-making process when choosing the local-application method of extinguishment. The concentration of carbon dioxide may still be a hazard to building occupants.

The effectiveness of a local-application system is dependent upon many variables. The hazard must be identifiable, and the combustible materials must be compatible for extinguishment by a carbon dioxide system. Discharge nozzles are intended to avoid splashing and the entrainment of air. Although normally installed inside buildings, local-application systems can also be placed outdoors. When used in this manner, shielding is provided to minimize disruption of the discharging agent by the wind.[5]

DIP TANKS

In a local-application system for a **dip tank**, nozzles discharge their agent directly at the localized fire hazard. Only liquid discharge is effective. Sector valves allow the application of carbon dioxide onto select areas of the tank. The entire fire hazard is covered in carbon dioxide without actually filling the room or enclosure to a predetermined concentration.

Dip tank ■ A tank containing flammable liquid used for dipping, coating, or stripping an object.

Total-Flooding Extinguishing Systems

Total-flooding extinguishing systems (see Figure 7.3) protect enclosures from exposed solid material fires as well as flash fires involving burning liquid and gas. They work by discharging a predetermined percentage of gas for a required "soaking" period of time. The enclosed space containing the fire hazard must therefore be reasonably leakproof. System design includes the automatic closing of openings (doors, windows, louvers, etc.) prior to or simultaneously with the introduction of carbon dioxide. If forced-air ventilation exists inside the space, shutdown of the system and/or closure of shutters, dampers, and vents should also occur upon activation of a total-flooding system. Also, any leakage that does occur should be compensated for by the fire protection designer through the application of additional carbon dioxide into the enclosure, as noted in *NFPA 12: Standard on Carbon Dioxide Extinguishing Systems*.

These systems consist of an in-house supply of carbon dioxide permanently connected to stationary piping. Nozzles are attached to this piping and are positioned so as to not disrupt burning materials when discharging their agent. They are also arranged so that the effects of discharge do not damage or destroy the equipment or materials being protected. In total-flooding systems, the nozzles are generally located in the upper level of the enclosure. If the flooding area has a high ceiling, additional nozzles may have to be installed at approximately one-third of the room height.

Integrity of Enclosure

Noted ahead are some design features, evaluations, tests, inspections, and installations that a fire protection engineer would include and/or recommend in order to help ensure that the enclosure with a carbon dioxide total-flooding system will contain the extinguishing agent

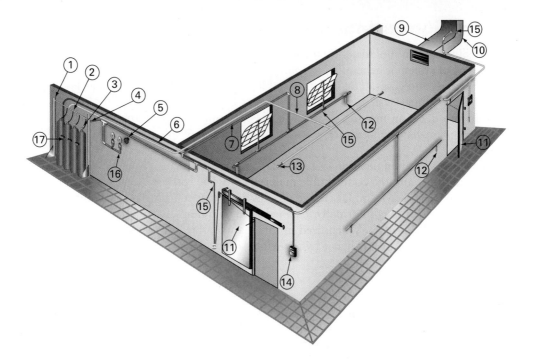

1. Cylinder Framing
2. Cylinder Manifold
3. Pressure-Operated Discharge Head
4. Flexible Connection
5. Alarm Gong
6. Remote-Control Cable (1/16-in.) (1.6-mm), run in 3/8-in. (9.5-mm) galvanized pipe or conduit with corner pulleys at all changes in directions; no bends or offsets allowed
7. Carbon Dioxide Piping
8. Actuator Tubing, run in 1/2-in. (13-mm) conduit
9. Self-Closing Weight-Operated Damper
10. Air-Exhaust Duct
11. Fire Door
12. Nozzle
13. Actuator on Ceiling
14. Remote-Control Pull Box
15. Pressure-Operated Trip to Release Self-Closing Damper Fire Doors, and Windows
16. Pressure-Operated Switches to Sound Alarm and Slow Down Fan
17. Pneumatic-Control Head with Local Manual Control

FIGURE 7.3 Carbon dioxide total-flooding system

for the required duration of time needed to successfully put out the fire hazard. System requirements may include the following:

- Install automatic closure devices for all openings.
- Measure openings that cannot be covered for computations regarding supplemental agent.
- Test the air-holding integrity of the enclosure by using a fan with a calibrated pressure-reading device.
- Securely fasten and seal acoustical ceiling tiles.
- Inspect and seal wall joints and floor openings that may be avenues for escaping agent.
- Inspect the walls of the enclosure for rigidity and integrity. The positive pressure caused by a carbon dioxide release can blow out lightweight walls and windows. Wall systems may have to be reinforced.
- Inspect the heating, ventilation, and air-conditioning (HVAC) system. It is important that detectors are installed to isolate the HVAC system before and during the activation of the carbon dioxide system. Detectors will be used to close off supply air into the space automatically via the activation of a duct damper. Return air will also be controlled by detectors through the use of dampers.

Minimum Carbon Dioxide Concentrations for Extinguishment

Note that theoretical minimum concentration percentages are not intended to be used as design concentrations. They are reference numbers gathered through testing under laboratory conditions. In general, most enclosures do not meet the integrity standards of the laboratory, and therefore the minimum design concentration percentage is used. The lowest minimum design concentration percentage for the group of materials listed in Table 7.2[6] is 34%. This figure holds true in most AHJs. For materials that are not given, the minimum theoretical carbon dioxide concentration will be obtained from a recognized source or determined by testing.

SURFACE FIRES AND DEEP-SEATED FIRES

Total-flooding systems are used for both **surface fires** and **deep-seated fires**. For surface fires—of liquid fuels, for example—a minimum design concentration of 34% carbon dioxide by volume is mandated. Most surface burning and open flaming will stop when the concentration of carbon dioxide in the air reaches approximately 20%. There is, however, a considerable safety factor built into minimum carbon dioxide concentrations required by NFPA 12.

Deep-seated fires require greater amounts of carbon dioxide by volume to extinguish effectively. Minimum design concentrations generally begin well above the percentages required for surface fires. For instance, 50% design concentration is used for hazards involving energized electrical equipment. Archives, records, and bulk paper storage require a 60% minimum design concentration. A **baghouse** type of dust collector needs a 75% minimum design concentration.[7] Deep-seated hazard systems are discharged at a slower flow rate and for a longer duration of time to counter the characteristics of a smoldering fire. The design concentration is achieved within seven minutes.

For deep-seated fires, total-flooding systems must also allow time for the heat that is stored in the fuel to dissipate in order to prevent reflash. Smoldering must also be eliminated. To accomplish this, the concentration of carbon dioxide is held for a longer "soaking" period to permit adequate dissipation of built-up heat. NFPA 12 requires at least a 20-minute "soaking" time.[8]

Surface fire ■ A fire on the outside or exterior of a substance.

Deep-seated fire ■ A fire burning below the surface of combustible materials (mulch and peat moss, for example), as contrasted with a surface fire.

Baghouse ■ An air pollution–control device used at power plants, steel mills, pharmaceutical facilities, food manufacturers, chemical producers, and other industrial companies.

EXTENDED DISCHARGE SYSTEM

An extended discharge system will provide compensatory amounts of carbon dioxide to account for avenues of gas leakage inside the enclosure. Without such a system, in some cases minimum design concentration percentages would not be attained. This detached assembly of stored carbon dioxide provides a supplementary amount of extinguishing agent through a separate selector valve upon completion of the primary discharge.

Hand Hose-Line Extinguishing Systems

Although not a substitute for local and total-flooding systems, hand hose lines can supplement them. They can be connected by means of permanent piping to a storage supply of carbon dioxide. Hand hose lines also may be used off of a standpipe outlet where they are supplied by containers of carbon dioxide mounted upon mobile units that can be moved quickly and coupled to a standpipe inlet. Hand hose lines can be used by in-house fire brigade members or firefighters. The hazard being protected should be within a reasonable distance of both the permanent piping and standpipe outlet systems.

TABLE 7.2 — Minimum Carbon Dioxide Concentrations for Extinguishment

MATERIAL	THEORETICAL MINIMUM CO_2 CONCENTRATION (%)	MINIMUM DESIGN CO_2 CONCENTRATION (%)
Acetylene	55	66
Acetone	27*	34
Aviation gas grades 115/145	30	36
Benzol, benzene	31	37
Butadiene	34	41
Butane	28	34
Butane-l	31	37
Carbon disulfide	60	72
Carbon monoxide	53	64
Coal or natural gas	31*	37
Cyclopropane	31	37
Diethyl ether	33	40
Dimethyl ether	33	40
Dowtherm	38*	46
Ethane	33	40
Ethyl alcohol	36	43
Ethyl ether	38*	46
Ethylene	41	49
Ethylene dichloride	21	34
Ethylene oxide	44	53
Gasoline	28	34
Hexane	29	35
Higher paraffin hydrocarbons	28	34
Hydrogen	62	75
Hydrogen sulfide	30	36
Isobutane	30*	36
Isobutylene	26	34
Isobutyl formate	26	34
JP-4	30	36
Kerosene	28	34
Methane	25	34
Methyl acetate	29	35
Methyl alcohol	33	40
Methyl butene-l	30	36
Methyl ethyl ketone	33	40
Methyl formate	32	39
Pentane	29	35
Propane	30	36
Propylene	30	36
Quench, lube oils	28	34

Note: The theoretical minimum CO_2 extinguishing concentrations in air for the materials in the table were obtained from a compilation of Bureau of Mines, Bulletins 503 and 627. Minimum CO_2 concentrations for the materials in the table are for deep-seated fires.
*Calculated from accepted residual oxygen values.

Source: "Minimum Carbon Dioxide Concentrations for Extinguishment" from NFPA 12-2012 Standard on Carbon Dioxide Extinguishing Systems. Copyright © 2010. Used by permission of National Fire Protection Association. This reprinted material is not the complete and official position of the NFPA on the referenced subject, which is represented only by the standard in its entirety.

FIGURE 7.4 Carbon dioxide portable fire extinguisher with conical applicator nozzle.
Source: Ronald R. Spadafora

Portable Fire Extinguishers

Carbon dioxide portable fire extinguishers are most effective on liquid/gas and electrical fires. They are normally ineffective on ordinary combustible fires, because although carbon dioxide excludes oxygen it does not significantly cool the burning materials. When the carbon dioxide disperses, Class A commodities can reignite upon exposure to atmospheric oxygen. Because CO_2 disperses quickly, these extinguishers are only effective from 3 to 8 ft. (0.9 to 2.4 m). Carbon dioxide is stored as a compressed liquid in the portable extinguisher; upon release it expands and transforms into dry ice "snow," cooling the surrounding air. It may cause frostbite upon contacting human skin in some cases (see Figure 7.4).

 CO₂ PORTABLE FIRE EXTINGUISHER MISHAP

A 20-year-old man was admitted to a hospital burn care unit with frostbite involving his left foot. His frostbite injury resulted from contact with the discharge from a CO_2 fire extinguisher. While he was playing football, the patient sprained his ankle and complained of severe pain. His teammates used the CO_2 fire extinguisher as a cooling spray. The carbon dioxide gas spray was sustained for a period of 30 to 45 seconds. Immediately after the exposure, the skin of his foot went white and cold. Afterwards, he initially felt numbness, followed by increasing foot pain.

Upon arrival at the hospital, two hours later, the patient was immediately taken to a cleaning tank. His foot was irrigated with sterilized water for 15 minutes. The initial physical examination demonstrated second-degree frostbite of the left foot. After 24 hours, the top portion of the left foot was inspected and blisters were present. All hemorrhagic blisters were punctured and debrided under sedation. Burn areas were dressed every day and healed gradually over two weeks[9] (see Figure 7.5).

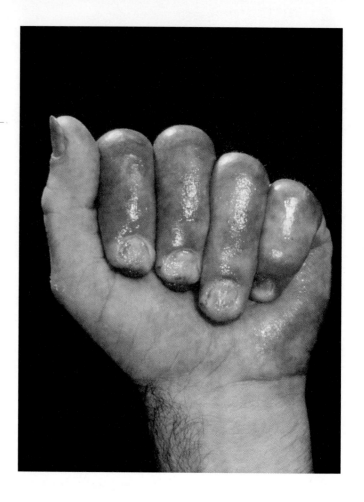

FIGURE 7.5 Second-degree frostbite of the hand sustained as a result of exposure to a carbon dioxide fire extinguisher gas spray for a period of 30 to 45 seconds.
Source: SIU Biomed Com/Custom Medical Stock Photo

Examining the Risks

The risk involved with the use of carbon dioxide systems is based on the fact that the amount of extinguishing agent needed to suppress fires and protect the contents of the enclosure is many times greater than the lethal concentration for humans. For instance, the minimum design concentration to suppress a kerosene fire is 34%. This concentration of carbon dioxide can produce convulsions, unconsciousness, and death within a very short period of time. In addition, carbon dioxide storage areas are often relatively small compared to the protected enclosures. Inadvertent discharges or leakages into these storerooms will also produce carbon dioxide concentrations much higher than the lethal level. The consequences of exposure happen quickly, and therefore there is little or no margin for error. A look at Table 7.3[10] reveals that as little as 2% of carbon dioxide in an enclosed space can have a detrimental effect on the body.

Air normally contains about 21% oxygen, 78% nitrogen, and 1% other gases, principally argon. Large amounts of carbon dioxide will displace oxygen in a space and lead to asphyxiation. In general, a minimum concentration of 15% oxygen in the air is needed to support flaming combustion. Smoldering combustion, however, can continue in an atmosphere with as little as 3% oxygen. Detrimental physical symptoms can prevent building occupants from escaping when oxygen levels drop below 17%. When oxygen is diminished by carbon dioxide to below 10%, marked impairments can occur. In areas in which the oxygen levels fall below 10%, the danger of victims falling unconscious is all too real.[11] Symptoms associated with oxygen deficiency are shown in Table 7.4.[12]

TABLE 7.3 Acute Health Effects of High Concentrations of Carbon Dioxide (with Increasing Exposure Levels of Carbon Dioxide)

CONCENTRATION (% CARBON DIOXIDE/AIR)	TIME	EFFECTS
2%	Several hours	Headache, dyspnea upon mild exertion
3%	One hour	Dilation of cerebral blood vessels, increased pulmonary ventilation, and increased oxygen delivery to the tissues
4%–5%	Within a few minutes	Mild headache, sweating, and dyspnea at rest
6%	1–2 minutes	Hearing and visual disturbances
	< 16 minutes	Headache and dyspnea
	Several hours	Tremors
7%–10%	A few minutes	Unconsciousness or near unconscious
	1.5 minutes–1 hour	Headache, increased heart rate, shortness of breath, dizziness, sweating, rapid breathing
10%–15%	1+ minute	Dizziness, drowsiness, severe muscle Twitching, and unconsciousness
17%–30%	< 1 minute	Loss of controlled and purposeful activity, unconsciousness, convulsions, coma, and death

United States Environmental Protection Agency, "Carbon Dioxide as a Fire Suppressant: Examining the Risks."

Total-flooding carbon dioxide systems are designed so that human exposure to the agent does not occur during a fire suppression incident. Discharge time delays and alarms are required, as denoted in NFPA 12, to prevent such occurrences. Activation of a single detector will initiate an audible and visual alarm capable of being perceived above ambient light and noise levels. This is intended to prevent building occupants from entering the protected area and allow them to safely exit. A signal will also be relayed to a fire alarm control unit (FACU), and auxiliary contacts are operated for air-conditioning shutdown. Upon completion of a time delay, the system will cause a discharge alarm to be activated prior to releasing carbon dioxide. During release, auxiliary contacts for emergency

TABLE 7.4 Symptoms of Oxygen Deficiency

PERCENT OF OXYGEN IN THE AIR	HUMAN SYMPTOMS
20.9%	Typical oxygen in air: no symptoms
19.5–16%	Any form of exertion can cause increased heart/breathing rate and impaired judgment/perception/muscular coordination
16–12%	At rest: increased heart/breathing rate, dizziness, headache, fatigue, faulty judgment
10–6%	Nausea, vomiting, unconsciousness, ashen face, blue lips
<6%	Convulsions, breathing cessation, cardiac arrest

Note: The Occupational Health and Safety Administration (OSHA) notes 19.5 percent oxygen as the level below which an oxygen-deficient atmosphere exists.
Based on Texas State University—San Marcos, *Safety Manual*, p. 6-9.

powering off of all electrical equipment will be activated. This shutdown excludes lighting and emergency circuits for life safety. Alarm bells, horns, and horn/strobe appliances are commonly installed. Strobes are positioned outside and above each exit door from the protected area.

In occupied buildings, carbon dioxide extinguishment systems are also configured so that a second independent detector activation is required to initiate discharge. In this design, the sequence of operations starts with the activation of a detector providing an alarm but not initiating time-delay discharge. A second detector activation is needed for the discharge countdown to begin.

A manual pull station or release switch provides a means of physically discharging the carbon dioxide. They are installed at each exit from the protected hazard. An advisory sign identifying the activation device and its function will be nearby. Manual release of a carbon dioxide system will bypass the time-delay function and will also cause shutdown switches to operate as if the system had activated automatically.[13]

A review of accidental deaths or injuries related to carbon dioxide use in fire protection reveals that fatal accidents most often occur during maintenance and testing. Inadvertently activating the system due to a lack of adequate safety procedures to prevent such discharges and failure to adhere to safety protocols are major causes for fatalities. The risk associated with the use of carbon dioxide for fire protection in enclosures is not well understood by the maintenance workers who perform functions on or around carbon dioxide systems. Evacuation is particularly difficult once discharge begins because of reduced visibility (fogging), the loud noise of activation, and the disorientation resulting from the physiological effects of carbon dioxide. Precautionary measures must be mandated to ensure that these personnel follow strict guidelines regardless of the type of work being performed.[14]

Firefighting Operations

First responders encountering carbon dioxide in any of its three phases should be thoroughly familiar with the associated hazards. There are several conditions under which extreme danger to personnel and equipment may exist. The following information is provided for formulating procedures and guidelines to prevent incidents that can lead to serious injury or death.

- Always wear the proper personal protective equipment (PPE). The use of gloves and eye protection when handling equipment containing vapor, liquid, and solid carbon dioxide is essential to staying safe. Skin, mouth, or eye contact with solid carbon dioxide can lead to lesions, corneal burn, or more serious injury from deep freezing of the tissues. Contact between exposed skin and cold piping or carbon dioxide vapor can also cause frost burns. Liquid carbon dioxide discharging from a container produces high-velocity abrasive particles, in addition to being cold, which can cause eye and skin injury.
- Carbon dioxide monitoring should be carried out by firefighters before entering any confined space or low area in which this gas may have accumulated. The carbon dioxide should be removed by ventilation fans to a negligible concentration and a positive pressure, self-contained breathing apparatus (SCBA) must be donned before entering.
- Do not attempt to remove anyone exposed to high concentrations of carbon dioxide without using proper PPE and rescue equipment. Would-be rescuers account for a large percentage of confined-space fatalities.
- If a person has inhaled large amounts of carbon dioxide and is exhibiting adverse effects, move the victim to fresh air at once. If breathing has stopped, perform

- artificial respiration. Administer oxygen to the victim. Keep the affected person warm and at rest. Fresh air and assisted breathing is appropriate for all cases of overexposure to gaseous carbon dioxide.
- Carbon dioxide in the gaseous state is not easily detected due to its physical properties. Do not rely solely on measuring the oxygen content of the air, because high concentrations of carbon dioxide can be dangerous even with adequate oxygen to support life.
- The physiological effects of carbon dioxide are unique, because it is a product of normal metabolism. The respiratory control system maintains carbon dioxide pressure at a relatively high level (50 mm Hg) in the arterial blood and tissue fluids. This maintains the acidity of the tissue and cellular fluids at the proper level for essential metabolic reactions and membrane functions. Changes in the normal carbon dioxide tissue pressure can be damaging. The effects produced by low and moderate concentrations of carbon dioxide are, however, reversible.
- Liquid carbon dioxide in a hose or pipe flows like water. When the pressure is reduced below 75.1 psia (518 kPa abs), the liquid changes into a mixture of vapor and carbon dioxide in solid form. Solid carbon dioxide (dry ice) when formed in a pipe or hose may create a "plug" that prevents depressurization and thereby creates a safety hazard. This can lead to hose or pipe rupture. As the dry ice sublimes, the gas pressure buildup can forcibly eject the plug. A dry ice plug can be pushed out from any open end of a hose or pipe with enough force to cause serious injury.
- Depressurization of a liquid carbon dioxide system can result in low-temperature liquid carbon dioxide and/or the formation of dry ice that causes the container, piping, and hoses to be brittle. Many materials safe to use at normal liquid carbon dioxide temperatures may fail if stressed when subjected to dry ice temperatures.
- Liquid carbon dioxide trapped between two closed valves will increase in pressure as it warms and expands. This pressure can exceed what the piping and hoses can withstand, leading to rupture of hose or piping, with possible serious injury and property damage. To prevent trapped liquid from becoming a hazard, all liquid carbon dioxide piping and transfer lines should be equipped with pressure relief devices located in all parts of the system (valves, check valves, and pumps) in which this phenomena can occur.
- Although low-pressure storage containers are generally equipped with refrigeration systems to maintain the liquid temperature at 0°F (−18°C), power failures can occur. Without backup power, temperature and pressure will rise, and the liquid carbon dioxide will expand, creating undue stress to the container.
- Carbon dioxide cylinders are constructed from steel or aluminum with a minimum rated service pressure of 1,800 psi (12,411 kPa). Aluminum cylinders subjected to heat or fire should not be reused. Temperatures in excess of 350°F (177°C) will irreversibly change the properties of the aluminum, and the cylinder must be condemned.
- The manufacturing of solid dry ice produces static electricity charges (>100,000 volts). This may lead to a discharge of the static electricity to any grounded object or person. Carbon dioxide "snow" or solid dry ice in combustible environments should be carefully evaluated.
- Liquid carbon dioxide should not be used for inerting combustible atmospheres because of the extremely high static charges produced during the formation of dry ice. Gaseous carbon dioxide can be used for inerting combustible atmospheres without the risk of generating static charges.

Appropriate warning signs should be placed at the entrance to confined areas in which high concentrations of carbon dioxide gas can accumulate. A typical warning is shown in Figure 7.6.

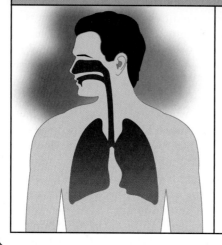

FIGURE 7.6 Warning sign used at every entrance to a space protected by CO_2 systems that include a wintergreen odorizer

Low-Pressure and High-Pressure Storage Systems

LOW PRESSURE

If storage area space is a factor, a low-pressure system is typically installed (see Figure 7.7). This eliminates the need to store dozens of high-pressure cylinders throughout the facility. General criteria for whether to install a low-pressure system or a high-pressure system is based upon the amount of product required to protect the hazards present. In general, when more than 4,000 lb. (1,814 kg) of carbon dioxide is needed a low-pressure system should be installed.[15] Low-pressure tank sizes, however, range from 500 lb. (227 kg) of storage capacity to tens of thousands of pounds of liquid carbon dioxide.

Reserve supplies are obtained by increasing the size of the storage tank. Both the main supply and a reserve supply are incorporated into the low-pressure system storage tank. It is also possible to design for simultaneous discharge into several hazards with a low-pressure system. There is no need for manifolds and valves as would be needed in high-pressure systems. With a low-pressure system, nearly all of the liquid in the storage container is effective for local application firefighting. This is not the case with high-pressure systems. When using high pressure for local application, at least 40% additional liquid is required.

The insulated carbon dioxide tank is a refrigeration unit. The temperature is kept at 0°F (−18°C), and the pressure is relatively low at above 300 psi (2,068 kPa); (see Figure 7.2 for temperature and pressure carbon dioxide liquid requirements). A compressor circulates refrigerant through coils located at the top of the tank. Tank pressure is controlled by condensation of carbon dioxide vapor.

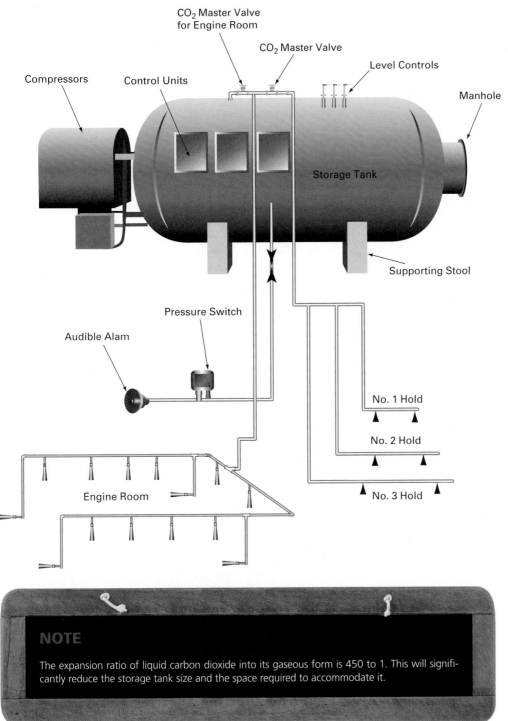

FIGURE 7.7 Carbon dioxide low-pressure storage system

> **NOTE**
>
> The expansion ratio of liquid carbon dioxide into its gaseous form is 450 to 1. This will significantly reduce the storage tank size and the space required to accommodate it.

A low-pressure system can protect a myriad of medium- to large-size total-flooding application hazards as well as local applications. It can also supply hand hose-line reels from a single storage container. Less hardware is required compared to a high-pressure system. Overall system cost for extinguishing agent, equipment, and installation is therefore reduced. A low-pressure system is also able to cover hazards at distances greater than 500 ft. (152 m) from the storage unit.

HIGH PRESSURE

A high-pressure system (see Figure 7.8) lends itself to covering small hazards with individual steel or aluminum cylinder storage banks located throughout the facility. These cylinders are also designed to store liquid carbon dioxide. They are kept at ambient room temperature. At 70°F (21°C), the internal pressure inside the cylinders reaches 850 psi (5,861 kPa). Storage cylinders are designed, tested, and filled to US Department of Transportation (DOT) specifications. The maximum filling density permitted is equal to 68% of the weight of water that the container can hold at 60°F (16°C). Cylinders are commonly designed to hold 5, 10, 15, 20, 25, 35, 50, 75, and 100 lb. (2.3, 4.5, 6.8, 9.1, 11.3, 15.9, 22.7, 34, and 45.4 kg) of carbon dioxide.[16] Each cylinder has its own individual release valve.

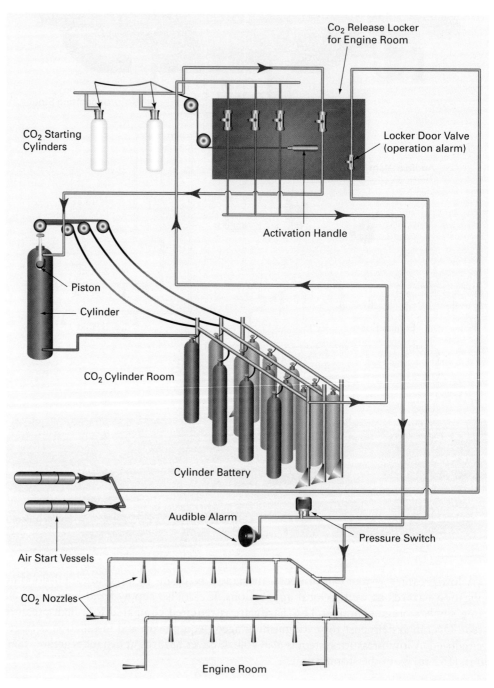

FIGURE 7.8 Carbon dioxide high-pressure storage system, ANSUL

Because the maximum capacity of a high-pressure cylinder is 100 lb., most carbon dioxide high-pressure systems consist of multiple cylinders that are manifolded together to provide the required design amount. Multiple hazard protection by a single bank of high-pressure cylinders is often limited by the distance between the hazard and the storage. With high pressure, less than 30% of the discharge is dry ice particles, lessening the effectiveness of local-application and hose-line applications.

In order to provide a reserve capability in a high-pressure system, a duplicate bank of cylinders must be connected to a common system discharge pipe. A manual switch commonly is provided to put the reserve bank of high-pressure cylinders online.

CHAPTER REVIEW

Summary

Carbon dioxide is used throughout the world for non-water-based fire extinguishing systems. It extinguishes fire by physically attacking all three elements of the fire triangle: fuel, heat, and oxygen. Fire extinguishment applications include local application, total flooding, hand hose lines, and portable fire extinguishers. There are risks involved with the use of carbon dioxide systems, however, because the amount of extinguishing agent needed to suppress fires is many times greater than the lethal concentration for humans. Design safeguards are implemented to prevent injury and death to occupants in areas in which carbon dioxide is discharged. First responders encountering carbon dioxide should be thoroughly familiar with its hazards. Proper precautions should include wearing appropriate PPE, using positive pressure, self-contained breathing apparatus (SCBA), monitoring air, and using fans for ventilation.

Review Questions

1. What is the global warming potential (GWP) of carbon dioxide?
2. In what phase is carbon dioxide when stored inside a Dewar?
3. Carbon dioxide portable fire extinguishers are most effective on what type of fires?
4. The effectiveness of a local-application system is dependent upon what variables?
5. How do total-flooding extinguishing systems protect enclosures from fire?
6. In a low-pressure system, the insulated tank stores carbon dioxide in its liquid form at what temperature and pressure?
7. High-pressure carbon dioxide systems use steel or aluminum cylinder storage banks that are designed, tested, and filled to what agency's specifications?
8. What should first responders do when encountering a person who has inhaled large amounts of carbon dioxide and is exhibiting adverse effects inside a building?

Endnotes

1. United States Environmental Protection Agency, "Carbon Dioxide as a Fire Suppressant: Examining the Risks," *Report EPA430-R-00-002* (Washington, DC: USEPA, 2000).
2. Asia Industrial Gases Association (AIGA), "Carbon Dioxide," AIGA 068/10 Globally Harmonized Document, Based on CGA G-6, 7th Edition (Singapore: Asia Industrial Gases Association, 2009).
3. James V. Dunford, Jon Lucas, Nick Vent, Richard F. Clark, and F. Lee Cantrell, "Asphyxiation Due to Dry Ice in a Walk-In Freezer," *The Journal of Emergency Medicine* 36 no. 4 (2009): 353–356.
4. Constance Cooper, "Carbon Dioxide Blamed for Pooler McDonald's Death," last modified September 15, 2011, http://savannahnow.com/news/2011-09-14/carbon-dioxide-blamed-pooler-mcdonalds-death#.T8pM9FJrGuI.
5. Norb Makowka, "Why Carbon Dioxide (CO_2) in Fire Suppression Systems?" National Association of Fire Equipment Distributors (NAFED), www.nafed.org/resources/library/whyCO2.cfm.
6. National Fire Protection Association (NFPA), *NFPA 12: Standard on Carbon Dioxide Extinguishing Systems* (Quincy, MA: NFPA, 2011).
7. Ibid.
8. Ibid.
9. Celalettin Sever, Yalcin Kulahci, Fatih Uygur, and Cihan, "Frostbite Injury of the Foot from Portable Fire Extinguisher," *Dermatology Online Journal* 15 no. 9 (2009): 10.
10. United States Environmental Protection Agency (EPA), "Carbon Dioxide as a Fire Suppressant: Examining the Risks," *Report EPA430-R-00-002*, Washington, DC: EPA, February 2000.

11. Canadian Standards Association (CSA), *Standard Z94.4-02: Selection, Use, and Care of Respirators* (Rexdale, Ontario: CSA, 2002).
12. Texas State University—San Marcos, *Safety Manual* (San Marcos, Texas: Texas State University, 2011).
13. Robert T. Wickham, "Review of the Use of Carbon Dioxide Total-Flooding Fire Extinguishing Systems," www.epa.gov/ozone/snap/fire/co2/co2report2.pdf.
14. Robert M. Gagnon, *Design of Special Hazard and Fire Alarm Systems*, 2nd ed. (Clifton Park, NY: Delmar Cengage Learning, 2008).
15. Thomas J. Wysocki, "Carbon Dioxide and Applications Systems," *Fire Protection Handbook*, 17th ed. (Quincy, MA: NFPA, 1991).
16. Ibid.

Additional References

Ross Bynum, "CO_2 Caused Ga. McDonald's Soda Fountain Death," www.boundtreeuniversity.com/print.asp?act=print&vid=1123521.

Josh, Cable, "NIOSH Report Details the Dangers of Carbon Dioxide in Confined Spaces," http://ehstoday.com/news/ehs_imp_37358.

Chemetron Fire Systems, "Carbon Dioxide Fire Suppression—Flexographic Printing—Paper, Film, & Foil Conversion," *Printing Industry Bulletin* 0215 (1996).

Chemetron Fire Systems, "Carbon Dioxide Fire Suppression—Heat Treating Facilities, Part 2: Open Quench Tanks," *Industrial Processes Bulletin* 0505 (1996).

Chemetron Fire Systems, "Carbon Dioxide Fire Suppression—Protection of Wet Benches," *Industrial Processes Bulletin* 0510 (1998).

Fire Department, City of New York, "Carbon Dioxide (CO_2) Incident," *Pass It On Program* 8 (2011).

Raymond Friedman, "Theory of Fire Extinguishment," *Fire Protection Handbook*, 17th ed. (Quincy, MA: NFPA, 1991).

James R. Gill, Susan F. Ely, and Zhongxue Hua, "Environmental Gas Displacement: Three Accidental Deaths in the Workplace," *American Journal of Forensic Medicine & Pathology* 23, no. 1 (2002): 26–30.

International Organization for Standardization (ISO), *Standard 6183: Fire Protection Equipment—Carbon Dioxide Extinguishing Systems for Use on Premises—Design and Installation, for CO_2 Fire Suppression Systems*" (Geneva, Switzerland: ISO, 1990).

Frank Leeb, "The Dangers of Carbon Dioxide in Fast Food-Type Occupancies," *With New York Firefighters (WNYF)* 1 (2012): 26–28.

National Fire Protection Association (NFPA), *NFPA 704: Standard System for the Identification of the Hazards of Materials for Emergency Response* (Quincy, MA: NFPA, 2012).

National Institute for Occupational Safety and Health (NIOSH), *NIOSH Pocket Guide to Chemical Hazards*, Pub. No. 2005-149 (Cincinnati, OH: NIOSH (DHHS), 2006).

Jonathan L. Scott, David G. Kraemer, and Randal J. Keller, "Occupational Hazards of Carbon Dioxide Exposure," *Journal of Chemical Health and Safety* 16, no. 2 (2009): 18–22.

Joseph Z. Su, Andrew K. Kim, George P. Crampton, and Zhigang Liu, "Fire Suppression with Inert Gas Agents," *Journal of Fire Protection Engineering* 11, no. 2 (2001): 72–87.

United States Department of Labor, Occupational Safety & Health Administration (OSHA), "Evacuation Plans and Procedures (Fixed Extinguishing Systems)," www.osha.gov/SLTC/etools/evacuation/fixed.html.

WorkSafeBC, "No Warning with Deadly Lack of Oxygen in Air," *WorkSafe Bulletin*, WS 06 03 (2006).

Torbjørn Laursen, "Overview of Toxicity/Effectiveness Issues" (presentation, Halon Alternatives Working Conference, Albuquerque, NM, May 11–13, 1993).

CHAPTER 8

Inert Gas and INERGEN Fire Protection Systems

Source: Ronald R. Spadafora

KEY TERMS

Inert gas, *p. 138*   **Hypoxia,** *p. 148*   **Hypercapnia,** *p. 148*

OBJECTIVES

After reading this chapter, the reader should be able to:

- Discuss the physical properties of inert gases
- Analyze inert gas fire extinguishment applications
- Be familiar with total-flooding application systems that use inert gases
- Evaluate the hazards of using inert gas
- Understand the extinguishing atmosphere of INERGEN
- Review the agent's physical properties
- Describe the physiological effects of INERGEN
- List and describe the various application systems
- Understand the design of the extinguishing system

Introduction

Inert gases (IG) are applied as total-flooding agents. (The IG designation comes from the National Fire Protection Association [NFPA] and International Organization for Standardization [ISO] standards.) Similar to carbon dioxide suppression systems, the displacement of oxygen, high noise levels created by nozzle discharges, and rapid cooling are also a life safety design concern if the agent is to be discharged into an occupied space. The most commonly used inert gases are argon and nitrogen. They reduce the concentration of oxygen below the level required for combustion. The extinguishing properties for all inert gases are similar for Class A, B, and C fires. In some instances, argon is suitable for Class D

Inert gas ■ A gas that is nontoxic and does not react with other substances. Common inert gases used for fire extinguishment are nitrogen and argon.

FIGURE 8.1 Nitrogen high-pressure steel cylinders. *Source:* Ronald R. Spadafora

fires. Inert gases are not subject to thermal decomposition when used in extinguishing fires and therefore form no toxic combustion byproducts. A large quantity of agent contained in high-pressure steel cylinders generally is required. This factor corresponds to design challenges relating to the size and load-carrying capacity of the storage space (see Figure 8.1).

Inert gases are typically stored at high pressure—2,900 to 4,351 psi (19,995 to 29,999 kPa)—in gas cylinders. In some cases, however, they may be stored as liquefied gases (see Figures 8.2 and 8.3). Inert gases normally will require a system that is robust to withstand

FIGURE 8.2 Liquid nitrogen used for inerting at an LNG facility. *Source:* Ronald R. Spadafora

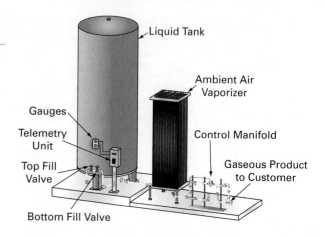

FIGURE 8.3 Typical liquid storage system used for argon and nitrogen

the high pressures involved, and the hardware required for this will be similar to that for carbon dioxide systems. The component gases of IG-55 (argon and nitrogen) are blended so as to have a density similar to that of air.

Overview

Inert gas agents are environmentally safe, because they are natural components in the atmosphere. Systems are designed to reduce the ambient oxygen concentration in a protected space to between 14% and 10%. At this oxygen level, the atmosphere is still breathable for occupants but will not support combustion. Occupants, however, will have a limited time frame to exit the area. Although inert gases are not subject to thermal decomposition, they can asphyxiate. Health considerations must be recognized and addressed. Inert gases are odorized for added detection. Relatively large amounts of agent being infused into a room require special safety feature installations. The concentrations of inert gases required for fire extinguishment generally range from 34% to 52% by volume.

Argon

Argon is an inert gas ideal for protecting areas containing valuable equipment that could be damaged by water. Argon is not electrically conductive, noncorrosive, and does not produce any toxic byproducts. It leaves no residue after discharge, so cleanup and business interruption is minimal after a fire. This inert gas is suitable for use on Class A, B, C, and some D fires (the single-molecule structure (Ar) is stable during high-heat scenarios common in metal fires such as those in magnesium and lithium).

Argon is commonly stored as a compressed gas in high-pressure steel cylinders. A typical total-flooding system is initiated upon fire detection. Odorization using 3 to 10 ppm of pyridine helps warn occupants of an accidental argon buildup of any severity. An audible prealarm is activated prior to discharge. The distribution system funnels the inert gas from the cylinders that are connected to a common manifold through piping. Valves for each individual hazard being protected by an argon system are calibrated to supply the proper amount of agent calculated by the design professional. This will decrease oxygen to a specified level. Through evenly spaced nozzles, argon gas is released, lowering the oxygen level to a percentage that will not support flaming combustion. In enclosures with high ceilings, multiple levels of nozzles may be installed to ensure more effective distribution of agent. After the flame is extinguished, it may be necessary to maintain the concentration for an extended period of time until hot surfaces and embers have cooled down in order to avoid reignition.

Factors that determine the amount of agent required include the volume of the enclosed area, leakage, and the type or amount of combustible material being protected.

Design concentration is normally obtained within sixty seconds. The absence of "fogging" demonstrated during carbon dioxide release allows occupants to see their way to the means of egress from the space.

The most significant hazard associated with an argon/oxygen mixture is the inhalation of an oxygen-deficient atmosphere. High concentrations of argon gas released into a poorly ventilated area or space will result in a depletion of oxygen. Persons breathing such an atmosphere may experience symptoms that include headaches, dizziness, drowsiness, fatigue, nausea, vomiting, and unconsciousness. In some circumstances, death can occur.

Applications for argon fire protection systems include paper storage, switch gear rooms, cell phone sites, telecommunications, areas with flammable liquids and paints, computer and server rooms, museums, libraries, data archives, magnesium and lithium processing, electrical generator enclosures, medical facilities, and art galleries. Argon is also effective in specialized spaces in which objects are shielded or difficult to reach. These areas include tool-making machinery, silos, turbines, transformers, textile machines, and engine rooms on ships.

PHYSICAL PROPERTIES

Argon is an odorless, tasteless, and colorless gas. It is completely undetectable by the human eye. Argon is also heavier than air (specific gravity of 1.38). It is nonflammable. Argon has approximately the same solubility in water as oxygen and is 2.5 times more soluble in water than nitrogen. Additional information is shown in Table 8.1.

Environmental Issues

There is no concern of ozone depletion or global warming from inert gases. They also provide protection from the potential harmful effects that water and chemical powders can have on machinery, equipment, and electrical wiring. Areas in which inert gas systems could provide a feasible alternative to halons include telecommunications facilities, computer rooms, data centers, banks, military installations, pipeline pumping stations, petrochemical plants, pharmaceutical facilities, control rooms, transformer and switchgear rooms, archives, museums, paper storage, cultural heritage, flammable liquid hazards, laboratories, and shipboard machinery spaces.

TABLE 8.1	Physical Properties
Chemical name	Argon
Chemical formula	Ar
Denomination according to NFPA 2001	IG-01
Molecular weight	39.9
Critical temperature	−188.14°F (−122.3°C)
Critical pressure	710.68 psi (49,00 kPa)
Maximum filling pressure	2,900/4,351 psi (19,995/29,995 kPa)
Typical design concentration for heptane	48.8%
Design concentration for Class A surface fires (NFPA)	38.7%
No Observed Adverse Effect Level (NOAEL)	43%
Lowest Observed Adverse Effect Level (LOAEL)	52%
Ozone-depletion potential (ODP)	0
Global warming potential (GWP)	0

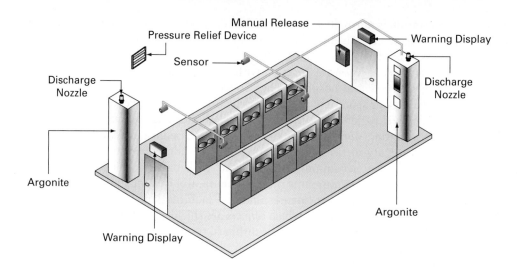

FIGURE 8.4 IG-55 (Argonite) total-flooding system protecting a computer room

CLEAN AGENT

Inert gases are naturally occurring and have zero ozone-depletion potential (ODP) and zero global warming potential (GWP). IG-01 (Argotec) is made up of 100% argon gas. IG-55 (Argonite) is an equal blend of argon and nitrogen (see Figure 8.4). Both IG-01 and IG-55 extinguishing agents are electrically nonconductive and leave no residue. Argon is considered a clean agent, and it complies with *NFPA 2001: Standard on Clean Agent Extinguishing Systems*.[1] When used under total-flooding application design, the ambient atmosphere's level of oxygen is reduced from 21% to between 12% and 14%. Between these percentages, the effect of argon on occupants inside the enclosed space is minimal, and medical research has stated that there are no short- or long-term influences on humans.[2]

Liquid Argon

Liquid (cryogenic) argon gas extinguishing systems have been developed for the protection of hazards requiring large amounts of extinguishing agent. Argon in liquid form has a density 850 times greater than that of gaseous argon. It allows for an alternative to carbon dioxide low-pressure extinguishing systems. In addition, liquid argon technology features compact, space-saving storage capability. System design can have a lone tank supplying multiple areas. Storage of liquid argon is at a temperature of −225°F (−143°C) and a relatively low pressure of 290 psi (1,999 kPa; see Figure 8.5). Liquid argon requires only small pipe diameters, even for large discharge areas.

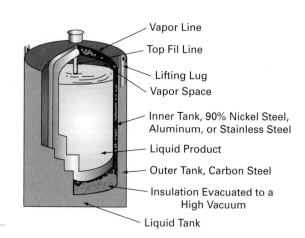

FIGURE 8.5 Cutaway of a typical cryogenic liquid storage tank

SPECTRUM NATURALS CONFINED SPACE INCIDENT

Two workers at Spectrum Naturals, a natural foods company in Petaluma, California, lost their lives in a confined-space incident involving argon gas. One of the workers was cleaning an empty tank used to store flaxseed oil when he reportedly lost consciousness. A coworker then entered the container in an attempt to rescue him and was also asphyxiated.

According to investigators on the scene, argon gas was pumped into the tank to displace oxygen in order to prevent the oil from spoiling. The tank was eventually emptied of oil, and air was forced into the tank and the vents opened to remove the argon gas. Apparently, not all of the argon gas was purged. Argon, because it is heavier than air, remained in significant amounts at the bottom of the tank.

The Petaluma Fire Marshal stated that his crew was on the scene within minutes of the call coming in, but it took about 20 minutes for them to retrieve the two workers from the confined space. A firefighter wearing a self-contained breathing apparatus (SCBA) had to be lowered into the tank with a rescue harness to remove the two men, who were pronounced dead at a local hospital.[3]

WHY INERT ATMOSPHERES IN CONFINED SPACES ARE DANGEROUS

California's Occupational Safety and Health Administration (Cal-OSHA) defines a confined space as a work area that meets one or more of the following conditions:

- Is large enough for a person to enter
- Has limited entry and exit points
- Is not meant for continuous employee occupancy
- Has the potential to be filled with toxic or oxygen-deficient air
- Has certain physical safety hazards

Health researchers from the University of California at Berkeley performed a study that analyzed 530 US worker deaths from 1992 to 2005 that were due to toxic or oxygen-deficient atmospheres in confined spaces. They also obtained data from urban fire departments in California about fire crew arrival times and estimates of the length of time it would take to complete a rescue of a victim from a confined space. Results revealed that the time needed for confined-space rescue operations ranged from 48 to 173 minutes. The study stated that even though the arrival time of the first engines averages five to seven minutes firefighters need time to evaluate and control the hazard before they enter the confined space.[4]

FIREFIGHTING PROCEDURES

Lessons learned from the Spectrum Naturals incident reinforce the importance of using oxygen meters prior to entry into a confined space and the wearing of positive pressure, SCBA by workers designated to enter the area. Before entry into a confined space, firefighters must first test the internal atmosphere with meters and other calibrated direct-reading instruments for oxygen content, flammable gases, and toxic air contaminants. If hazardous conditions are detected, the need for forced air ventilation is evident prior to entry.

Forced air ventilation is directed into the immediate area in which the rescue firefighters will be conducting operations, and the space must be periodically tested as necessary to ensure ventilation is preventing the reoccurrence of a hazardous atmosphere. When work is being conducted inside tanks or confined spaces in which inert gas has been used for purging, requirements must include continuous ventilation to help ensure a safe environment.[5]

Nitrogen

Nitrogen (N_2) gas is manufactured from ambient air. It is not regulated under any governmental environmental statutes. Nitrogen is designated as IG-100. It is a colorless, odorless, tasteless, and electrically nonconductive gas.

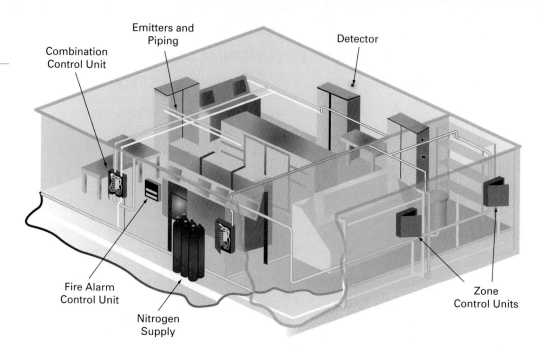

FIGURE 8.6 IG-100 (nitrogen) total-flooding fire suppression system

In the 1980s, scientists discovered that the Earth's stratospheric ozone layer was being destroyed by manmade chemical substances released into the atmosphere. Chlorofluorocarbons (CFCs), along with other chlorine- and bromine-containing compounds (such as halon extinguishing agents), were among the worst ozone-depleting substances. Total-flooding fire suppression systems containing nitrogen gas began to replace Halon 1301 fire extinguishing agent. Nitrogen gas systems consist of detection and control devices, cylinders containing the inert gas, discharge manifold, valves, piping, pipe hangers, and nozzles. The high pressure allows the cylinders to be stored remotely from the hazard being protected. More than one hazard can be guarded by a single bank of nitrogen cylinders. In addition, a wide range of reusable cylinder sizes can be employed. Cylinder discharge valves are operated by electromechanical solenoid, pneumatically, or a combination of both.

Nitrogen fire extinguishing systems reduce the oxygen concentration within the hazard area from 21% in air to below 15%. IG-100 is stored in high-pressure cylinders in a compressed gas form. Fire extinguishing systems using this agent are designed for cylinder pressures of 2,900/4,351 psi (19,995/29,995 kPa; see Figure 8.6). Nitrogen cylinders should be stored upright in a well-ventilated, secure area protected from harsh environmental conditions. Storage area temperatures should not exceed 125°F (52°C).

In addition, nitrogen can be used as either a backup to or a substitute for an instrument air system. If it is used as a backup supply in case of air compressor failures, it is common to find a nitrogen supply connected to an air supply by means of isolation valves. There is the constant danger that the released nitrogen may accumulate in poorly ventilated areas and enclosures, presenting a serious asphyxiation risk.

Nitrogen is safe to use in occupied spaces (within the parameters of design oxygen levels) and maintains excellent visibility in the room of discharge to facilitate egress. The human body is naturally conditioned to breathe high amounts of this gas. Nitrogen systems protect machinery, museums, telecommunications equipment, computer rooms, electrical installations, archives, and logistics control areas.

Nitrogen is mainly found in the atmosphere, but it is also found in the Earth's crust, in organic form (plants), and in mineral form. In gaseous form, nitrogen is neutral and does not sustain life. The physical properties of nitrogen are listed in Table 8.2.

TABLE 8.2	Physical Properties
Chemical name	Nitrogen
Chemical formula	N_2
Denomination according to NFPA 2001	IG-100
Molecular weight	28.02
Critical temperature	−232.6°F (−147°C)
Critical pressure	493 psi (3,399 kPa)
Typical design concentration for heptane	34.6%
Design concentration for Class A surface fires (NFPA)	37.2%
NOAEL	43%
LOAEL	52%

UNION CARBIDE'S TAFT/STAR PLANT TRAGEDY

On March 27, 1998, one worker was killed and another severely injured when they were asphyxiated by nitrogen that was venting through a large, open pipe. The men were working at Union Carbide's Taft/Star plant in Hahnville, Louisiana. The plant manufactures industrial chemicals, and the accident occurred in the plant's ethylene oxide production unit that was undergoing a major maintenance and equipment removal project. The work left a pipe, which was connected to a processing unit, open to the air at one end. Two workers were inspecting the inside of the open pipe end to determine the effectiveness of an earlier cleaning effort. They performed their tasks from the exterior of the opening as well as just inside the pipe in a crouched position. While the men worked, nitrogen gas was being used to purge air and moisture from the unit and protect the chemicals inside.

During their pipe inspection, the two workers used a black light, which causes grease, oil, and other contaminants to glow in the dark. Two additional contractors assisted by holding up a dark plastic sheet to shield the opening from the sunlight that made it difficult to see the glow. However, the sheet created a dangerous enclosure in which the nitrogen gas could gather and displace oxygen. The workers were unaware that this condition had been generated. The contractors holding the sheet became concerned when they had not heard from the workers inside the pipe for approximately 15 minutes. They called into the pipe without hearing a response. The sheet was pulled away, and they discovered one worker unconscious and the other in a bewildered state. One worker was pronounced dead on arrival at a hospital. The second victim was hospitalized in critical condition but survived. The US Chemical Safety Board (CSB) investigated the incident and found that inadequate confined space warnings and entry procedures were at the root of the tragedy.

The evening prior to the deadly accident, one of the two victims had directed the nitrogen-purging operation. Apparently, he did not realize that dangerous amounts of nitrogen were flowing through open valves to the pipe end where the incident occurred, which was about 150 ft. (46 m) and several floors away from the origin of the nitrogen. The CSB investigation report noted that it is important to look beyond the immediate task and anticipate secondary hazards that may not be obvious.

In this situation, there was a failure to evaluate the risks to the workers downstream from the flow of the gas caused by the nitrogen purge. The dangers at the open pipe end were also not recognized. The investigation concluded that a hazard evaluation likely would have led to the posting of warning signs at the pipe opening. This action may have prevented loss of life.

After the accident, managers at the plant stated that they were unaware that the workers were going to perform a black light inspection of the pipe, which led to the creation of a temporary enclosure. However, the CSB investigation found that the open pipe end itself presented a hazard. A worker simply leaning over or into the open pipe would be in danger of being overcome depending upon the wind and other factors.

The CSB investigation noted that human senses cannot detect excess levels of nitrogen. If the nitrogen had a chemical odorant added, the workers most likely would have realized they were in danger before it was too late. Odorizing nitrogen would have provided an extra measure of safety, the CSB determined, in addition to appropriate confined-space procedures, hazard evaluations, and warning signs.[6]

> **NOTE**
>
> Sewer workers have been overcome by gases without even entering confined spaces. Pent-up gases rise up upon removal of manhole covers. The lack of oxygen coupled with high levels of nitrogen can quickly lead to asphyxiation. A more recent nitrogen gas incident involved the death of a Middletown, Ohio, public works maintenance employee, which occurred on May 7, 2010. The worker was wiping the edge of a sewer manhole cover clean in preparation for inserting a remote camera as part of a routine check when he was overcome and fell into the opening. Investigative air testing inside the confined space of the manhole revealed nitrogen gas content in excess of 90%. During rescue operations, three Middletown firefighters were also overcome. However, they were treated at local hospitals, recuperated, and were released.[7]

CSB SAFETY BULLETIN

In 2003, the CSB issued a Safety Bulletin pertaining to the ongoing problem of nitrogen asphyxiation. Nitrogen asphyxiation caused 80 deaths and 50 injuries in industrial settings between 1992 and 2002, the bulletin said. Deaths and injuries occurred at chemical plants, food-processing facilities, laboratories, medical facilities, and other sites. The majority of accidents occurred during work in or near confined spaces, and many incidents were caused by the failure to detect an oxygen-deficient environment.

The Safety Bulletin identified a number of good practices to prevent nitrogen-related injuries, including comprehensive worker-training programs, warning systems, continuous ventilation, atmospheric monitoring, and planning for emergency rescue operations. Copies of the bulletin are available from the CSB website.

FIREFIGHTING PROCEDURES

The walls of inert gas cylinders will weaken when exposed to fire or high heat. As pressure builds up within the cylinders, they may eventually burst or fail in a violent manner if their maximum safe working pressure is exceeded. Firefighters should, if safety permits, make every effort to extinguish any fire involving or threatening inert gas cylinders. Apply water spray from hose lines directly on to any affected cylinders in an attempt to cool them down. Firefighters carrying out this task should use all available shielding/cover. Defensive strategy and tactics must be employed when an advanced fire is encountered for an unknown period of time. During severe fire conditions, use extreme caution and consider the use of unmanned large caliber streams to apply water onto the hazard. Cylinders may be thrown great distances when not contained within a structure or building. Establish an exclusion zone to ensure firefighter safety.

Firefighter Safety Concerns

There are many safety hazards associated with the use of inert gases. Oxygen is the only gas which supports life. If the concentration of inert gases increases within a space or enclosure, the oxygen concentration in the air decreases. This situation can lead to a dangerous condition that can threaten occupants' lives. When the oxygen concentration drops below 10%, the risk of death is real unless immediate rescue and recovery operations are conducted. Some key points for firefighters to understand follow:

- Because inert gases are odorless (an odorant may, however, be added), colorless, and tasteless, they are insidious (giving no warning signs of the danger of asphyxiation). This risk potential can also be found outdoors in the vicinity of gas leaks and vent exhausts.

- The use of inert cryogenic liquids such as nitrogen or argon poses additional hazards. Cryogenic liquids are extremely cold and can cause serious frostbite burns on contact with the skin. Once vaporized, they generate large volumes of gas. Cold gas will flow and collect in low-lying spaces, such as basements and elevator shafts.
- Workplaces often use compressed inert gas to power equipment (such as pneumatic jackhammers). Pneumatically operated tools and instruments vent continuously. Proper air monitoring with oxygen-concentration meters should be performed.
- Removal of argon or cryogenic nitrogen vapors from large vessels, trenches, and subgrade areas can be difficult due to the relatively high density of the gas compared with air. Purging of these gases should be performed from the bottom of the space.
- Oxygen-deficiency fatalities involving inert gases happen unexpectedly. All personnel who work with or are exposed to inert gases must be given initial and periodic awareness training in respect to the hazards involved.
- A confined space in which there is the potential for an inert gas/oxygen-deficient atmosphere poses a serious risk to life. A victim collapsed inside such a space for any length of time is very likely to be dead. First responders must first evaluate risk versus reward before attempting rescue/body-recovery operations.

INERGEN

INERGEN® was developed as a direct result of the ban on Halon 1301 production in 1993 due to its negative effects on the Earth's ozone layer. INERGEN is environmentally friendly and made exclusively of gases that humans breathe: nitrogen, argon, and carbon dioxide. Upon discharge, INERGEN returns to the atmosphere in its natural state and poses no threat to the ozone layer. INERGEN is designed so as not to be banned by future legislation. INERGEN is an odorless, colorless, and nontoxic gas. It is nonflammable, nonexplosive, and is not an oxidizer.

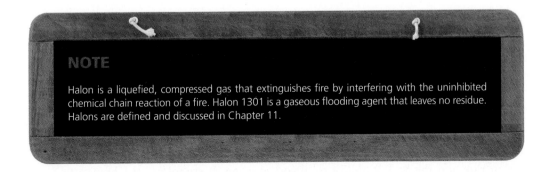

NOTE
Halon is a liquefied, compressed gas that extinguishes fire by interfering with the uninhibited chemical chain reaction of a fire. Halon 1301 is a gaseous flooding agent that leaves no residue. Halons are defined and discussed in Chapter 11.

INERGEN extinguishing systems are designed to protect against surface-burning fires for Class A, B, and C hazards. INERGEN is mainly used in total-flooding fire suppression systems (see Figure 8.7). Electrically nonconductive and noncorrosive, it is especially suited for the protection of electrical and electronic equipment. INERGEN extinguishes fire by lowering the oxygen content well below the level (15%) that supports combustion. In addition, when introduced into an enclosure it creates a mixture of gases that will allow occupants to breathe in a reduced-oxygen environment.[8] Because INERGEN agent is composed of atmospheric gases, it does not pose toxicity problems associated with chemically derived halon alternative agents. INERGEN systems are environmentally friendly and safe to use in enclosed areas where people may be present. This agent is zero ODP, zero GWP, and has a zero atmospheric lifetime (ATL). INERGEN systems use a fixed-nozzle agent-distribution network. They are designed and installed in accordance with NFPA 2001.

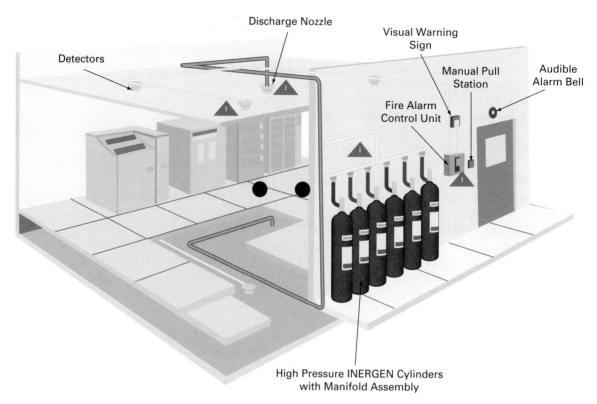

FIGURE 8.7 INERGEN total-flooding system

PHYSIOLOGY OF INERGEN

Nitrogen and argon are physiologically inert at normal atmospheric pressures. The sustained breathability of the extinguishing atmosphere created by INERGEN is directly related to the degree of **hypoxia** (decreased oxygen) and **hypercapnia** (increased carbon dioxide).[9]

INERGEN actually enhances the body's ability to assimilate oxygen. The normal atmosphere in a room contains approximately 21% oxygen and less than 1% carbon dioxide. During the discharge of INERGEN, the oxygen content is reduced below the level most ordinary combustibles need to burn. The carbon dioxide content, however, will increase to above 3%. The increase in the carbon dioxide enhances a person's respiration rate and the body's ability to absorb oxygen. It stimulates the human body to breathe deeply and rapidly (hyperventilation) to compensate for the lower oxygen content of the atmosphere. The brain, therefore, continues to receive the same amount of oxygen as it would in a normal atmosphere! The design engineer must, however, take into consideration situations in which occupants cannot escape the area of discharge. Detectors that activate alarms, sirens/horns, and strobes are employed to initiate a speedy evacuation. These components work in conjunction with a short time delay prior to release of the agent.

It must be noted, of course, that it is always recommended that occupants evacuate the space when fire is detected. Enhanced respiration in a fire scenario also means an increase in the inhalation of combustion byproducts that are harmful. Nevertheless, the use of INERGEN minimizes the intensity of fire and the production of carbon monoxide (which is a primary cause of death during fire incidents) and other toxic gases. INERGEN

Hypoxia ■ A pathological condition in which the body as a whole (generalized) or a region of the body (tissue) is deprived of adequate oxygen supply.

Hypercapnia ■ A condition in which there is too much carbon dioxide in the blood. It triggers a reflex that increases breathing and access to oxygen.

also does not produce a heavy fog upon discharge, as do carbon dioxide systems, so escape routes can remain visible.

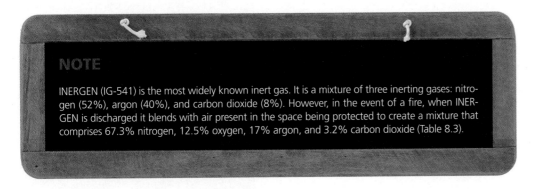

> **NOTE**
>
> INERGEN (IG-541) is the most widely known inert gas. It is a mixture of three inerting gases: nitrogen (52%), argon (40%), and carbon dioxide (8%). However, in the event of a fire, when INERGEN is discharged it blends with air present in the space being protected to create a mixture that comprises 67.3% nitrogen, 12.5% oxygen, 17% argon, and 3.2% carbon dioxide (Table 8.3).

DESIGN AND OPERATION

INERGEN is stored in high-strength alloy steel cylinders constructed and tested in accordance with applicable US Department of Transportation (DOT) specifications. Cylinder-charging pressure is 2,175 psi (14,996 kPa) at 70°F (21°C). Large systems with multiple cylinders incorporate a manifold assembly. The release of the agent into the hazard area is accomplished through manual pull stations or automatic (electric or pneumatic) actuators. Selector valves are used to direct the flow of INERGEN to a single hazard or multiple hazard systems. A pressure reducer is installed within the distribution system. A network of piping distributes the gas to discharge nozzles. Each nozzle is designed with orifices to use the stored pressure from the cylinders and to deliver a uniformed, reduced-turbulence flow rate of the agent. Discharge patterns of 360° or 180° are standard. Additional equipment includes control units, detection devices, alarms, door and window closures, pressure trips and switches, and ventilation controls (see Figure 8.8).

INERGEN, like all inert gas systems, has overpressure vents fitted to protect the room or area from damage during a discharge. As extinguishing agent is discharged into a room, it needs to be able to vent to the open atmosphere to protect the walls, ceilings, and doors from damage. During discharge, positive pressure starts to build up within the room. Vents allow this pressure to dissipate.

APPLICATIONS

INERGEN systems are used for suppressing fires in areas that are normally occupied, areas in which an electrically nonconductive agent is desired, and places where cleanup of

TABLE 8.3	How INERGEN Works		
GAS	AIR %	INERGEN %	AIR AND INERGEN % MIXED DURING DISCHARGE
Nitrogen	78.0%	52%	67.3%
Oxygen	20.9%	NA	12.5%
Argon	1.0%	40%	17.0%
Carbon dioxide	0.03%	8%	3.2%

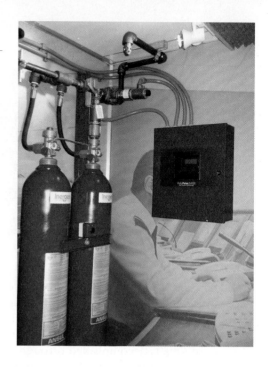

FIGURE 8.8 INERGEN system component display.
Source: Ronald R. Spadafora

other extinguishing agents would present a problem. Typical hazard areas protected by INERGEN fire suppression systems include:

- Data processing
- Computers
- Tape storage
- Cables
- Subflooring
- Industrial
- Process control rooms
- Robotic systems
- Automated production lines
- Power-generating equipment
- Drying ovens
- Laboratories
- Printing machines
- Flammable liquid storage
- Commercial and institutional
- Vaults
- Document storage areas
- Museums/art galleries
- Libraries
- Archives
- Medical diagnostic rooms
- MRI scanners
- Aviation areas
- Flight simulators
- Air traffic control centers
- Marine/offshore/naval
- Communications

- Telephone exchanges
- Radio/television stations
- Communications centers
- Satellite stations
- Remote cellular sites
- Telecommunications/switchgear rooms

ABRAHAM LINCOLN PRESIDENTIAL LIBRARY AND MUSEUM

The Abraham Lincoln Presidential Library and Museum (opened in 2005) is the largest and best attended of all US Presidential Libraries in the country. Located in Springfield, Illinois, it houses thousands of priceless artifacts, including books, manuscripts, monographs, writings, and other historical materials within the 200,000 ft.^{2} (18,581 m^2) structure.

An INERGEN fire suppression system protects two occupied spaces: the Treasures Gallery and the Viewing Vault. The system is designed to control and extinguish a fire without causing any collateral damage to archived material. Both areas are open to the public, which is a significant reason that INERGEN was selected.

The INERGEN system has a control unit with cross-zoned photoelectric and ionization smoke detectors. Cross-zoning with these two types of detectors requires two independent signals to be received by the control unit prior to going into the alarm stage. This design lessens the chance for false discharge.[10]

FIREFIGHTING PROCEDURES

Firefighters should use the following points of information gathered from material safety data (MSD) sheets to help formulate successful strategy and tactics when extinguishing fires and mitigating emergencies involving INERGEN release. This knowledge will also assist in the treatment of victims and firefighters who may be exposed to the agent.

- Although INERGEN is a nonirritating gas, eyes exposed to INERGEN should be flushed with water for a minimum of 15 minutes.
- If INERGEN is inhaled, remove the victim to fresh air. If breathing is difficult, give oxygen. If the victim is not breathing, administer artificial respiration.
- Although gas cylinders are equipped with pressure- and temperature-relief devices, they should be removed from high-temperature areas or fires to avoid possible rupture.
- Appropriate personal protective equipment (PPE) for the incident should be worn.
- During a fire, the oxygen level may be reduced below safe levels, and the combustion products formed by the fire are likely to be hazardous to health. Use positive pressure, SCBA, as required by the Occupational Safety and Health Administration (OSHA).[11]

CHAPTER REVIEW

Summary

Argon and nitrogen are used as fire extinguishing agents in total-flooding applications. They put out fire by displacing oxygen in the space they are designed to protect. These gases are odorless, colorless, and tasteless. They provide no warning signs of the dangers of asphyxia. Good safety practices are important to prevent inert gas–related injuries for both fire extinguishing systems and inerting applications. These measures should include occupant and worker educational and training programs, warning signs, adequate ventilation, atmospheric monitoring, and planning for emergency rescue operations.

INERGEN is a fire suppression agent composed of three naturally occurring gases found in the air we breathe: nitrogen, argon, and carbon dioxide. It is used to protect enclosed areas where there is a need for quick suppression of fire and people may be present. This mixture of gases actually enhances human safety and escape potential when discharged into the hazard area.

Review Questions

1. Define the term inert gas.
2. What is the range of inert gas concentration required for fire extinguishment in total-flooding applications?
3. When using inert gas systems, what is the design ambient oxygen concentration range in the protected space?
4. What should first responders do prior to attempting to rescue an unconscious victim inside a confined space?
5. Why is argon suitable for use on some Class D fires?
6. Why are inert gases considered clean agents?
7. How does the California Occupational Safety and Health Administration define a confined space?
8. What type of storage system design do fire extinguishing systems using nitrogen (IG-100) employ?
9. In 2003, the CSB issued a Safety Bulletin pertaining to the ongoing problem of nitrogen asphyxiation. Between 1992 and 2002, what was the primary cause of death and injury occurring during work in or near confined spaces?
10. What three gases make up the fire suppression agent INERGEN?
11. When INERGEN is discharged, it creates a mixture with air in the space being protected containing approximately what percentage of oxygen?
12. What effect does the increased percentage of carbon dioxide into the area of INERGEN discharge have on humans?
13. During the discharge of INERGEN, positive pressure starts to build up within the room. What design feature is installed to protect the walls, ceilings, and doors from damage?
14. What first aid measure should be taken for a victim whose eyes have been exposed to INERGEN?

Endnotes

1. National Fire Protection Association (NFPA), *NFPA 2001: Standard on Clean Agent Extinguishing Systems* (Quincy, MA: NFPA, 2012).
2. Steven Zumdahl and Donald J. Decoste *Chemical Principles*, 7th ed. (Belmont, CA: Brooks/Cole, 2013).
3. Sandy Smith, "Two Die in Confined Space Incident at California Natural Foods Company," http://ehstoday.com/news/ehs_imp_35332.
4. Sarah Yang, "On-Site Worker Rescue Plan Urged for Confined Spaces," http://newscenter.berkeley.edu/2012/02/13/confined-spaces.
5. US Department of Labor (DOL), Occupational Safety and Health Administration (OSHA),"Permit-Required Confined Spaces," *OSHA Confined-Space Standard, 29 CFR 1910.146* (Washington, DC: US DOL-OSHA1993).

6. US Chemical Safety and Hazard Investigation Board, *Summary Report: Nitrogen Asphyxiation-Union Carbide Corporation, Hahnville, Louisiana, March 27, 1998. Report No. 98-05-I-LA* (Washington, DC: Chemical Safety and Hazard Investigation Board, 1999).
7. Ansul Incorporated "The Physiology of INERGEN Fire Extinguishing Agent," White Paper 1005 (Marinette, WI: Ansul Incorporated, 1995).
8. Ibid.
9. Steve Bennish, "Nitrogen Is Implicated in Middletown Sewer Worker's Death," *Dayton Daily News*, May 14, 2010, www.daytondailynews.com/news/news/local/nitrogen-is-implicated-in-middletown-sewer-workers/nNCrq.
10. ANSUL Incorporated, "Presidential Library Guarded By INERGEN System," www.tycofsbp.com/downloads/LincolnCaseStudy.pdf
11. US Department of Labor (DOL), Occupational Safety and Health Administration (OSHA), "Permit-Required Confined Spaces," *OSHA Confined-Space Standard, 29 CFR 1910.146* (Washington, DC: US DOL-OSHA1993).

Additional References

ANSUL Incorporated, "INERGEN Material Safety Data Sheet," www.efire.info/MSDS/Inergen.pdf.

ANSUL Incorporated, "Summary: UL Fire Test Results—INERGEN Fire Extinguishing Agent," www.firedot.com/fire_test_results.htm.

European Industrial Gases Association (AISBL), "Hazards of Inert Gases and Oxygen Depletion," www.eiga.org/index.php?id=173&tx_abdownloads.

Gielle Srl, "Argon Fire Suppression Systems: Argon Gas Extinguishing," www.fm200.biz/argon.htm.

Gielle Srl, "INERGEN Fire Suppression Systems," www.fm200.biz/inergen.htm.

CHAPTER 9

Dry Chemical Fire Suppression Systems

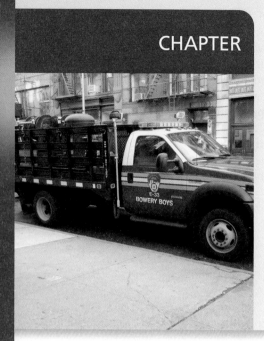

Source: Ronald R. Spadafora

KEY TERMS

Acidic, *p. 155*
Alkaline, *p. 155*
Free radicals, *p. 156*
Pyrophoric, *p. 156*
Decrepitate, *p. 158*

OBJECTIVES

After reading this chapter, the reader should be able to:

- Review the physical properties of dry chemical
- Understand how dry chemical agent extinguishes fire
- Examine the different dry chemical agents and their uses
- Analyze system components and functions
- Review the methods of dry chemical agent application

Introduction

Dry chemical is an extinguishing agent that is a powder mixture. Applications include fixed total-flooding and local systems, hand hose lines, mobile systems, and portable fire extinguishers. There are five basic varieties of dry chemical extinguishing agents. Dry chemical extinguishing agents have been used since the start of the twentieth century. Early in the development of dry chemical agents, sodium bicarbonate was found to have greater effectiveness on flammable liquid fires than other chemicals being used at the time, and it is still widely used today. In the 1950s and 1960s, new dry chemical agent types and manufacturing processes were introduced that provided even greater effectiveness on Class B and Class C fires.

Underwriters Laboratories Inc. (UL) issued the first UL Listing for a dry chemical fire extinguishing system unit in the mid-1950s. Subsequently, the National Fire Protection Association (NFPA) adopted *ANSI/NFPA 17: Standard for Dry Chemical Extinguishing Systems,* which specifies that installation system units are UL Listed for use in a variety of fire suppression applications. Dry chemical systems are also required to comply with *NFPA 33: Standard for Spray Application Using Flammable or Combustible Materials* and *UL 1254: Standard for Pre-Engineered Dry Chemical Extinguishing System Units.*

Overview

The first dry chemical agents developed were borax (sodium borate) and sodium bicarbonate based. In the 1960s, this dry chemical was modified to make it compatible with protein-based low-expansion foams so that it could be used as a dual-attack agent. Subsequently, multipurpose (monoammonium phosphate-based) and "Purple K" (potassium bicarbonate-based) dry chemical were developed. Later on, "Super K" (potassium chloride-based) dry chemical was made in an effort to create a high efficiency, protein-foam compatible agent. At the end of the 1960s, scientists in Great Britain created another high-performance agent, known as urea-potassium bicarbonate-based dry chemical.[1]

The terms "regular dry chemical" and "ordinary dry chemical" refer to dry chemical agents that are Listed for use on Class B and Class C fires. Dry chemical agents are exceptionally efficient in the extinguishment of flammable liquid fires. They have limited use, however, on fires involving energized electrical equipment, because the residue they leave behind can have detrimental corrosive effect on cables and wiring. Regular dry chemical can in some instances extinguish flash surface fires involving ordinary (Class A) combustibles, but water will have to be applied for the total extinguishment of deep-seated, smoldering fires. "Multipurpose dry chemical" refers to dry chemical agents that are Listed for use on Class B, Class C, and Class A fires.

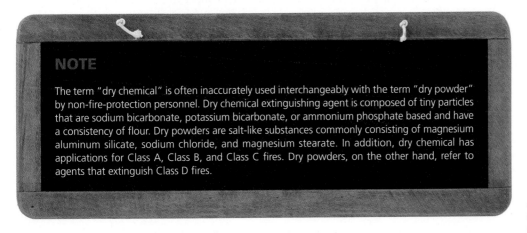

NOTE

The term "dry chemical" is often inaccurately used interchangeably with the term "dry powder" by non-fire-protection personnel. Dry chemical extinguishing agent is composed of tiny particles that are sodium bicarbonate, potassium bicarbonate, or ammonium phosphate based and have a consistency of flour. Dry powders are salt-like substances commonly consisting of magnesium aluminum silicate, sodium chloride, and magnesium stearate. In addition, dry chemical has applications for Class A, Class B, and Class C fires. Dry powders, on the other hand, refer to agents that extinguish Class D fires.

Physical Properties

Common additives (magnesium and calcium stearates, tricalcium phosphate, and silicones) mixed into dry chemical agents improve their storage, flow, and water repellency characteristics. Dry chemicals are stable at both low and high temperatures. Some additives, however, may melt and cause sticking. An upper storage temperature limit of 120°F (49°C) is recommended. Under fire conditions, active ingredients disassociate or decompose. Mixing dry chemical agents must be avoided. For example, mixing **acidic** monoammonium phosphate dry chemical with sodium bicarbonate, which has an **alkaline** base (like most other dry chemical agents), will result in an undesirable reaction that releases

Acidic ■ Chemicals or substances with the property of an acid are said to be acidic. The lower pH value means a higher acidity and thus a higher concentration of hydrogen ions in the solution.

Alkaline ■ Substances with a relatively low concentration of hydrogen ions in the solution and thereby a pH value of greater than 7.

carbon dioxide gas. Extinguisher shells have been known to lose their strength and explode because of this positive-pressure gas generation. Mixing dry chemical agents can also cause caking (hardening) of the powder and lead to poor performance as well as create destructive internal hardware problems.

Ingredients used in dry chemical agents are nontoxic. Large-quantity discharge of dry chemical agent in fixed total-flooding applications can, however, cause temporary breathing difficulties during and immediately after discharge and interfere with visibility. Dry chemical particle size ranges from less than 10 micrometres (μm) up to 75 μm (1 μm = 0.000039 in.). Particle size has an effect on extinguishing efficiency. In general, the greater the surface area to mass ratio, the more efficient the particle. Best results are obtained by a heterogeneous mixture with a median particle size in the range of 20 to 25 μm.

Although considered nontoxic, exposure to dry chemical through eye and skin as well as inhalation should be avoided. Eye and skin contact for short periods of time may cause irritation. First aid measures include immediately flooding the eye with copious amounts of water for at least 15 minutes while holding the eye open. Wash affected skin areas with soap and water. Inhalation victims may complain of irritation of the respiratory tract. They can exhibit a transient cough and shortness of breath. Move exposed persons to fresh air. If the victim has difficulty breathing, obtain medical attention immediately.[2]

Varieties

SODIUM BICARBONATE

Sodium bicarbonate is the original dry chemical agent. It has an alkaline base and is mixed with other ingredients to improve flow characteristics and expelling properties. It also has a silicon supplement in order to impart excellent water repellency characteristics. Sodium bicarbonate is still very popular because it provides a highly reliable, cost-effective, and user-friendly extinguishing medium. Known as "regular" or "ordinary" dry chemical, it is used on Class B and Class C fires. It extinguishes fire by interrupting the chemical chain reaction between heat, fuel, and oxygen. By so doing, it halts the production of fire-sustaining **free radicals**, thereby extinguishing the fire. When discharged into a fire, sodium bicarbonate heats up and releases a cloud of carbon dioxide. This gas drives oxygen away from the fire, creating a smothering action. Water vapor is also released, which has a cooling effect on the fire. This agent is not generally effective on Class A fires; when it is expended, the cloud of carbon dioxide gas and water vapor dissipates quickly, and if the fuel is still sufficiently hot it can reignite. Sodium bicarbonate is less effective than Purple-K for Class B fires.

Free radicals ■ Molecules that contain at least one unpaired electron.

FOAM-COMPATIBLE SODIUM BICARBONATE

Although magnesium and calcium stearate additives waterproof dry chemical agent, they will also act to destroy the foam blanket created by protein (animal)-based foams. Therefore, foam-compatible sodium bicarbonate–based dry chemical was developed for use with protein foams for fighting Class B fires. It uses silicone as a waterproofing agent. Silicon has no harmful effect on protein foam. Foam-compatible sodium bicarbonate–based dry chemical's effectiveness is identical to regular dry chemical.

MET-L-KYL

Pyrophoric ■ A substance that will ignite spontaneously when exposed to the air.

MET-L-KYL (a product name of Ansul, Incorporated) is a specialty variation of sodium bicarbonate. It is used on metals that are **pyrophoric** liquids, such as hydrazine and triethylaluminum. In addition to sodium bicarbonate, it also contains silica gel particles that absorb unburned fuel, thereby insulating it from contact with air. It is also effective on standard Class B fuels.[3]

MONOAMMONIUM PHOSPHATE

Monoammonium phosphate is also known as "multipurpose" or "ABC" dry chemical agent. It can be used on Class A, B, and C fires. It receives its Class A rating from the agent's ability to melt and flow at 350°F (177°C) to produce a thin, oxygen-sealing residue that smothers ordinary (Class A) combustible fires. This agent is more corrosive (acidic) than any of the other dry chemicals and has a pH value* in the range of 4 to 5. Therefore, the removal of multipurpose dry chemical agent residue will require specific cleaning solutions. Unprotected metal surfaces exposed to heat and multipurpose dry chemical agent may at a later time display surface corrosion if they were not properly cleaned.

Multipurpose dry chemical is coated with water-repellant silicone. In a fire, it will decompose to release carbon monoxide, carbon dioxide, sulfur dioxide, and oxides of phosphorous and ammonia. In general, however, the decomposition of dry chemical agents occurring during discharge into a fire is limited, and from a health perspective such decomposition is considered extremely minimal compared to the toxic decomposition products commonly produced from burning materials.

POTASSIUM BICARBONATE

Potassium bicarbonate (a.k.a. Purple-K) is a violet (purple)-colored powder developed by the US Naval Research Laboratory (NRL) and is used on Class B and Class C fires. NRL discovered that the salts of potassium were more effective on flammable liquid fires than the salts of sodium. Potassium bicarbonate is, therefore, more effective on Class B fires than sodium bicarbonate. It is especially valuable as a fire extinguishing agent for flammable liquid pool fires and oil spray fires. Although effective on electrical fires, it will leave an alkaline residue that may be difficult to remove. If moisture is present, this residue can corrode or stain surfaces on which it settles. Therefore, it should not be used in installations where relays and delicate electrical contacts are present.

Purple-K is the preferred dry chemical agent of the oil and gas industry (see Figure 9.1). It is also useful as a dual-attack agent in combination with alcohol-resistant firefighting foams such as alcohol-resistant aqueous film-forming foam (AR-AFFF) and alcohol-resistant film-forming fluoroprotein foam (AR-FFFP). Various additives are mixed with Purple-K base ingredients to enhance storage, flow, and water-repellency characteristics. Silicones are commonly used to coat Purple-K particles, making them free-flowing and resistant to the caking effect of moisture.

When discharged into a fire, Purple-K extinguishes active flaming by breaking the chemical chain reaction. Decomposition of Purple-K creates an opaque, gaseous cloud in the combustion zone. This layer limits the amount of heat that can be radiated back to the seat of the fire. Less fuel vapor is thereby produced because of reduced radiant heat.

Purple-K has a few drawbacks, however. It does not have a cooling capability nor does it produce a lasting inert atmosphere above the surface of a flammable liquid. Application may not result in complete extinguishment if ignition sources (hot metal surfaces or electrical arching) are present. Purple-K is also not effective for extinguishing fires involving Class A combustibles.

---

*The pH scale ranges from 0 to 14 and is used to measure how acidic or alkaline a substance is. Pure water is considered neutral and therefore has a pH of 7 (in the middle of the scale). A pH less than 7 is acidic, and a pH greater than 7 is alkaline. Each whole pH value below 7 is ten times more acidic than the next higher value, and each pH value above 7 is ten times more alkaline than the next lower whole value.

FIGURE 9.1 Dry chemical (Purple-K) local application system using "thruster" nozzles (one is shown with a dust cap in the lower right corner of the photo) protecting against a Class B hazard. *Source:* Ronald R. Spadafora

POTASSIUM CHLORIDE

Potassium chloride (a.k.a. Super-K) dry chemical is a metal halide salt composed of potassium and chlorine. In its pure state, it is odorless and has a white or colorless crystal appearance. It was created in the 1960s, prior to Purple-K, in an effort to make a high-efficiency, protein-foam-compatible dry chemical. Super-K is more effective than sodium bicarbonate–based dry chemical when used in portable and wheeled fire extinguisher applications. It was never as popular as other dry chemical agents, however, because as a salt it was very corrosive. This agent fell out of favor with the introduction of Purple-K dry chemical in the late 1960s, which is more effective and much less corrosive. Super-K is rated for Class B and Class C fires.

UREA-POTASSIUM BICARBONATE

Urea-potassium bicarbonate (a.k.a. Monnex®) is a grey powder dry chemical agent used on Class B and Class C fires. It is effective against fire involving alcohols, ketones, and esters. It is more effective than all other dry chemicals due to its ability to **decrepitate** in the combustion or flame zone. This phenomenon creates a larger surface area for free radical inhibition. This type of dry chemical is popular for high-risk applications in petrochemical, oil, and gas facilities. Monnex also has extensive military and aviation applications.

Decrepitate ■ Break up into small particles with enhanced surface area when heated.

Extinguishing Properties

Dry chemical agent primarily extinguishes fire by preventing the chemical reactions involving heat, fuel, and oxygen. It also halts the production of fire-sustaining free radicals. When discharged directly into the fire area, dry chemicals cause very rapid flame extinguishment. Monoammonium phosphate residue creates an oxygen barrier

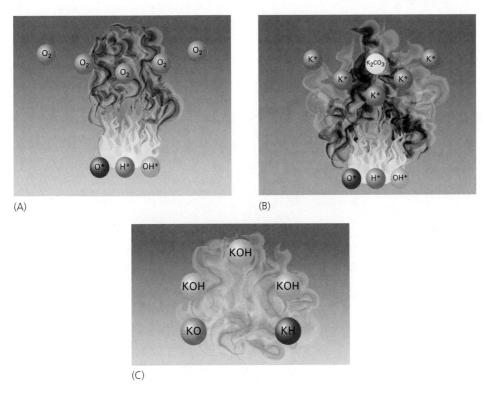

FIGURE 9.2 Purple-K dry chemical extinguishing agent interrupts the chemical chain reaction between heat, fuel, and oxygen. A) Chemical chain reaction of fire forming free radicals (O*, H*, and OH*). B) Purple-K dry chemical extinguishing agent is applied onto the fire forming the free radical (K*) from the disassociation of K2CO3. C) Stable compounds (KOH, KH, and KO) are formed causing the fire to be extinguished.

around Class A combustibles. When dry chemical is decomposed upon application onto a fire, water vapor and carbon dioxide are generated and provide some cooling action. The solid particles of dry chemical agent form a barrier between the fuel and radiant heat being created by the fire. This action shields heat from returning to the fuel for continued ignition. However, these supplementary dry chemical extinguishment characteristics are considered insignificant when compared to its chemical chain-breaking attributes.

INTERRUPTION OF THE CHEMICAL CHAIN-BREAKING REACTION

The principle way dry chemical agent extinguishes fire is through the interruption of chemical chain-breaking reactions. Application of dry chemical prevents the free radicals hydrogen (H) and hydroxyl (OH) from linking up with oxygen in the air, generating heat energy, and sustaining the reaction. It is suggested that the decomposition of dry chemical forms new compounds (NaOH, NaH, KO, and KOH) when reacting with free radicals, which are stable products. Heat production is drastically reduced, thereby terminating the generation of new free radicals. This action extinguishes fire without having to deplete oxygen (see Figure 9.2)[4].

System Components

Dry chemical extinguishing systems are intended for use in situations in which prompt extinguishment is required and sources for reignition are not present. Sodium bicarbonate, potassium bicarbonate, and ammonium phosphate are used as an extinguishing agent

FIGURE 9.3 Pre-engineered, skid-mounted dry chemical delivery system at an LNG facility.
Source: Ronald R. Spadafora

for fixed total-flooding and local-application systems as well as hand hose lines and portable extinguishers. Design calculations for fixed systems are performed to ensure that a sufficient amount of agent is stored inside the container. These calculations are based upon the hazard presented and the expected fire. A single hazard normally does not require a reserve supply of agent. Dry chemical extinguishing systems protecting multiple hazards, however, do require an additional supply.

Pre-engineered dry chemical fire extinguisher systems consist of a container holding the dry chemical extinguishing agent, expellant gas storage tank(s), manual and/or automatic control hardware, piping or tubing to deliver the agent, and nozzles to effectively disperse the agent onto the fire area (see Figure 9.3). Pressure drop through lengths of piping, tees, and elbows must be calculated and dealt with by the design engineer in order to provide adequate flow rate to the nozzles. The minimum rate of flow for total-flooding systems is vital due to the fact that dry chemical will not extinguish a fire if applied too slowly. Too high a rate of discharge for local-application systems may result in uneven discharge of dry chemical before extinguishment is accomplished. Expellant gas is also needed to mix with the dry chemical in order to "fluidize" the agent (allowing it to flow easily through the piping) for distribution onto the hazard. Gas cylinders generally contain nitrogen or dry air, but in smaller extinguishing systems carbon dioxide may be used. Volume and storage pressure for the expellant gas is determined by the gas used and the requirements of the system.

Containers in which dry chemical agent and expellant gas are stored together are equipped with a fill opening, a valve with a discharge outlet, an expellant gas-charging inlet, and a pressure gauge. In systems in which dry chemical storage tanks and expellant gas cylinders are separate, the gas is admitted to the tank to first fluidize the dry

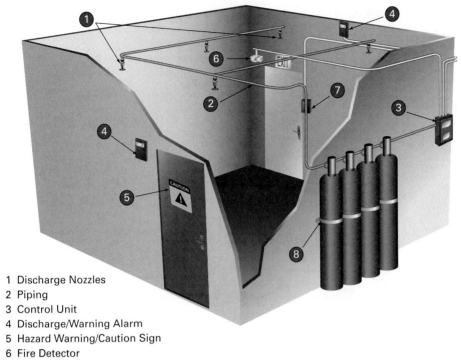

1 Discharge Nozzles
2 Piping
3 Control Unit
4 Discharge/Warning Alarm
5 Hazard Warning/Caution Sign
6 Fire Detector
7 Manual Discharge Station
8 Dry Chemical Agent and Pressurization Cylinders

FIGURE 9.4 Dry chemical system components

chemical agent while the pressure builds up equally throughout the volume of the tank before the agent is released. Application of dry chemical agent is from nozzles mounted on the side of the hazard and/or overhead (see Figure 9.4). The installation requirements for a pre-engineered fire extinguishing system are specified by the system equipment manufacturer. The manufacturer also provides references including the quantity of extinguishing agent, pipe size/length, number and type of fittings, nozzle type, and installation locations.

Sequence of Operation

In Figure 9.5, the numerals in circles correspond to the numbered list ahead. This diagram will aid the reader's understanding of dry chemical extinguishment system operational design and the sequence of operation events.

1. *Detection:* A detector (smoke, heat, or flame) senses a fire condition.
2. *Transmission:* An electronic signal is sent from the detector via wiring to the fire alarm control unit.
3. *Analysis:* The fire alarm control unit interprets the signal as a fire condition.
4. *Alarm:* A warning signal is sounded.
5. *Shutdown:* Equipment, including the heating, ventilation, and air-conditioning (HVAC) system, is shutdown.
6. *Manual pull-station activation:* The system may be activated manually via pull station.
7. *Manual push-button activation:* The system may be activated manually via a push-button.

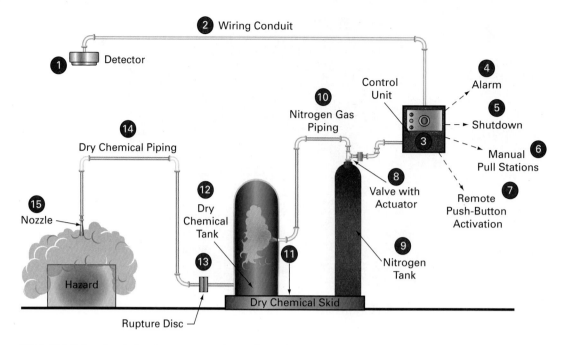

FIGURE 9.5 Dry chemical system sequence of operation Robert M. Gagnon, *Design of Special Hazard and Fire Alarm Systems*, 2nd ed. (Clifton Park, NY: Delmar Cengage Learning, 2008), 312.

8. *Nitrogen actuator alert:* The signal is delivered to an actuator on the nitrogen (expellant) tank.
9. *Nitrogen actuation:* The nitrogen cylinder valve is opened.
10. *Nitrogen release:* Nitrogen under pressure is piped to the dry chemical storage tank.
11. *Fluidization:* Nitrogen mixes with the agent upon entry into the dry chemical tank.
12. *Pressure increase:* Pressure inside the dry chemical tank builds up.
13. *Rupture disc bursts:* Increased pressure inside the tank causes the rupture disc located at the dry chemical storage tank discharge outlet to burst.
14. *Agent flow:* Fluidized dry chemical and nitrogen gas mixture flows through system piping under pressure.
15. *Nozzle application:* The nozzle applies dry chemical extinguishing agent to the fire.

Total-Flooding and Local-Application System Methods

Pre-engineered, fixed dry chemical extinguishing system units employ two basic methods to achieve fire suppression: total flooding and local application. Either method or a combination of both may be used, depending on the end-use application. Total-flooding applications discharge dry chemical into an enclosed space. This enables the extinguishing system to build up the necessary concentration of dry chemical within a short period of time to ensure extinguishment. The quantity of agent and flow rate must be sufficient to provide a fire-suppression concentration throughout all parts of the area being protected. Upon completion of discharge, the agent quickly settles to low areas of the space, and its presence within the combustion zone diminishes.

Total-flooding systems have a supply of dry chemical permanently connected to fixed piping. Nozzles are arranged to discharge dry chemical into an enclosed area onto a

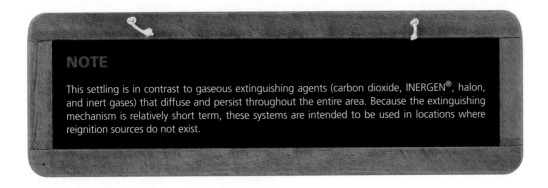

NOTE

This settling is in contrast to gaseous extinguishing agents (carbon dioxide, INERGEN®, halon, and inert gases) that diffuse and persist throughout the entire area. Because the extinguishing mechanism is relatively short term, these systems are intended to be used in locations where reignition sources do not exist.

designated hazard. These systems are used where there is a permanent enclosure about the hazard that is capable of allowing the required concentration of agent to remain in the area of protection. Leakage of dry chemical from this space must be insignificant. With a total-flooding system, the rate of application will be such that the design concentration in all parts of the enclosure is obtained within 30 seconds.

Local-application protection is a different method than total flooding in that the dry chemical is applied directly onto the surface of the fire source. Discharging a dry chemical locally is effective only in applications for which the fire risk is isolated or all fire hazards are protected simultaneously. Local-application systems are primarily used for the extinguishment of fires in flammable or combustible liquids and gases, where the combustibles are not enclosed or the enclosure does not conform to the requirements for total flooding (see Figure 9.6). The hazard includes all areas that can become coated by liquids or solids. Areas subject to spillage, leakage, dripping, splashing, or condensation must be protected. Additional materials and equipment to be safeguarded are stock, drainboards, hoods, and ducts that can extend the fire outside the protected area.

FIGURE 9.6 Close-up of an automotive filling station canopy with dry chemical nozzles. Note the gas sensor at the center of the photo.
Source: Ronald R. Spadafora

FIGURE 9.7 Remote manual emergency fuel shutoff device at an automotive filling station.
Source: Ronald R. Spadafora

 FIRE TESTS—FLAMMABLE LIQUIDS

Pre-engineered dry chemical extinguishing system units are used to protect localized flammable liquid fires. These units are designed to discharge automatically upon activation from fire detection systems or manually discharge the extinguishing agent either overhead or horizontally across the fuel surface. Additional safety features incorporate remote manual emergency fuel shutoff switches (see Figure 9.7). Tests are conducted with nozzles positioned at the maximum and minimum height above a flammable liquid surface.

Fire tests use a deep steel pan with areas referenced in the manufacturer's installation instructions. The pans are filled with 2 in. (51 mm) of heptane. The liquid is ignited and allowed to burn freely for at least 30 seconds, after which the extinguishing system is actuated. To comply with UL 1254, the test fire must be completely extinguished.[5] Fire tests for extinguishing system units intended to protect automobile service filling stations are evaluated in a similar manner as local-application systems. In lieu of using a single pan of heptane, however, small cans of heptane are located around the perimeter of the protected area to determine the coverage space for each discharge nozzle.

Hand Hose-Line Method

This type of method uses a storage supply of dry chemical and an expellant gas with one or more hose lines to discharge the agent onto the hazard. Mobile and wheeled units provide a way for firefighters and fire brigade members to bring the agent along great distances to the hazard (see Figure 9.8). Hose stations are also used. This application connects the hose to the agent storage container via piping. Large amounts of agent can be provided through this method to protect major fire hazards.

Portable Fire Extinguishers

Dry chemical extinguishers expel a finely powdered, nonconductive agent that, on contacting flame, releases many times its volume in nontoxic fire extinguishing gases. The agent is modified to make it free flowing, but portable extinguishers still require nitrogen gas as a propellant stored in the same chamber as the agent itself or a carbon diox-

FIGURE 9.8 Fire brigade apparatus with dry chemical extinguishing agent capability. *Source:* Ronald R. Spadafora

FIGURE 9.9 A stored-pressure dry chemical ABC portable fire extinguisher. *Source:* Ronald R. Spadafora

ide cartridge in a separate chamber (see Figure 9.9). A continuous squeeze of the handle will discharge a full 30 lb. (14 kg) container in a little more than 10 seconds. These extinguishers, depending on the agent, can be used on Class A, Class B, and Class C fires. Dry chemical agent, however, is difficult to remove from motors and generator windings. These portable extinguishers are effective on flammable liquid fires in vats and pools as well as spilled fires on floors.

Incipient fire may initially be extinguished, but heated combustible vapors contacting hot surfaces will cause reignition. Dry chemical extinguishers have an advantage over CO_2 extinguishers because they leave a coating on the extinguished material, reducing the likelihood of this occurrence. Although dry chemical agent is nontoxic, the discharge of a portable fire extinguisher in a confined space can cause a sudden reduction of visibility, which may temporarily hinder escape and rescue efforts.[6] If there is a possibility that personnel may be exposed to a dry chemical discharge, positive-pressure, self-contained breathing apparatus (SCBA) should be used.

As with all types of portable fire extinguishers, dry chemical should be inspected regularly. Is the shell undamaged and in good condition? Are all external parts serviceable? Check to make sure it has proper container pressure (gauge in green area), the correct volume of extinguishing agent (weighing), and has been tested and maintained according to authority having jurisdiction (AHJ) requirements. The bottom of the container can be hit with a rubber mallet to prevent the dry chemical agent from hardening.

Application Techniques

Purple-K portable dry chemical extinguishers have shown a superior extinguishing capability on alternative fuel fires. Methanol, for example, is used extensively as an alternate fuel for gasoline- and diesel-powered vehicles and equipment. Purple-K has also been highly effective on gasoline, diesel, and compressed gas fires.[7] The dry chemical container is made of heavy-gauge steel welded construction, holding approximately 30 lb. (14 kg) of agent. The discharge range is (19–20 ft.; 5.7–6.0 m) allowing the operator to stand back from the fire a good distance, thereby enhancing safety. The following dry chemical agent application techniques are used by professional firefighters for the extinguishment of incipient fires involving Class B combustibles.

FLAMMABLE LIQUID SPILL FIRE

Approach the fire from the upwind side. Hold the nozzle at a 45-degree angle to the ground. Stand back approximately 10 ft. (3 m) away from the front edge of the fire. Direct the dry chemical agent 6 in. (152 mm) ahead of the flame edge. Direct the stream using a rapid, side-to-side sweeping motion. Each sweep of stream must be slightly wider than the near edge of the fire. Extinguish the small fire at the back edge of the spill by aiming directly at the base of the flame.

THREE-DIMENSIONAL (DIP TANK), FLAMMABLE LIQUID, GRAVITY-FED FIRE

Approach the fire from the upwind side. Hold the nozzle at a 45-degree angle to the ground. Stand back approximately 10 ft. (3 m) away from the front edge of the fire. Make a very slow sweep with the dry chemical agent across the front edge of the flammable liquid spill. Ensure that the spill is completely extinguished before moving forward. Do not close the extinguisher nozzle. Raise the dry chemical stream to the front lip of the dip tank. Extinguish the fire in the dip tank. Direct the stream up the dip tank tray, pushing the fire towards the fuel source. Finally, extinguish the fuel source.

FLAMMABLE LIQUID PRESSURE FIRE

Approach the fire from the upwind side. Stand back approximately 15 ft. (5 m) away from the front edge of the spill fire area. Direct the dry chemical stream at the source of the fuel. Move the dry chemical stream down toward the escaping fuel. Extinguish remaining flammable liquid spill using a rapid, side-to-side sweeping motion. In some cases, the fire should not be extinguished unless the fuel supply can be shut off.

FLAMMABLE GAS PRESSURE FIRE

Approach the fire from the upwind side. Direct the dry chemical stream at the fuel source. Move the dry chemical stream up toward the escaping gas. Continue to move the dry chemical stream upward until the fire is extinguished. In some cases, the fire should not be extinguished unless the fuel supply can be shut off.

LIQUEFIED PETROLEUM GAS (BROKEN FLANGE) FIRE (IN CONJUNCTION WITH WATER STREAMS)

Approach the fire from the upwind side. Open hose streams first, and adjust nozzles to the fog position. Direct the dry chemical stream at the source of the fuel. If a spill fire is present, move the dry chemical stream down toward the escaping fuel to the spill on the ground. With a side-to-side sweeping motion, extinguish the spill fire with the dry chemical agent. Cool the piping and surroundings with water. Close the supply valve to stop the flow of fuel. Dry chemical agent and water fog streams can be highly effective in this type of scenario. Water acts as a coolant and heat shield, and the dry chemical agent provides rapid flame extinguishment. Fog nozzles can also be used to enhance the reach of dry chemical agent.

FIGURE 9.10 FDNY dry chemical (Purple-K) unit. *Source:* Ronald R. Spadafora

FDNY Purple-K Units

The Fire Department of New York (FDNY) currently has completely self-contained, firefighting Purple-K mobile units. A typical Purple-K unit is a truck with a platform body. It carries 1,000 lb. (454 kg) dry chemical units on skid-mounted modules. The agent is expelled by two nitrogen cylinders into two rubberized hose reels. Specialized nozzles are attached to the hose for discharging the agent at the rate of five pounds per second per line. Each nitrogen cylinder has a manual valve installed for individual activation. A fully loaded system will supply both hand lines for approximately two minutes.[8] In addition, these units carry portable, 30 lb. (14 kg) stored-pressure dry chemical fire extinguishers (see Figure 9.10).

These units were originally designated for alternate-fuel (CNG) vehicle responses. Today, their function has been expanded to include all incidents for which dry chemical agent can be effective (LNG, LPG, alcohols, transformers, aircraft fuels, electrical substations, gasoline filling stations, and pressurized gas fires).

CHAPTER REVIEW

Summary

The most commonly used dry chemical fire extinguishing agents are sodium bicarbonate, potassium bicarbonate, and monoammonium phosphate (multipurpose). The principle way in which dry chemical agent extinguishes a fire is by interrupting the chain reaction; the chemicals prevent the union of the free radical particles and oxygen during the combustion process. Dry chemicals can be used to extinguish Class A, Class B, and Class C fires. Dry chemical agents are applied through fixed piping systems, hand hose lines, and portable fire extinguishers.

Review Questions

1. What are the basic varieties of dry chemical extinguishing agent?
2. What dry chemical agent is Listed for use on Class B, Class C, and Class A fires?
3. Why should firefighters avoid mixing dry chemical agents?
4. What is the primary extinguishing characteristic of dry chemical extinguishing agent?
5. Why do most dry chemicals contain metal stearates?
6. Why is potassium bicarbonate (a.k.a. Purple-K) more effective on flammable liquid fires than sodium bicarbonate?
7. What benefit can be achieved when dry chemical is applied in conjunction with water streams expelled through fog nozzles?
8. Why would a firefighter hit the bottom of a dry chemical portable fire extinguisher with a rubber mallet?

Endnotes

1. Walter M. Haessler, *The Extinguishment of Fire* (Quincy, MA: NFPA, 1974).
2. Kidde, "Material Safety Data Sheet—Commercial ABC Dry Chemical (Fire Extinguishing Agent)," last modified February 28, 2011, www.kidde.com/Documents/msds%20abc%20english.pdf.
3. Ansul, Incorporated, "Fire Protection for Metal Alkyls," www.ansul.com/AnsulGetDoc.asp?FileID=8091.
4. J. Craig Voelkert, *Fire and Fire Extinguishment—A Brief Guide to Fire Chemistry and Extinguishment Theory for Fire Equipment Service Technicians* (J. Craig Voelkert, publisher, 2009).
5. Underwriters Laboratories Inc., *UL 1254: Standard for Pre-Engineered Dry Chemical Extinguishing System Units* (Northbrook, IL: UL, 1992).
6. James Angle, David Harlow, William Lombardo, Craig Maciuba, and Michael Gala, *Command 1A: Command Operations for the Company Officer*, California Edition (Clifton Park, NY: Delmar, Cengage Learning, 2012).
7. Fire Department of the City of New York (FDNY), "Purple-K Dry Chemical Extinguisher," *Training Bulletin*, September 1, 2010.
8. James S. Griffiths, *Fire Department of New York—an Operational Reference*, 9th ed. (New York, NY: FDNY Foundation, 2012).

Additional References

Badger Fire Protection, "Dry Chemical Extinguishing Agent Exposure and Agent Clean Up," *Technical Bulletin* #103-0110, www.badgerfire.com/utcfs/ws-603/Assets/T103-0110.pdf.

Fire Department of the City of New York (FDNY), "Purple-K Dry Chemical Extinguishers, Addendum 1 (Purple-K Unit)," *Training Bulletin*, February 4, 2004.

James D. Lake, "Chemical Extinguishing Agents and Application Systems," in *Operation of Fire Protection Systems*, ed. Arthur C. Cote (Quincy, MA: NFPA, 2003), 601–613.

Kerry Bell and Kenneth W. Zastrow, "Pre-Engineered Chemical Extinguishing Systems," www.nafed.org/resources/library/Prechem.cfm.

National Fire Protection Association (NFPA), *NFPA 17: Standard for Dry Chemical Extinguishing Systems* (Quincy, MA: NFPA, 2009).

National Fire Protection Association (NFPA), *NFPA 33: Standard for Spray Application Using Flammable or Combustible Materials* (Quincy, MA: NFPA, 2011).

Scott Dornan, *Industrial Fire Brigade: Principle and Practice* (Sudbury, MA: Jones and Bartlett, Inc., 2008).

CHAPTER **10**

Combustible Metal Suppression Agents

Source: IEP Technologies

KEY TERMS

Explosion, *p. 170*
Alloy, *p. 171*
Deflagration, *p. 172*
Fines, *p. 176*
Chips, *p. 176*

Turnings, *p. 176*
Ribbon, *p. 177*
Shavings, *p. 177*
Allotrope, *p. 178*
Swarfs, *p. 182*

Halogens, *p. 183*
Ingots, *p. 185*
Detonation, *p. 186*
Thermal imaging camera, *p. 191*

OBJECTIVES

After reading this chapter, the reader should be able to:

- Identify the five components of the dust explosion pentagon
- List the physical properties of combustible metals
- Explain the uses of Class D extinguishing agents
- Describe the four categories of combustible metals
- Understand explosion prevention and protection systems
- Discuss the guidelines and procedures of workers and first responders

Introduction

Combustible dusts are finely ground organic or metal particles that present an **explosion** hazard under certain conditions when suspended in air. A dust explosion can cause loss of life, serious life-changing injuries, and destruction of buildings. The US Chemical Safety and Hazard Investigation Board (CSB) identified 281 combustible dust incidents between 1980 and 2005 that led to the deaths of 119 workers, injured 718, and extensively damaged many

Explosion ■ A rapid release of combustion energy.

industrial facilities. A wide variety of materials will burn when in dust form and can potentially generate an explosion. Industry examples include food (sugar, spice, starch, and flour), grain, tobacco, plastics, wood, paper, rubber, textiles, pesticides, pharmaceuticals, dyes, coal, and metals (aluminum, iron, magnesium, and zinc).[1]

IMPERIAL SUGAR FACTORY EXPLOSION

On February 7, 2008, a series of sugar-dust explosions at the Imperial Sugar manufacturing facility in Port Wentworth, Georgia, resulted in 14 worker fatalities.

In addition, 36 workers were treated for serious burns and injuries. The explosions and subsequent fires destroyed the sugar-packing buildings, silos, bulk train car-loading area, and parts of the sugar-refining process areas.

The CSB determined that the first dust explosion started in the enclosed steel-belt conveyor located below the sugar silos. Steel cover panels on the belt conveyor allowed explosive concentrations of sugar dust to accumulate inside the enclosure. An unknown source ignited the sugar dust, causing a violent explosion. It lofted sugar dust that was able to accumulate on the floors and elevated horizontal surfaces, propagating more dust explosions throughout the buildings. Secondary dust explosions heaved concrete floors and resulted in the collapse of brick walls.

The CSB investigation identified sugar- and cornstarch-conveying equipment as not designed or maintained to minimize the release of sugar dust into the work area. Inadequate housekeeping practices were also blamed for significant accumulations of combustible sugar dust on the floors and elevated surfaces throughout the packing buildings. This dust accumulated above the minimum explosion concentration (MEC) inside the newly enclosed steel-belt assembly under the silos. The CSB determined that an overheated bearing in the steel-belt conveyor most likely ignited a primary dust explosion and that the 14 fatalities were most likely the result of the secondary explosions and fires.[2]

Under certain circumstances, nearly all metals will burn when exposed to the atmosphere. Combustible metals (Class D materials), however, refer to certain metals that ignite quite easily. Important factors in assessing the combustibility of metals include size, shape, quantity, and **alloy**. Hot or burning metals may react violently upon contact with other materials, including some fire extinguishing agents. Burning metals generate temperatures much higher than Class A, Class B, and Class K fires. Class C fires, once de-energized, revert to the classification of fire matching what is actually burning. Class D fire extinguishing agents are used to control and extinguish combustible metal fires.

Alloy ■ A homogeneous mixture of two or more metals, the atoms of one replacing or occupying interstitial positions between the atoms of the other.

Life safety objectives for a facility in which combustible metals are located will be designed to maintain its structural integrity despite the effects of fire and explosion. This will provide the time necessary to evacuate, relocate, or shelter in place occupants not in the immediate vicinity of the explosion. The facility and processes will also be built to prevent fires, explosions, or releases of hazardous materials that can cause failure of adjacent sections of the structure, life safety systems, adjacent properties, and adjacent storage.

Overview

Similar to other fuels, combustible metal's surface area to mass ratio is inversely proportional to the energy required for ignition. Large blocks of metal with a low surface area to mass ratio require high amounts of heat energy over a long period of time to ignite. On the other hand, finely ground metal dust, with a high surface area to mass ratio, needs a relatively low amount of heat energy applied for a few seconds or less to ignite or explode.

Class D fires involve extreme temperatures and highly reactive fuels. Burning magnesium, for example, breaks down the water molecule (H_2O) into hydrogen and oxygen gas to accelerate the intensity of the fire and cause violent hydrogen gas and steam explosions. This conversion of water into hydrogen has a heat value (British thermal units per pound [Btu/lb.]) of about 2.8 times that of gasoline (assuming 100% conversion of the hydrogen in the water). This equates to flowing 42.8 gpm (162 L/min.) of gasoline onto the fire for every 100 gpm (379 L/min.) of water. In addition, the release of large amounts of oxygen, an oxidizer, will enhance the burning characteristics of the material.[3]

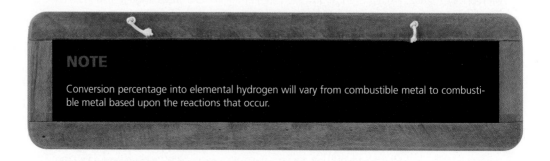

NOTE

Conversion percentage into elemental hydrogen will vary from combustible metal to combustible metal based upon the reactions that occur.

Dust Explosion Pentagon

Deflagration ■ A reaction in which the flame front travels into unburned material at subsonic speed.

A dust **deflagration** is complex. It requires not only the three components that make up the fire triangle but also that the fuel must be confined within an enclosure as well as easily dispersed within it. The five components necessary to initiate a dust deflagration can be referred to as the dust explosion pentagon.[4] They include fuel, ignition source, oxidizer, confinement and dispersion (see Figure 10.1).

An examination of these components follows.

FUEL

There are three major factors that pertain to fuel and its likelihood to burn. The fuel component consists of small, easily suspended particles. They do, however, have a relatively large surface area to mass ratio compared to larger particles and therefore are ignited readily. Moisture is another consideration. The more moisture dust clouds absorb, the more difficult they are to ignite. Moisture also reduces the severity of the deflagration. The final fuel factor is its Kst value. The Kst value is a measure of the relative energy of a dust explosion. The higher the Kst value, the more energy the dust deflagration will generate (see Table 10.1).

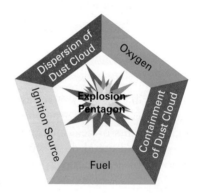

FIGURE 10.1 Dust explosion pentagon

TABLE 10.1 Kst Values for Common Metal and Nonmetal Dusts

DUST	SIZE (IN MICRONS)	Kst VALUE
Aluminum grit	41	100
Aluminum powder	22	400
Charcoal	29	117
Cotton	44	24
Magnesium	28	508
Milk powder	165	90
Rice starch	18	190
Silicon	10	126
Sulfur	20	151
Tobacco	49	12
Wood dust	43	102

NOTE

A dust cloud's explosive power is measured via the dust's deflagration index (rate of pressure rise). The National Fire Protection Association (NFPA) classifies dusts according to their explosiveness and corresponding Kst rating. Class 0 dusts (silica, for example) have a 0 Kst rating. They are not subject to deflagration. Any dust with a Kst value greater than 0 can burn. Class 1 dusts have weak explosion characteristics and ratings below 200 Kst. Class 2 dusts (strong explosion characteristics) are within a Kst range of 200 to 300. Class 3 dusts have very strong explosion characteristics and are rated above 300 Kst.

IGNITION SOURCE

The key factor to predicting a dust particle's likelihood to ignite is its minimum ignition energy (MIE). Dusts that have low MIE are more easily ignited than dusts that have higher MIE.

OXIDIZER

Reducing the oxygen content increases the amount of energy required to ignite the combustible dust and may prevent ignition. Determining the minimum oxygen content (MOC) is an important strategy used to prevent explosions in process systems by lowering the oxygen concentration to the point at which dust cloud deflagrations can no longer be supported.

CONFINEMENT

The minimum explosive concentration (MEC) is the minimum concentration of combustible dust and air that is needed for a deflagration to occur. The MEC is analogous to the lower flammable limit for gases/vapors in air.

DISPERSION

Dispersion of dust in air is yet another component of the dust pentagon. A dust layer depth of 1/32 in. (0.8 mm) or greater covering an area more than 10% above the floor may present a dust deflagration hazard for many combustible dusts. Accumulations of dust on floors and working surfaces of equipment as well as from structural building elements, ventilation duct work, and utility conduits may also provide a sufficient amount of fuel to propagate a dust deflagration.

Physical Properties

The physical properties of combustible metals vary widely. The following is a list of a few combustible metals that first responders may encounter during their careers in public service.

ALUMINUM

Aluminum (Al) has a specific gravity of 2.64 and melting point of 1,220°F (660°C). The surface of this metal is covered with a thin layer of oxide that helps protect it from corrosion by air. Normally, aluminum metal does not react with air. If the oxide layer is damaged, however, aluminum will burn in oxygen with a brilliant white flame. When large aluminum particles are heated, they can be ejected from burning surfaces by gas pressures formed during the combustion reaction. If sufficiently hot, they can continue radiating outside of the combustion zone as sparks. Aluminum powders have high Kst rates. Aluminum is used in products such as beverage cans, food containers, institutional foil, and electronics and electrical equipment (see Figure 10.2).

 DUMPSTER FIRE FATALITY

On the evening of Dec. 29, 2009, a fire company was dispatched for a dumpster fire at a foundry located in Wisconsin. The facility was in the business of melting aluminum to produce aluminum sand castings. The first arriving unit found a dumpster outside a structure emitting two-foot high bluish-green flames from the open top. There was a reddish-orange glow in the middle of the dumpster. A hose line was stretched, and a portable ladder was used to examine the contents of the dumpster. The dumpster was made of steel and measured approximately 17 ft. (5 m) long by 5 ft. (1.5 m) wide by 7 ft. (2 m) high. It contained a large amount of aluminum alloy shavings, foundry floor sweepings (consisting of metal particles), and several open-top 55-gallon (208 L) drums of slag (a byproduct of the casting process). The color of the flame indicated that metal cutting fluids and/or oils were burning. Approximately 700 gallons (2,650 L) of water was flowed into the dumpster with no effect on the fire. An additional 100 gallons (379 L) of firefighting foam was applied. During foam application, the contents of the dumpster started sparking. Subsequently, an explosion occurred within seconds, emitting shrapnel into the air. The explosion killed one firefighter and injured eight others.[5]

NFPA 484: Standard for Combustible Metals states that it is extremely important to gather situational awareness of possible hazards by identifying the combustible metals involved.[6] This can be simplified through recording the combustible metals being used or stored during unit building inspection. Visiting facilities where combustible metals are processed or stored can also provide the fire company with the physical state of the metals, their location relative to other combustible materials, and the amounts involved.

Manufacturing facilities that use combustible metals should implement measures such as container labeling to control risks to emergency responders from waste fires. This should be carried out by personnel enforcing fire code requirements in the authority having jurisdiction (AHJ). However, it is also the responsibility of all firefighters to make the AHJ aware of the hazard.

In lieu of a water-based extinguishing agent, it is recommended that the appropriate Class D extinguishing agent be used upon identifying the burning metal. When the quantity of the needed extinguishing agent is not readily available, the best approach may be to isolate the material as much as possible, protect exposed property, and allow the fire to burn itself out.

FIGURE 10.2 Recycling of waste from electronic and electrical equipment (WEEE), includes shredded and pulverized aluminum cables and components. *Source:* Andrew Pini/Getty Images

The National Institute for Occupational Safety and Health (NIOSH) investigation identified the following items as key contributing factors in the Wisconsin firefighter fatality incident:[7]

- Wet extinguishing agent applied to a combustible metal fire
- Lack of hazardous materials awareness training
- No documented site preplan
- Insufficient scene size-up and risk assessment
- Inadequate disposal and storage of materials

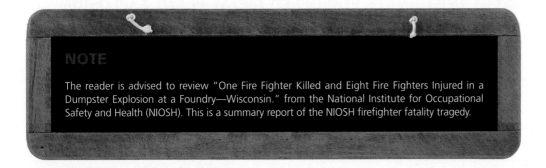

NOTE

The reader is advised to review "One Fire Fighter Killed and Eight Fire Fighters Injured in a Dumpster Explosion at a Foundry—Wisconsin." from the National Institute for Occupational Safety and Health (NIOSH). This is a summary report of the NIOSH firefighter fatality tragedy.

BERYLLIUM

Beryllium (Be) has a specific gravity of 1.85 and melting point of 2,400° F (1,316°C). In powder form, ignition may occur if the metal is heated to about 1,112°F (600°C). Burning occurs with an intense flame but can be extinguished by water. Beryllium powder is kept away from air and moisture and stored in tight containers, preferably under argon gas. Beryllium is toxic, and contact with skin or inhalation of dust or fumes should be avoided. Industries that use beryllium include aerospace, aircraft manufacturing, computer manufacturing, dental laboratories, telecommunications, foundries, and metal reclamation.

CAESIUM

Caesium (or cesium; Cs) has a specific gravity of 1.87 and melting point of 82°F (28°C). Caesium metal is highly reactive. It reacts explosively with water and ignites spontaneously with air. It has a range of applications in the production of electricity, electronics, and chemistry.

CALCIUM

Calcium (Ca) has a specific gravity of 1.55 and melting point of 1,542°F (839°C). Calcium is a soft alkaline earth metal that reacts with water to produce hydrogen gas. In powdered form, however, the reaction with water is extremely rapid, because the increased surface area of the powder accelerates the reaction. The flammability of calcium is dependent upon the moisture in the air. In a finely divided form, it will spontaneously ignite in air. Barium and strontium have similar fire characteristics. Calcium is used in manufacturing cement, mortar, lime, and limestone and aids in production in the glass industry.

HAFNIUM

Hafnium (Hf) has a specific gravity of 13.36 and melting point of 4,032°F (2,222°C). This metal burns with little visible flame but with a high heat release rate (HRR). It reacts with water to form hydrogen, which may ignite spontaneously. Hafnium **fines** are pyrophoric (ignite instantly upon exposure to air). Its hazards are similar to those of zirconium. It is used in the aerospace, electronics, and nuclear energy industries.

Fines ■ Crushed or powdered metal generated by such activities as grinding, sawing, cutting, and sanding.

IRON

Iron (Fe) has a specific gravity of 7.86 and melting point of 2,802°F (1,539°C). Iron presents a dust explosion hazard under favorable conditions. It can be easily ignited in the form of steel wool, fines, **turnings**, or **chips** containing oil. Iron is used primarily in structural engineering, maritime, automobile, machinery, and general industrial applications.

Turnings ■ Slender metal particles produced when shaping metal on a lathe.

Chips ■ Extremely sharp metal particles produced from cutting.

HOEGANAES POWDERED METAL PLANT EXPLOSION

The first of three 2011 combustible metal accidents occurred on January 31 at the Hoeganaes powdered metals plant in Gallatin, Tennessee. Fine particles of iron dust ignited while two workers were troubleshooting a problem with a bucket elevator. Both employees suffered burns and later died from their injuries. Two months later, on March 29, a similar flash fire accident burned another Hoeganaes worker. At a news conference in Tennessee on May 11, 2011, the CSB released laboratory test results on dust samples taken from the plant after the second accident. The testing demonstrated the combustibility of even small amounts of the iron dust when dispersed in air in the presence of an ignition source.

On May 27, just a few weeks after the CSB news conference, a hydrogen explosion erupted in the plant after the gas began leaking from a corroded furnace pipe. The blast shook loose iron-dust accumulations from the upper reaches of the building, which ignited and rained down on workers. The explosion and ensuing fire killed three workers and injured two others. The CSB found that the company did not require atmospheric testing for hydrogen or other explosive gases. Inspections conducted previously by the CSB found "significant quantities of accumulated metal dust on surfaces within close proximity to the incident locations and elsewhere throughout manufacturing areas, including elevated surfaces," according to a CSB statement. Additionally, laboratory tests showed that metal dust collected from various locations around the facility was combustible and capable of exploding when dispersed in air and confined.[8]

LITHIUM

Lithium (Li) has a specific gravity of 0.53 and melting point of 356°F (180°C). Ignition and burning occur when lithium is heated to its melting point. It reacts less vigorously than sodium with water or air and usually does not ignite. It reacts strongly, however, with organic compounds that contain chlorine, fluorine, and bromine. Lithium also reacts with the halogens, hydrogen peroxide, and sulfuric acid. In the presence of moisture, lithium reacts exothermically (liberates heat) with nitrogen at ordinary temperatures. When lithium is heated near its melting point, it ignites in air and burns with an intense, brilliant white flame. When mixed with nitric acid, it may explode on very light impact or friction. Lithium has many industrial uses. It goes into glasses, ceramics, pharmaceuticals, and aluminum and magnesium alloys.

MAGNESIUM

Magnesium (Mg) has a specific gravity of 1.74 and melting point of 1,202°F (650°C). Although its ignition temperature is near its melting point, ignition of some forms may occur at lower temperatures. When pulverized into a dust cloud or in **ribbon** form, magnesium can be ignited almost instantly. Loose **shavings** also ignite fairly readily. A compact pile of chips is less easy to ignite. Magnesium fines wet with oils may ignite spontaneously. Fines wet with acids, water, water-soluble oils, or oils containing fatty acids will generate hydrogen. They will also react with chlorine, bromine, iodine, and oxidizing agents. In powder form, explosive mixtures with air are generated that can be ignited by a spark. Magnesium has applications in cases in which weight reduction is important (aerospace and missile construction). Magnesium components are also widely used in industry and agriculture.

Ribbon ■ A metal particle that is less than 0.1 in. (3.2 mm) in two dimensions or less than 0.05 in. (1.3 mm) in a single dimension.

Shavings ■ Shredded metal particles produced as a result of machining.

MANGANESE

Manganese (Mn) has a specific gravity of 7.43 and melting point of 2,246°F (1,230°C). When in the form of a dust cloud, it can be ignited at 840°F (449°C) in air. Manganese is used in many industrial metallurgical processes. It is of particular importance in the creation of alloys of steel and aluminum.

MAGNESIUM SHAVINGS FIRE

On the morning of April 23, 2012, Portland, Oregon, firefighters responded to reports of a fire involving magnesium shavings at a commercial (engraving) occupancy. Four-foot (1.2 m) flames were reported from neighboring businesses. Initial size up revealed that magnesium shavings and plates were on fire. Firefighters arriving on the scene used dry chemical and CO_2 portable fire extinguishers in an attempt to douse the flames.

The fire was difficult to extinguish, and their attempts were unsuccessful. A hazardous materials response team (HMRT) was called to the scene to provide combustible metal fire extinguishers. As a safety precaution, neighboring businesses and civilians were evacuated to a safe location. Bags of sand that were available at a place of business near the fireground were used by firefighters to suppress the active flaming. Firefighters maintained a presence at the scene long after the sand was applied. Once it was determined safe to pull the combustible metal from the sand, portable extinguishers were used to fully extinguish the fire.[9] This case study underscores the need for the fire service to preplan firefighting procedures within facilities that manufacture, use, or store combustible metals. The use of standard portable fire extinguishers (water, dry chemical, or carbon dioxide) on incipient combustible metal fires as well as the application of water through hose lines on advanced fires is typically not appropriate. At this fire, the availability of dry sand in the building allowed firefighters to smother the fire and exclude oxygen. In time, the burning metal cooled down to a point at which combustible metal portable fire extinguishers could be used to attain complete extinguishment.

FIGURE 10.3
Magnesium reacting to water application.
Source: Bill Greenblatt UPI Photo Service/Newscom

> **NOTE**
>
> Some vehicles use magnesium to construct engine blocks and power train parts. Large amounts of water applied by firefighters using hose lines will serve to absorb the heat from the magnesium fire. Water flow must be continuous until the material is no longer burning and completely cool. A magnesium fire not fully extinguished and allowed to flare up due to insufficient water will require much more water to successfully put out. The officer in charge may opt to take a defensive position—protecting exposures around the vehicle and letting the fire burn out on its own. It should also be noted that in addition to engine blocks, automotive factories are using magnesium to manufacture various car parts: steering column mounts, transmission covers, exterior cowling, and dash and seat frames (see Figure 10.3).
>
> Manufacturing and recycling facilities for combustible metals pose a different hazard to firefighters. Combustible metals in minute sizes burn at much higher temperatures, 5,000°F (2,760°C) for magnesium, and present an explosion hazard when water is applied to this burning material.

PHOSPHOROUS

Allotrope ■ A variant of a substance consisting of only one type of atom. It is a new molecular configuration with new physical properties. Substances that have allotropes include carbon, oxygen, sulfur, and phosphorous.

Phosphorous (P) has a specific gravity (white) of 1.82 and a melting point (white) of 111°F (44°C). Phosphorus has three main **allotropes**: white, red, and black. White phosphorus is poisonous and can spontaneously ignite when it comes in contact with air. Therefore, white phosphorus must be stored under water. It is normally used to produce phosphorus compounds. Red phosphorus is formed by heating white phosphorus to 482°F (250°C) or by exposing white phosphorus to sunlight. Red phosphorus is nonpoisonous and is not as dangerous as white phosphorus, although frictional heating can convert it to white phosphorus. Red phosphorus is used in safety matches, fireworks, and pesticides. Black phosphorus is also formed by heating white phosphorus. It is the least reactive form of phosphorus and has no significant commercial uses. Phosphorus is widely found in the fertilizer industry. It is also extensively used in the production of metal alloys and as an additive to industrial oils.

PLUTONIUM

Plutonium (Pu) has a specific gravity of 19.84 and melting point of 1,182°F (639°C). In a finely divided form, it is pyrophoric. It has an ignition temperature of 1,112°F (600°C) and burns in a similar manner as uranium. Steam generated when water is applied to a plutonium fire significantly increases radiologic contamination. Therefore, it is normally allowed to burn itself out. It is used in nuclear weapons and nuclear power stations.

POTASSIUM

Potassium (K) has a specific gravity of 0.86 and melting point of 145°F (63°C). It emits hydrogen gas when introduced to water. Potassium also burns vigorously in oxygen. It is violently reactive with sulfuric acid and most halogens. It will detonate in contact with liquid bromine. Its primary use is in agriculture and horticulture as a fertilizer.

RUBIDIUM

Rubidium (Rb) has a specific gravity of 1.53 and melting point of 103°F (39°C). This soft, silvery metal reacts violently with water and has also been known to ignite spontaneously in air. It burns with a yellowish-violet flame. Rubidium's compounds have various chemical and electronic applications.

SILICON

Silicon (Si) has a specific gravity of 2.33 and melting point of 2,588°F (1,420°C). A silicon dust cloud can be ignited in air at 1,425°F (774°C). Under certain conditions, pure silicon metal dust has been shown to be highly explosive. Pure silicon is used to produce wafers used in the semiconductor industry, in electronics, and in photovoltaic (solar) applications.

SODIUM

Sodium (Na) has a specific gravity of 0.97 and melting point of 208°F (98°C). When found in laboratories, sodium metal is most often used to remove excess water from flammable liquids during distillation. It is also used in the reduction of organic esters. Sodium metal reacts violently with water to produce highly flammable hydrogen gas and sodium hydroxide, a caustic, corrosive solid. The reaction is exothermic and can cause hydrogen gas to autoignite. This action can create a severe fire or explosion. When sodium is finely divided, it ignites spontaneously in air. Molten sodium reacts with most gases and liquids except the noble gases (helium, neon, argon, krypton, xenon, and radon) and nitrogen. Solid sodium reacts strongly with water, alcohol, some hydrocarbons, halogens (fluorine, chlorine, bromine, iodine, and astatine), acidic oxides, sulfuric acid, mercury, and certain alloys of lead, tin, zinc, and bismuth. Sodium compounds have a wide range of uses. Various industries that use sodium include petroleum, chemicals, soaps, textiles, and paper.

STRONTIUM

Strontium (Sr) has a specific gravity of 2.6 and melting point of 1,386°F (752°C). When water is applied to this metal, it releases hydrogen readily but normally without ignition. The major use of strontium is in the production of color television tubes. It is also used in the manufacture of ceramics and specialty glass.

TANTALUM

Tantalum (Ta) has a specific gravity of 16.62 and melting point of 5,425°F (2,996°C). Under the right conditions (particle size, dispersion, and ignition source), tantalum presents a moderate dust explosion hazard. About half the tantalum consumed each year is used within the electronics industry, mainly as wire for capacitors.

THALLIUM

Thallium (Tl) has a specific gravity of 11.85 and melting point of 572°F (300°C). Thallium compounds are extremely toxic when heated to decomposition. Firefighters must wear positive-pressure, self-contained breathing apparatus (SCBA) with full protective clothing to prevent contact with skin and eyes. Flammable hydrogen gas is also emitted. Thallium is used for making low-melting-point specialty glass for highly reflective lenses.

THORIUM

Thorium (Th) has a specific gravity of 11.6 and melting point of 3,090°F (1,699°C). Thorium is a metal that is both pyrophoric and radioactive. It burns rapidly, releasing dense, white, toxic fumes. Thorium may be transported in a molten form. As a dry powder, it has a low ignition temperature. Thorium powder is shipped under helium or argon gases in special containers. It can explode from heat or contamination. Ignition has occurred due to chemical reaction between finely divided thorium and water at ordinary temperatures. Thorium is used in the production of a wide array of products and processes, including ceramics, carbon arc lamps, strong alloys, and in mantles for lanterns.

TIN

Tin (Sn) has a specific gravity of 7.29 and melting point of 450°F (232°C). When this metal is pulverized into a dust cloud, it can be ignited in air at 1,165°F (629°C). The largest single application of tin is in the manufacture of tinplate. Over 90% of the world production of tinplate is used for containers (tin cans). Other applications of tinplate include fabrication of signs, batteries, toys, gaskets, and containers for pharmaceuticals, cosmetics, fuels, and numerous other commodities.

TITANIUM

Titanium (Ti) has a specific gravity of 4.54 and melting point of 3,000°F (1,649°C). Titanium burns in air and is the only element that burns in nitrogen. It is slow to react with air and water, however, because it forms a passive and protective oxide coating that prohibits it from further reaction. Titanium powder can form an explosive suspension in air. Titanium is commonly employed in the aerospace industry. It also is used to make pipes for nuclear, oil, and chemical industry applications.

TITANIUM STORAGE AND RECYCLING FACILITY EXPLOSIONS

Three firefighters were injured while operating at a huge fire at a south Los Angeles titanium storage and recycling facility in July, 2010 (see Figure 10.4). At this incident, several firefighters noticed a bright, white-hot fire, white sparks, bluish-green hues of flame, and white smoke. They did not, however, recognize that what they saw could be indicative of burning combustible metals. The fire department did not suspect that combustible metals were present until after the first explosion and the discovery of the placard indicating that oxidizers were in the structure. Once identified, command directed water away from areas of suspected burning combustible metals.

The fire, which was discovered shortly before midnight, set off several explosions that hurled burning chunks of titanium debris into the neighborhood. The airborne material also damaged several fire engines. The explosions were the result of water being applied onto burning titanium. Strategy and tactics were adjusted; the use of water was stopped for several hours, allowing the burning combustible metal to burn itself out and become more manageable.

Large amounts of water and firefighting foam were then applied from a defensive position via unmanned ladder pipes in an attempt to protect surrounding properties. This fire progressed to a point at which the entire building was engulfed in flames. Four other businesses were destroyed. More than 200 firefighters, working throughout the night into the early morning hours, were needed to bring this fire under control. Two firefighters suffered burn injuries, and another firefighter suffered a concussion-type injury to his ear from one of the explosions. All three firefighters were taken to local hospitals.[10]

Well into the incident, a few concentrated areas remained on fire. Copious amounts of water were directed to these areas to extinguish the flames. This caused another explosion. Fortunately, no one was hurt. The titanium that was involved in the second explosion had developed a protective crust during the fire. This phenomenon contributed to the shaped-charge effect when the molten metal under the crust came in contact with the water. The development of a protective crust is normal in combustible metal fires. It limits open flames emitting from the combustible metal. Therefore, it is very important not to disturb combustible metal fires once in their advanced stage of burning.

Firefighters should allow the crust formation to control and extinguish the fire and ensure that water is not directed onto the combustible metal. Contributing factors related to the injuries at this fire included the unrecognized presence of combustible metals and unknown building contents as well as the use of traditional fire-suppression tactics. Key NIOSH recommendations are as follows:

- Ensure preincident plans are updated and available to responding fire crews
- Ensure firefighters are rigorously trained in combustible metal fire recognition and tactics
- Ensure that policies are updated for the proper handling of fires involving combustible metals
- Ensure that first arriving personnel and fire officers look for occupancy hazard placards on commercial structures during size up
- Ensure that all firefighters communicate fireground observations to the IC (Incident Commander).
- Ensure that firefighters wear all personal protective equipment when operating in an environment immediately dangerous to life and health
- Ensure that an Incident Safety Officer is dispatched on the first alarm of commercial structure fires
- Ensure that collapse/hazards zones are established on the fireground[11]

FIGURE 10.4 South Los Angeles titanium storage and recycling facility fire.
Source: Mike Meadows/AP Photo

URANIUM

Uranium (U) has a specific gravity of 18.7 and melting point of 2,069°F (1,132°C). This metal is harmful due to both its toxicity and its radioactivity. Exposure to uranium increases the risk of getting a variety of cancers. It is a dense, radioactive metal. When finely powdered, it ignites spontaneously and explosively in air. Uranium is a highly reactive metal and reacts with almost all the nonmetallic elements and many of their compounds. The generation of hydrogen gas through oxidation/corrosion of uranium metal by its reaction with water can create a flammable atmosphere during transport and storage operations. Although most people think uranium is extraordinarily rare, it is in fact more abundant than familiar elements such as mercury and silver. It is most commonly used in the nuclear power industry to generate electricity.

ZINC

Zinc (Zn) has a specific gravity of 7.13 and melting point of 786°F (419°C). As a dust cloud, it can be ignited at 1,112°F (600°C). When zinc is in the form of dust and in contact with moisture, alkaline solutions, or acetic acid, it will heat spontaneously to ignition. Large pieces of zinc are difficult to ignite, but once ignited it will burn vigorously. Acids liberate hydrogen on contact with zinc. Zinc is used to make die castings for use in the electrical, automotive, and hardware industries.

ZIRCONIUM

Zirconium (Zr) has a specific gravity of 6.51 and melting point of 3,326°F (1,830°C). When this metal is in a fine particulate form, it is pyrophoric. Particles can be ignited in nitrogen gas above 986°F (530°C) and in carbon dioxide above 1,040°F (560°C). Particles may also form explosive mixtures with oxidizing materials. When in a dry powder form, zirconium or zirconium–copper alloys in glass containers may explode by impact or friction if the container breaks. Zirconium ignites more easily than magnesium. Zirconium powder is commonly handled wet to make it more difficult to ignite. Once ignited, however, it will burn violently. Zirconium is used extensively by the chemical industry, in which corrosive agents are employed. Zirconium is also used as an alloying agent in steel and as a component in surgical appliances, photoflash bulbs, explosive primers, and lamp filaments.

Extinguishing Agent Applications

An incipient fire will be ringed with a dam of dry sand or with dry material that will not react with the metal being extinguished or with a Listed Class D extinguishing powder in accordance with the manufacturer's instructions. Application of a dry extinguishing agent will be conducted to avoid any disturbance of the combustible metal which could cause a dust cloud. The use of pressurized extinguishing agents is not permitted on a combustible metal powder fire or chip fire unless applied carefully so as not to disturb it. Only Listed Class D extinguishing agents or those tested and shown to be effective for extinguishing combustible metal fires are permitted.

Combustible metals present a fire or explosion hazard when finely divided into dust (420 micrometre [μm] or smaller in diameter). The degree of hazard will depend upon the type of combustible dust and processing methods used. Metal particles (chips, fines, shavings, ribbons, **swarfs**, turnings, powders, etc.) produced from cutting or machining vary in ease of ignition and rapidity of burning, depending on their size and geometry.

Fires involving large quantities of combustible metals may be impossible to extinguish if not caught in the Incipient phase. The heat produced by a combustible metal fire will often make it impossible to get close enough in order to apply the extinguishing agent. In this situation, the best approach may be to isolate the material as much as possible, if it can

Swarfs ■ Fine metallic particles removed by a cutting tool.

be done safely. The fire can be allowed to burn itself out naturally to minimize hazards to personnel and losses to surrounding property.

There are several common types of agents that can be used on numerous kinds of combustible metal fires, and there are other types of agents that are employed for specific kinds of metal fires. Moreover, there are important differences in the way various Class D agent fire extinguishers are operated. Firefighters should receive training on these appliances.

Categories

There are four general combustible metal categories. Alkali metals (lithium, sodium, potassium, rubidium, caesium, and others) should be considered the most dangerous by first responders, because they are water reactive under both fire and nonfire conditions. They are located in Group I of the Periodic Table of Elements. Alkaline earth metals (magnesium, barium, radium, beryllium, calcium, strontium, and others) are almost as reactive as the alkali metals. They react easily with the **halogens** (fluorine, chlorine, bromine, iodine, and astatine) and with water. The alkaline earth metals can be found in Group II of the periodic table. The third category consists of the transitional metals (hafnium, niobium, titanium, tantalum, zirconium, and others). These metals are between Groups III to XII in the periodic table. They have an incomplete inner electron shell and serve as transitional links between the most and the least electropositive metals. Metals (aluminum, tin, lead, and others) located in Groups XIII to XV in the periodic table are in the fourth category of combustible metals. Table 10.2 provides a noninclusive listing of some of the most commonly encountered combustible metals and the categories in which they belong.

Halogens ■ A group in the periodic table consisting of five chemically related elements: fluorine, chlorine, bromine, iodine, and astatine.

TABLE 10.2 Categories of Combustible Metals

ALKALI	ALKALINE EARTH	TRANSITIONAL	OTHER
Lithium	Magnesium	Hafnium	Aluminum
Sodium	Barium	Niobium	Tin
Potassium	Radium	Titanium	Lead
Rubidium	Beryllium	Tantalum	
Caesium	Calcium Strontium	Zirconium	

Alkali Metals

Alkali metals are all highly reactive. Because of this, they are usually stored immersed in mineral oil or kerosene. They are water reactive under nonfire conditions and should be placarded with the ₩ symbol under *NFPA 704: Standard System for the Identification of the Hazards of Materials for Emergency Response*.

Applying water to alkali metals under both fire and nonfire conditions will result in extreme reactions and the formation of hydrogen gas. The reaction is so vigorous in nature that the hydrogen gas produced during the reaction catches fire. Potassium and the liquid alloy of potassium and sodium are the most reactive, whereas lithium is the least reactive. Lithium is the only alkali metal that reacts slowly with water. If inhaled, the dust of alkali metal oxides can cause damage to mucous membranes and upper respiratory tracts. Contact with the skin may result in burns.

To extinguish fires and contain alkali metals, MET-L-X®, Lith-X®, NA-X®, anhydrous dry soda ash, dry sand, inert gas (argon and nitrogen), sodium chloride, copper powder,

FIGURE 10.5 MET-L-X portable fire extinguisher.
Source: Ronald R. Spadafora

and powdered graphite are commonly used. MET-L-X is a dry powder composed of a salt base with a polymer for sealing purposes. It has other additives to make it free flowing. MET-L-X hardens over the burning material to form a crust that excludes oxygen and absorbs heat. It is used on sodium, potassium, sodium–potassium alloy, and magnesium fires. In addition, it can also control/extinguish small fires involving zirconium and titanium. MET-L-X can be discharged through scoop applications, hose lines, portable handheld fire extinguishers, wheeled units, stationary units, and piped systems (see Figure 10.5).

Lith-X is a compound with a graphite base. It also contains additives to render it free flowing so that it may be used in extinguishers. Lith-X fire extinguishing agent is used for lithium fires. It will also, however, extinguish magnesium, sodium, and potassium fires. Lith-X will contain/extinguish zirconium, titanium, and sodium–potassium alloy fires as well. Unlike MET-L-X, it does not harden (crustover) to extinguish combustible metal fires. Instead, it conducts heat away from the burning material to dissipate heat.

NA-X is a sodium carbonate–based dry powder for use on most Class D fires involving sodium, potassium, and sodium–potassium alloy materials. It is an extinguishing agent used where stress corrosion of stainless steel needs to be minimized. Various additives are incorporated into the powder to make it easily fluidized for use in pressurized extinguishers. It also contains an ingredient that allows it to soften and crust over the exposed surface of burning metal.

The following additional combustible metal extinguishing agents have multiple applications. They may be used on alkali metals as well as those of other categories.

- Powdered sodium carbonate (soda ash) extinguishes combustible metal fires by excluding ambient oxygen and thereby smothering the material. Soda ash can be used on the alkali metals as well as aluminum, magnesium, and titanium.
- Argon is optimally suited for all four categories of combustible metal fires.
- Finely granulated sodium chloride is a popular combustible metal extinguishing agent. It can be applied by a shaker, scooper, or shovel. Sodium chloride is suitable for sodium and potassium fires. It can also be used on magnesium, titanium, aluminum, and most other metal fires. Powder that is thermoplastic is added to the mixture

along with metal stearates for waterproofing and anticaking ingredients to form the extinguishing agent. When it is applied to the fire, the salt acts like a heat sink, dissipating heat from the fire. Sodium chloride also forms an oxygen-excluding crust to smother the fire. The plastic additive melts and helps the crust maintain its integrity until the burning metal cools below its ignition temperature.

- Finely powdered graphite can be applied by a shaker or shovel onto sodium and potassium fires. For fires in finely powdered reactive metals, application via a long-handled scooper is preferred over the use of a portable extinguisher blast that could stir up the powder and cause a dust explosion. Graphite both smothers the fire and conducts away heat. It can also be used on lithium fires burning on a level surface and magnesium, titanium, aluminum, and zirconium combustible metal fires.
- Finely powdered copper liquidized by compressed argon is another extinguishing agent for lithium fires. It smothers the fire, dilutes the fuel, and conducts away heat. It is capable of clinging to dripping molten lithium on vertical surfaces.
- Cast-iron borings cool down combustible metal fires. This extinguishing agent is used on magnesium fires.

As a last resort, dry sand may be used to smother a metal fire. It is generally applied with a long-handled shovel to avoid the operator receiving flash burns. Sand that has absorbed moisture, however, may result in a steam explosion, spattering burning molten metal in all directions. Dry sand can generally be used to completely smother and cover any small metal fire that occurs. This Class D extinguishing agent is used on all four categories of combustible metal fires. Table 10.3 provides a brief overview of the compatibility of Class D extinguishing agents with some of the most common combustible metals encountered by first responders.

TABLE 10.3 Combustible Metal Fire Extinguishing Agents

EXTINGUISHING AGENT	PHOSPHORUS	SODIUM	LITHIUM	ALUMINUM	MAGNESIUM	TITANIUM METALS
Met-L-X	Yes	Yes	No	Yes	Yes	Yes
Lith-X	Yes	Yes	Yes	No	No	No
Copper powder	Yes	Yes	Yes	Yes	No	No
Dry sand	Yes	Yes	Yes	Yes	Yes	Yes
Soda ash	Yes	Yes	Yes	Yes	Yes	Yes
Sodium chloride	Yes	Yes	Yes	Yes	Yes	Yes
Argon	Yes	Yes	Yes	Yes	Yes	Yes
Nitrogen	Yes	No	No	No	No	No

Alkaline Earth, Transitional, and Other Combustible Metals

Large-mass (bars, **ingots**, castings, and plates) alkali earth and transitional metals are extremely difficult to ignite. They will normally self-extinguish when the heat source is removed. Combustible metal fires involving alkali earth and transitional metals require an inert atmosphere for extinguishment. This is normally impractical during most fire operations. Applying other agents will only contain the fire until the metal has oxidized to a point at which it cools down below its ignition temperature. This may take a long period of time and may not meet the needs of first responders during situations (inside the fuselage of a plane, for example) in which time is of the essence and rapid extinguishment is needed.

Ingots ■ Solid pieces of metal that have been formed into particular shapes (such as bricks) so that they are easy to handle or store.

Explosions

An explosion is a rapid release of combustion energy. It can be visualized as a fast-traveling fire that creates a pressure increase when in an enclosed area. Many metals can create an explosion hazard when transformed into byproducts and small particles during manufacturing. A deflagration and a **detonation** are two types of explosion reactions. A deflagration is a rapid high-energy release that propagates through a gas or across the surface of a material at a subsonic (slower than the speed of sound) rate. A detonation involves a supersonic flame front accelerating through a medium. Explosion suppression systems may be unable to control or extinguish a fire accelerating at this rate.

> **Detonation** ■ A supersonic flame front accelerating through a medium.

Explosion Prevention and Protection Systems

Managing explosion risks involves identifying the hazards and implementing prevention and protection methods. Explosion-suppression systems are designed in accordance with *NFPA 69: Standard on Explosion Prevention Systems*, to protect enclosures from rupture due to overpressurization from a deflagration. Type of processing, vessel size and shape, duct design, sources of ignition, and product hazard are just a few factors that need to be considered by the fire protection engineer during the design phase of explosion-prevention and protection systems.

Metal-dust explosions can be devastating. They have unique deflagration characteristics, making them difficult to defend against. Metal-dust explosions can produce flame temperatures of more than 6,332°F (3,500°C), with resulting high explosion pressures and rapid flame speeds. Success or failure of explosion suppression relies upon the speed of system activation. Overall speed relates to detection and mechanical response of the equipment.

A deflagration-suppression system can be used to protect processing equipment (pulverizes, grinders, mills, and dust collectors), storage equipment (pressure tanks), material-handling equipment (conveyors), and laboratory equipment (hoods). Design features pertaining to positioning of detection devices and location/quantity of suppressant cylinders are based upon the time required for detection, discharge pattern, suppressant efficiency, explosive nature of the combustible material, and the physical characteristics of the enclosure being protected. Deflagration suppression systems are especially valuable under circumstances that make venting impractical. These include facilities where:

- Equipment to be protected is located indoors
- Combustible metal has a high explosive potential
- There is insufficient area on the equipment for adequate exterior venting
- Process combustible metal vapors are toxic and cannot be vented into the air
- Potential exists for high flame propagation

Explosion-suppression systems for combustible metal hazards inside enclosures consist of cylinders filled with extinguishing agent, a detection system, and a fire alarm control unit (FACU). Suppressants include the clean agents, halon, and dry chemicals that are pressurized with an inert gas, generally nitrogen. Agent containers may be installed directly on the vessel they are protecting, on a wall inside a room or area, or at ceiling level (see Figure 10.6).

Extinguishing agents are introduced into an enclosed space to bring to a halt the combustion process. Containers have control valves that allow the agent to be released one millisecond from initiation. Activation can start with the firing of a low-power explosive device (squib). The extinguishing agent reacts with the deflagration at the Incipient phase thereby protecting the enclosed vessel or container (see Figure 10.7).

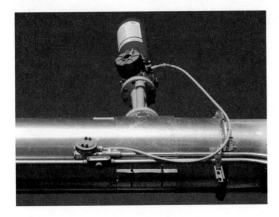

FIGURE 10.6 Chemical isolation. Explosion suppressant agent cylinder on duct. *Source:* IEP Technologies

The explosion-suppression system agent is selected based upon a number of factors. Suppression most often physically controls explosions by absorbing energy produced by the combustion reaction and/or chemically participating in the combustion reaction. Rock dust (calcium carbonate) is an example of a suppressant agent that is primarily a physical agent. It prevents explosion propagation by absorbing thermal and radiant energy produced by combustion. This absorption competes with the heating of the unburned fuel particles. Agents such as dry chemicals (sodium bicarbonate and monoammonium phosphate) extinguish flame by both physical and chemical mechanisms.

Life safety is always a priority. Exposure of the agent to building personnel and occupants can determine the kind of extinguishing agent used as well as the distance workers will be from the vessel being protected from rupture. Design fire protection engineers must ask themselves, "how much agent will be needed to successfully suppress an explosion?"

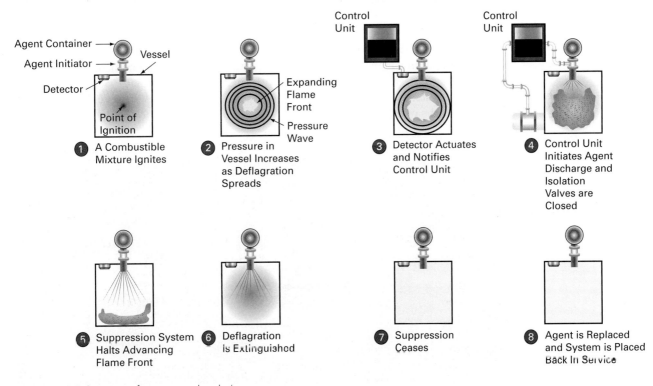

FIGURE 10.7 Sequence of a suppressed explosion

Chapter 10 Combustible Metal Suppression Agents

Detection Devices and Fire Alarm Control Units

Explosion-suppression systems use pressure sensors to identify a potential explosion event. Pressure sensors detect rapid increases in pressure in one millisecond. The FACU receives the detection signal from the pressure sensors and then sends a release signal to the cylinder/container holding the extinguishing agent. Pressure sensors are ideal for dust hazards found in enclosed spaces. If venting techniques are being employed, however, pressure sensors may not be suitable.

Explosion-suppression systems can also use the following detection devices, which can identify a hazard quickly, send a signal to the FACU to initiate an alarm, and activate the suppression system:

- Ultraviolet flame detectors
- Products of combustion (smoke) detectors
- Infrared spark/ember sensors

NOTE

Ultraviolet flame detectors and infrared spark ember sensors, when activated inside ductwork funneling metal dust to a storage container, trigger the opening of water spray nozzles installed a calculated distance downstream. The spray nozzles are designed to create a fine mist of water to cool and extinguish the spark before it reaches the dust collector (see Figure 10.8).

An FACU for an explosion-suppression system has multifaceted applications. It monitors operational integrity and provides for automatic control of equipment and transmission of an audible alarm. It allows for both manual and automatic activation of the suppression system. An uninterruptible power supply/battery backup provides emergency power when the main power supply fails.

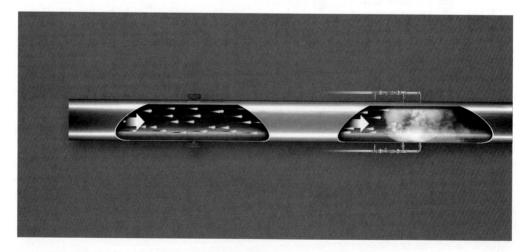

FIGURE 10.8 Infrared spark/ember sensor system protecting ductwork. Used by permission of GreCon Inc.

Alternatives and Enhancements to Explosion Suppression Systems

Techniques used to prevent explosions are also important features of explosion prevention and protection systems. Guarding against explosions can enhance overall effectiveness and reduce risk. Some of the methods used follow:

- Controlling oxidant concentration is performed by introducing inert gas (nitrogen, for example) into a volume of combustible dust. The addition of inert gas acts to purge the container of oxygen. The objective is to reduce the level of oxygen in the protected space below what is required for combustion. Nitrogen is the inert gas of choice for aluminum, whereas helium or argon is required for magnesium.
- Controlling combustible concentration of dust through ventilation or air dilution is another method used to limit the risk of explosion. This technique limits the amount of hazardous dusts permitted inside an enclosure, thereby reducing the risk of ignition.
- If overpressurization is a major concern, some vessels can be constructed using stronger metals and tighter seals to resist anticipated pressures. Containment techniques require that a vessel be constructed with a strength that corresponds with the hazard. High pressures resulting from explosions are possible during material processing. Connected vessels and ducts must also be able to sustain these pressures.
- When it is not feasible to make a container strong enough to withstand predicted pressures, passive safety devices (pressure-release door latches, louvers, and vents) may be used to protect against rupture. There is no active attempt to suppress the deflagration during the venting process. Passive vents are more economical than using strengthening techniques on the vessel.

Explosion venting is one of the most common and effective forms of explosion protection. It provides overpressure protection from processing equipment explosion hazards by providing a planned pathway for the deflagration to escape outside the structure. The vent openings are sized (using Kst values) to allow the expanding gases to leave at a rapid rate. This feature limits the internal pressure in the vessel. Precautions must be in place, however, to prevent a worker, occupant, or passerby from walking in the vicinity of the relief blast and becoming seriously injured (see Figure 10.9).

Not all venting needs to allow the products of the deflagration into the atmosphere. It is often necessary to locate process equipment indoors. Flameless venting technology consists of an assembly with layers of high-temperature stainless steel mesh that absorb heat produced during the combustion reaction. Flame gases above 2,732°F (1,500°C) are

FIGURE 10.9 Explosion vents on dust collector.
Source: IEP Technologies

FIGURE 10.10 Flameless vent (center of photo) on a dust collector inside a building. *Source:* IEP Technologies

cooled down to less than 200°F (93°C) via energy transfer within the steel mesh filter inlet. This cooling reduces gas volume as it extinguishes the flame. This allows conventional venting to be done indoors with no release of flame. Flameless venting also ensures combustible dust retention; no dust particles pass through the assembly. In addition, the pressure rise and increased noise level are significantly reduced to negligible levels compared with exterior explosion venting techniques (see Figure 10.10).

Manufacturing and processing (buffing/sawing/polishing) of metal or metal alloys can produce combustible dust. Segregation and separation techniques are used to limit the extent of a dust explosion. The segregation method uses physical barriers that are erected to segregate dust deflagration hazard areas. These barriers are constructed to remain upright during a dust explosion. Dust explosion hazard areas may exist in all operating areas of facilities processing or handling combustible metal dusts.

Doors and openings are generally not permitted in the barriers unless they are normally closed and have the same or greater strength and fire resistance rating as that required of the physical barrier. Moreover, seals are installed at all floor, wall, ceiling, and partition penetrations. This provides a fire-resistive rating. Fixed dust-collection systems located in areas where dust is released may be used to limit the dust deflagration hazard area.

Explosion-isolation systems prevent the propagation of the deflagration from one component of the industrial/manufacturing process to another through the use of fast-acting valves. Mechanical explosion isolation uses valves that provide a physical barrier to prevent the spread of the explosion and resulting pressure in both directions through connecting ducts/pipes. They are designed to close nearly instantaneously upon detecting an explosion (see Figure 10.11).

FIGURE 10.11 High speed mechanical explosion isolation valve. *Source:* IEP Technologies

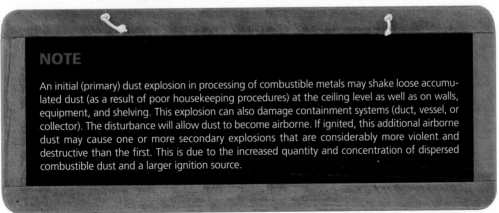

NOTE

An initial (primary) dust explosion in processing of combustible metals may shake loose accumulated dust (as a result of poor housekeeping procedures) at the ceiling level as well as on walls, equipment, and shelving. This explosion can also damage containment systems (duct, vessel, or collector). The disturbance will allow dust to become airborne. If ignited, this additional airborne dust may cause one or more secondary explosions that are considerably more violent and destructive than the first. This is due to the increased quantity and concentration of dispersed combustible dust and a larger ignition source.

Firefighting Operations and Emergency Procedures

Frequently, combustible metal fires will develop an oxide crust that will limit active flaming of the material. This crust should not be disturbed by hose streams. Firefighters must wait until the combustible metal is completely oxidized and extinguished. The time frame for this to occur is dependent upon the magnitude of the fire but could last more than 24 hours. Firefighters should make use of a **thermal imaging camera** to denote change in temperature during this process; even though there may not be signs of external combustion, the metal can still be extremely hot.

Unlike the other classes of fire, Class D fires require a unique application technique. The operator of the extinguishing agent device will not see a lot of flame or feel a lot of radiant heat in the early phases of the fire. If the fire involves magnesium, there will be intense light. Lithium will generate large volumes of dense smoke. Conversely, titanium or zirconium fires produce very little smoke. The inhibited nature of the fire or the lack of

Thermal imaging camera ■ A device that forms an image using infrared radiation. Used by firefighters to see through smoke, enabling them to find victims faster and identify the seat of the fire earlier.

smoke are not true indicators of the danger. Combustible metal fires are serious fires that should not be taken lightly. Firefighting foam should not be used to extinguish Class D fires. For every 100 gallons (379 L) of firefighting foam at 1%, 3%, or 6% concentration, there is 99%, 97%, or 94% water, respectively.[12]

If not properly identified and handled, combustible metal fires can present serious hazards to firefighters. Extreme temperatures generated by most combustible metals (5,000°F to 8,500°F, or 2,750°C to 4,704°C) result in rapid heat buildup and fire spread. The temperatures encountered with combustible metal fires will far exceed the specifications for standard personal protective equipment (PPE). Some burning metals can be fatal if their vapors are inhaled. Large fires or uncontained fires for which the ventilation system does not contain all of the fumes being generated by the combustible metal should be handled only by the fire department using positive-pressure, self-contained breathing apparatus (SCBA).

Fire departments should attempt to properly document the presence of combustible metals from information gathered on building inspections or routine response or gleaned from preincident plans or surveys. Seek out persons at the plant or facility who are familiar with the combustible metals and their hazards. Obtain material safety data (MSD) sheets to help better understand the characteristics of the metals, their hazards, and the appropriate firefighting and emergency procedures to deal with them.

Information obtained firsthand under nonemergency circumstances can be much more reliable than when it is obtained from facility workers during a fire. It should also be included in communication databases. This information can be invaluable when responding to the scene of the incident and on the fireground. An IC with knowledge of the presence of combustible metals at the fire scene can warn all members of the dangers and formulate the right strategy and tactics to ensure a more successful and safer operation.

CHAPTER REVIEW

Summary

Metal fires represent a unique hazard because first responders and workers who use these materials are often not aware of the characteristics of these fires and are not properly trained and prepared to fight them. In general, metal fire hazard potential exists when manufacturing and processing creates dust. Incipient fires are commonly extinguished using Class D agents. The most popular of these agents are sodium chloride and graphite powder. Fixed fire prevention and protection systems include methods and techniques such as using deflagration-suppression agents, controlling oxidant concentration, air dilution, strengthening containment vessels, explosion venting, physical barriers, isolation systems, and good housekeeping.

Review Questions

1. What are the five components of the dust explosion pentagon?
2. Regarding combustible dusts, what does the Kst value measure?
3. Why is pouring water on burning magnesium a bad idea?
4. List some extinguishing agents that should not be used on combustible metal fires.
5. What are the four general combustible metal categories?
6. What combustible metal category should be considered the most dangerous to first responders and why?
7. How do the two types of explosion reactions (deflagrations and detonations) differ?
8. What agent is used to control oxidant concentration inside a container holding a volume of combustible dust?
9. Explosion venting is one of the most common and effective forms of explosion protection. Briefly explain how it affords overpressure protection from processing equipment explosion hazards.

Endnotes

1. US Department of Labor, Occupational Safety and Health Administration, *Hazard Communication Guidance for Combustible Dusts*, OSHA 3371-08, 2009.
2. US Chemical Safety and Hazard Investigation Board, "Sugar Dust Explosion and Fire," *Investigation Report* Report No. 2008-05-I-GA, September 2009,
3. Kevin L. Kreitman, "Proper Handling of Combustible Metal Fires," *Fire Engineering* 161, no. 2 (2008): 115–116, 118.
4. Scott A. Stookey, "Dust Deflagrations: Recognizing and Regulating the Hazard," *Building Safety Journal* 4, no. 1 (2006): 14–20.
5. National Institute for Occupational Safety and Health (NIOSH), "One Fire Fighter Killed and Eight Fire Fighters Injured in a Dumpster Explosion at a Foundry—Wisconsin," *Fatality Investigation Report*, F2009-31 (Washington, DC: NIOSH, 2010).
6. National Fire Protection Association (NFPA), *NFPA 484: Standard for Combustible Metals* (Quincy, MA: NFPA, 2012).
7. National Institute for Occupational Safety and Health (NIOSH), "One Fire Fighter Killed and Eight Fire Fighters Injured in a Dumpster Explosion at a Foundry–Wisconsin," *Fatality Investigation Report*, F2009-31 (Washington, DC: NIOSH, 2010).
8. US Chemical Safety and Hazard Investigation Board, *Hoeganaes Corporation Fatal Flash Fires, Final Report* (Washington, DC: US Chemical Safety Board, 2012).
9. Portland Fire and Rescue, "Portland Fire and Rescue Respond to Commercial Fire at 2415 SE 10th Avenue" (press release, April 23, 2012),

www.portlandoregon.gov/FIRE/50972?&archive=2012-04&page=9.
10. John Gregory and John North, "Explosive Titanium Fire Injures Firefighters," http://abclocal.go.com/kabc/story?section=news/local/los_angeles&id=7554195.
11. CommandSafety.com, "NIOSH Report Addresses Operational Issues at Metal Recycling Facility Fire," http://commandsafety.com/2011/08/niosh-rcport-addresses-operational-issues-at-metal-recycling-facility-fire.
12. David Greene, "Combustible Metals: Where Fire Meets Hazmat," *Carolina Fire Rescue EMS Journal*, 25, no. 2 (2010): 5–6.

Additional References

Ansul, Incorporated, "Fire Protection for Metal Alkyls," 1998.

Ben Peetz, "Combustible Dust Fires and Explosions," *Fire Engineering* 165, no. 3 (2012): 59–72.

Lee Morgan and Tony Supine, "Five Ways New Explosion Venting Requirements for Dust Collectors Affect You," *Powder and Bulk Engineering* July (2008): 42–49.

Michael Garlock, "Explosion and Fire at Sugar Refinery Kill 14 Workers," *Firehouse* 33, no. 12 (2008): 70–72, 74–76.

National Fire Protection Association (NFPA), *NFPA 68: Standard on Explosion Protection by Deflagration Venting* (Quincy, MA: NFPA, 2007).

National Fire Protection Association (NFPA), *NFPA 69: Standard on Explosion Prevention Systems* (Quincy, MA: NFPA, 2008).

National Fire Protection Association (NFPA), *NFPA 654: Standard for the Prevention of Fire and Dust Explosions from the Manufacturing, Processing, and Handling of Combustible Particulate Solids* (Quincy, MA: NFPA, 2013).

National Fire Protection Association (NFPA), *NFPA 704: Standard System for the Identification of the Hazards of Materials for Emergency Response* (Quincy, MA: NFPA, 2012).

National Institute for Occupational Safety and Health (NIOSH), "Seven Career Fire Fighters Injured at a Metal Recycling Facility Fire—California," www.cdc.gov/niosh/fire/reports/face201030.html.

Robert M. Gagnon, *Design of Special Hazard and Fire Alarm Systems*, 2nd ed. (Clifton Park, NY: Delmar Cengage Learning, 2008).

CHAPTER 11

Halon Fire Suppression Systems

Source: Ronald R. Spadafora

KEY TERMS

Decommissioning, *p. 196*

Stratosphere, *p. 196*

Clean agent, *p. 201*

OBJECTIVES

After reading this chapter, the reader should be able to:

- Identify the recommendations and standards set by the Montreal Protocol
- Understand the physical and chemical properties of halons
- Be familiar with the US Army Corps of Engineers halon numbering system
- Review the two most significant types of halons
- Explain the uses and applications of Halon 1211 and Halon 1301

Introduction

Halon is a gas that suppresses fire by chemically interfering with the chemical chain reaction between combustible vapors and oxygen in the atmosphere. It is an effective fire extinguishing agent (even at low concentrations), easily recyclable, and chemically stable. Halon is rated for both Class B and Class C fires. It is also effective on Class A fires, although not as effective as water.

Halon has been used for fire protection throughout the 20th century. Because halon is a compound that damages the Earth's ozone layer, however, scientists have urged restriction of its use. Halon use during fire suppression, testing of equipment, servicing of equipment, and accidental release can still result in the release of halons into the environment. The two most popular, Halon 1211 and Halon 1301, both have applications for occupied areas.

Overview

In 1987, officials from many nations met in Montreal, Canada, to discuss the environmental impact that halons and other fluorinated hydrocarbons were having on the atmosphere. These risks spurred leaders to ratify the Montreal Protocol on Substances that Deplete the Ozone Layer (Montreal Protocol). Halon was deemed obsolete for fire protection purposes and has been eliminated, phased out, or replaced with alternative agents that are environmentally friendly.

This ban does not eliminate the use of halons as fire suppressants, but it does significantly affect the cost and availability. It is legal to continue to use an existing halon system and to purchase recycled halon to recharge a fire extinguishing system. Recycled halon can be purchased from many halon and fire protection equipment distributors or directly from owners who are **decommissioning** their halon systems. The production of halons, however, ceased in the United States on January 1, 1994, under the Clean Air Act.

Halon fire suppression systems are used for the protection of areas that contain sensitive or irreplaceable equipment or items that could be damaged or destroyed by other types of extinguishing systems that use water, foam, and dry chemical. This is primarily due to its ability to extinguish fire without the production of residues that could damage the assets being protected. Although Halon systems have one of the most expensive fire suppression system maintenance costs, they are still cost effective when compared with the loss of priceless artifacts, business continuity, or replacement of valuable equipment.

Decommissioning ■ A formal process to remove something from an active status.

Montreal Protocol

The ozone hole is not technically an opening in the sky where no ozone is present. It is actually a region of exceptionally depleted ozone in the **stratosphere**. Satellites provide daily images of the ozone layer over the Antarctic region. The ozone hole image in Figure 11.1 shows very low values (blue- and purple-colored area) centered over Antarctica on July 20, 2013.

Stratosphere ■ The second major layer of Earth's atmosphere, just above the troposphere and below the mesosphere. It is stratified in temperature, with warmer layers higher up and cooler layers farther down.

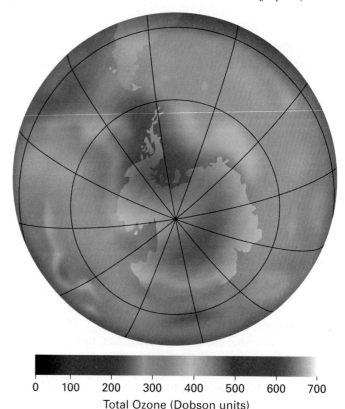

FIGURE 11.1 NASA's "Ozone Hole Watch," last updated July 20, 2013, shows the total ozone over the Antarctic pole. The purple and blue colors indicate where there is the least ozone, and the yellows and reds indicate where there is more ozone.

From a historic perspective, total column ozone values of less than 220 Dobson Units (a unit used to measure the columnar density of a trace gas in the Earth's atmosphere) were not observed prior to 1979. Over the Earth's surface, the ozone layer's average thickness is about 300 Dobson Units: a layer that is 0.1 in. (3 mm) thick.

From an aircraft field mission over Antarctica, it is known that a total column ozone level of less than 220 Dobson Units is a result of catalyzed ozone loss from chlorine and bromine compounds. For these reasons, the National Aeronautics and Space Administration (NASA) use 220 Dobson Units as the boundary of the region representing ozone loss. With daily snapshots of total column ozone, NASA can calculate the area on the Earth that is enclosed by a line with values of 220 Dobson Units.[1] The growing knowledge in atmospheric chemistry led to an enhanced understanding of the significant depletion of stratospheric ozone levels. Manmade chemicals such as chlorofluorocarbons (CFCs), halons, and a large number of industrial chemicals were verified as attackers of the Earth's ozone layer and recognized as ozone-depleting substances (ODS).

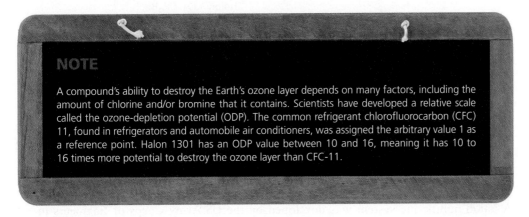

NOTE

A compound's ability to destroy the Earth's ozone layer depends on many factors, including the amount of chlorine and/or bromine that it contains. Scientists have developed a relative scale called the ozone-depletion potential (ODP). The common refrigerant chlorofluorocarbon (CFC) 11, found in refrigerators and automobile air conditioners, was assigned the arbitrary value 1 as a reference point. Halon 1301 has an ODP value between 10 and 16, meaning it has 10 to 16 times more potential to destroy the ozone layer than CFC-11.

The increase in ultraviolet (UV) radiation reaching the Earth's surface is a direct result of a compromised ozone layer. Negative consequences include disturbance of food chains (agriculture and fisheries) as well as biological diversity. Overexposure to UV radiation in humans can cause skin cancers and premature aging. It can also lead to eye damage (cataracts) and suppression of our immune system.

Physical and Chemical Properties

Halon has one or more of its hydrogen atoms replaced by atoms from the halogen series (fluorine, chlorine, bromine, or iodine). This substitution provides nonflammability and flame-extinguishing properties. In general, the halons are odorless and colorless, noncorrosive, nonreactive with water, and stable up to 900°F (482°C).

Halon 1211 (Bromochlorodifluoromethane) is a nonflammable, colorless gas with a sweet odor. Its chemical formula is CF_2BrCl (see Figure 11.2).

Halon 1211 has an ODP of 3. Additional properties can be found in Table 11.1.

TABLE 11.1	Properties of Halon 1211
PROPERTY	**HALON 1211**
Boiling point at 1 standard atmosphere (atm)	25°F (−4°C)
Vapor density (air = 1)	5.7
Vapor pressure at 70°F (21°C)	37.5 psi (259 kPa)

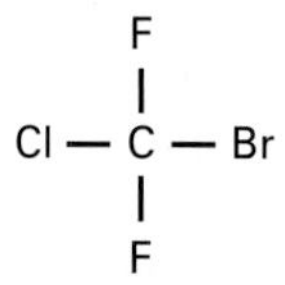

```
    F
    |
Cl—C—Br
    |
    F
```

FIGURE 11.2 Structural formula of Halon 1211

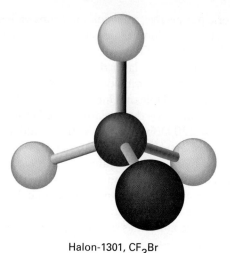

Halon-1301, CF$_3$Br

FIGURE 11.3 Structural formula of Halon 1301. Carbon atom = black, Fluorine atoms = green, Bromine atom = red.

Halon 1301 (Monobromotrifluoromethane) is a clear, colorless, liquefied gas with a slight ethereal odor. Its chemical makeup is CF$_3$Br (see Figure 11.3). Additional properties can be found in Table 11.2.

US ARMY CORPS OF ENGINEERS NUMBERING SYSTEM

A halon numbering system was developed by the US Army Corps of Engineers in the 1950s to describe the various halogenated hydrocarbons. This system eliminates the use of the long, confusing chemical names and formulas. The prefix term is always "Halon." The first digit in the number represents the number of carbon atoms in the compound molecule, the second digit stands for the number of fluorine atoms, the third digit corresponds to the number of chlorine atoms, the fourth digit signifies the number of bromine atoms, and the fifth digit notes the number of iodine atoms (if any). In this system, terminal zero digits are not expressed. Table 11.3 provides an example of how the system works.

Types and Applications

The configuration of the chemical formula of a halon determines how it is used. When fluorine is in a halogenated compound, for example, it increases its inertness and stability. The presence of other halogens, particularly bromine, adds to the fire extinguishing effectiveness of the compound.[2]

TABLE 11.2	Properties of Halon 1301
PROPERTY	**HALON 1301**
Boiling point at 1 atm	−72°F (−58°C)
Vapor density (air = 1)	5.1
Vapor pressure at 70°F (21°C)	450 psi (3103 kPa)

TABLE 11.3	US Army Corps of Engineers Numbering System
EXAMPLE (TEMPLATE)	
Halon-01234	where 0 = number of carbon atoms
	1 = number of fluorine atoms
	2 = number of chlorine atoms
	3 = number of bromine atoms
	4 = number of iodine atoms
Halon-1211 ($CBrClF_2$)	1 = carbon
	2 = fluorine
	1 = chlorine
	1 = bromine
Halon-1301 ($CBrF_3$)	1 = carbon
	3 = fluorine
	0 = chlorine
	1 = bromine

Halon 1211

Halon 1211 was first introduced as an effective gaseous fire-suppression agent in the early 1970s. Applications included protection of valuable objects inside museums and delicate electrical equipment in telephone communications centers. It was also widely used in the maritime industry to safeguard engine rooms on ships. Halon 1211 wheeled portable fire extinguishers began replacing CO_2 extinguishers for protection against Class B and Class C fires (see Figure 11.4).

Halon 1211 is stored and shipped as a liquid and pressurized with nitrogen gas. Pressurization is necessary, because the vapor pressure is too low to flow it properly to the fire area. It is discharged as a liquid and vaporizes into a cloud within a few feet of release. The range of the stream from a standard Halon 1211 portable fire extinguisher is 9 to 15 ft. (3 to 5 m). Upon release from its container, Halon 1211 inhibits fire chemically by interfering with the chain reaction between fuel vapors and oxygen. It decomposes

FIGURE 11.4 Wheeled Halon fire extinguishers.
Source: ANAM Collection/Alamy

principally into hydrogen fluoride and hydrogen bromide upon contact with flames or hot surfaces above 900°F (482°C). For effective firefighting purposes on surface fires involving Class A combustibles, a minimum concentration of 5% is recommended for total-flooding systems. Deep-seated fires involving Class B and Class C combustibles require much larger concentrations and extended holding times. Unfortunately, Halon 1211 is toxic to people when concentrations exceed 4%. This characteristic prevents its use as a total-flooding agent for occupied spaces.

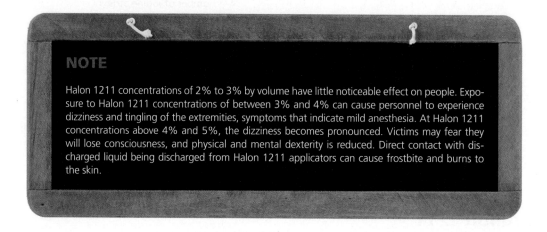

> **NOTE**
>
> Halon 1211 concentrations of 2% to 3% by volume have little noticeable effect on people. Exposure to Halon 1211 concentrations of between 3% and 4% can cause personnel to experience dizziness and tingling of the extremities, symptoms that indicate mild anesthesia. At Halon 1211 concentrations above 4% and 5%, the dizziness becomes pronounced. Victims may fear they will lose consciousness, and physical and mental dexterity is reduced. Direct contact with discharged liquid being discharged from Halon 1211 applicators can cause frostbite and burns to the skin.

LOCAL-APPLICATION SYSTEMS

Fixed local-application systems using Halon 1211 can be activated automatically or manually. These systems suppress fire by discharging the extinguishing agent in high concentrations over the burning hazard. In this type of application, neither the quantity of the agent nor the type, number, or arrangement of the applicators are sufficient to achieve total flooding of the enclosure.

SYSTEM COMPONENTS

Pre-engineered Halon 1211 local-application systems consist of components installed according to testing laboratory approval and Listing. Pressurized agent cylinders have valves designed to release their product automatically through detector activation or via a remote control manual release station. The fire alarm control unit (FACU) is connected between the fire detectors and cylinder release valves. The FACU may also sound alarms and shut off electrical power to the protected equipment. The piping system must be tested to ensure that it is securely fastened to prevent leakage during discharge. Halon nozzles are positioned to apply a high concentration of agent directly onto the hazard. Agent supply must be sufficient to flow for the required design time. Types of occupancies that are commonly protected by Halon 1211 in local-application systems include those that contain paint spray booths, dip tanks, oil-filled transformers, and printing machines.

FIGURE 11.5 Portable halon fire extinguisher (top) supplementing fixed halon extinguishing system.
Source: Coston Stock/Alamy

PORTABLE FIRE EXTINGUISHERS

Portable Halon 1211 fire extinguishers are normally found in locations containing sensitive equipment that are susceptible to damage from other, more traditional extinguishing agents, such as water and foam. In addition, they are often provided to supplement total-flooding Halon 1301 systems (see Figure 11.5).

HALON 1211 PORTABLE FIRE EXTINGUISHER INCIDENT

During an airline flight in December of 1998, an accidental release of all the contents of a Halon 1211 portable fire extinguisher occurred during handling, in the absence of a fire, in the rear part of the plane. Crew members were exposed while moving the defective fire extinguisher to a closed cabinet. Passengers were not involved in the incident. The estimated exposure time for the crew members was deemed to be less than five minutes. The level of exposure to crew members was considered to be high.

One exposed crew member was a 30-year-old, nonsmoking air hostess with no medical history of respiratory disease. On medical examination after the plane's arrival at the airport, she presented with signs of irritation of the upper respiratory tract. During the days following the incident, she experienced abnormal physical weakness on exertion and cough at night. She also stopped her sports activities because of respiratory symptoms. Twenty-eight months after the incident (April 2001), she still experienced cough at night, and lung function was unchanged.

Another exposed crew member was a 29-year-old, nonsmoking air hostess who also had no previous medical history of respiratory disease. Immediately after exposure, she presented with sore throat, nausea, and dizziness. During the days following the incident, she reported the presence of cough at night and dizziness. Twenty-eight months after the accidental exposure (March 2001), she still complained of cough.

A third exposed crew member was a 39-year-old steward who was an exsmoker with a medical history of allergies. He was the most severely exposed crew member. Immediately after the incident, he experienced malaise, wheezing, and limb numbness or weakness. In January 1999, he presented with shortness of breath on exertion.[3]

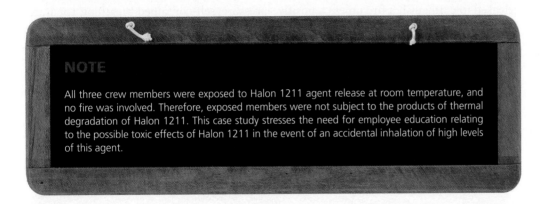

NOTE

All three crew members were exposed to Halon 1211 agent release at room temperature, and no fire was involved. Therefore, exposed members were not subject to the products of thermal degradation of Halon 1211. This case study stresses the need for employee education relating to the possible toxic effects of Halon 1211 in the event of an accidental inhalation of high levels of this agent.

ESSENTIAL APPLICATIONS

Recycling of Halon 1211 allows it to remain in use, although the availability of parts is limited to a few manufacturers. Halon 1211 is still widely used by the US military for "essential applications." An essential application is defined under the Montreal Protocol:

- There is imminent danger to human life in a place where human occupancy is essential and evacuation is not possible.
- There is imminent danger to human life and the continued operation is necessary to protect human life.
- The installation is essential to a community and to protect critical assets.
- A loss of critical equipment and/or its operation may have far reaching consequences.
- A situation in which no acceptable alternative means of fire protection exists.

Clean Agent Substitute for Halon 1211

Halotron® I is the replacement extinguishing agent for Halon 1211, although it requires a larger volume to get the same ratings as Halon 1211. Halotron I fire extinguishing agent is a US Environmental Protection Agency (EPA)-approved, **clean agent** alternative to ozone-depleting Halon 1211.

Clean agent ■ Defined by *NFPA 2001: Standard on Clean Agent Fire Extinguishing Systems* as an electrically non-conductive, volatile, or gaseous fire extinguishing agent that does not leave a residue upon evaporation.

This portable fire extinguisher discharges as a rapidly evaporating liquid that leaves no residue. The agent extinguishes Class C fires without conducting electricity back to the operator. It also provides good visibility during discharge without the risk of thermal or static shock to sensitive electrical equipment found in computer rooms and data-storage areas. Halotron I is also EPA approved for Class A and Class B combustibles.

Halon 1301

Fixed Halon 1301 fire-suppression systems are installed to provide a very high level of property protection with negligible secondary damage to crucial equipment and minimal disruption to resumption of operations. In a limited number of installations, Halon 1301 systems are installed to protect human life from fire, although in most cases they are only designed to protect facilities and equipment.

Halon 1301 is stored as a liquid in compressed gas cylinders pressurized with nitrogen to 600 psi (4,137 kPa) at 70°F (21°C) to provide an expellant pressure for the agent in excess of its normal vapor pressure. When released, Halon 1301 vaporizes into a gas with a density of approximately five times that of air. Equipment for Halon 1301 fire extinguishing systems is similar to that used for high-pressure carbon dioxide systems. Halon 1301 is the least toxic of the halon gases. The short discharge time (10 seconds maximum) also helps keep the products of thermal decomposition well below lethal concentrations. Personnel should not remain in a space in which Halon 1301 has been released to extinguish a fire unless a positive-pressure, self-contained breathing apparatus (SCBA) is used.

If Halon 1301 should inadvertently be released into a space in which *no fire exists*, personnel can be exposed to 5% to 7% concentrations for up to 10 minutes (depending upon the individual) without danger to their health. Halon 1301 can be considered a nontoxic and nonsuffocating extinguishing agent in this range. Enclosures should be evacuated, however, on system discharge. Vaporizing liquid discharged from Halon 1301 systems has a strong chilling effect on direct contact with human skin and can cause frostbite and burns. The liquid phase vaporizes rapidly when mixed with air, limiting this hazard to the immediate vicinity of the nozzle.[4]

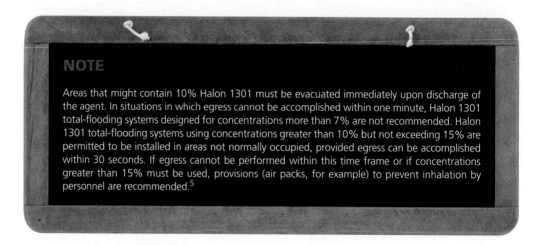

NOTE

Areas that might contain 10% Halon 1301 must be evacuated immediately upon discharge of the agent. In situations in which egress cannot be accomplished within one minute, Halon 1301 total-flooding systems designed for concentrations more than 7% are not recommended. Halon 1301 total-flooding systems using concentrations greater than 10% but not exceeding 15% are permitted to be installed in areas not normally occupied, provided egress can be accomplished within 30 seconds. If egress cannot be performed within this time frame or if concentrations greater than 15% must be used, provisions (air packs, for example) to prevent inhalation by personnel are recommended.[5]

Total-Flooding Systems

Halon 1301 is more commonly used in total-flooding systems for essential applications designed to protect enclosed hazards. Total-flooding systems are pre-engineered to provide a uniform fire extinguishing concentration of agent into a room or space. This allows the

FIGURE 11.6 Halon 1301 total-flooding system (token booth). *Source:* Ronald R. Spadafora.

agent to extinguish fire no matter where it is located. The enclosure must maintain this concentration for a designated period of time in order to facilitate permeation to all areas and ensure extinguishment of both surface and deep-seated fires. An effective concentration, if maintained correctly, will provide protection to allow emergency action by trained personnel. Deep-seated fires or a persistent source of ignition (electrical arcing) can, however, cause reignition of a fire after it is initially extinguished.

Halon 1301 total-flooding systems are installed in data-processing centers, storage vaults, electrical control rooms, electrical switchgear rooms, paper-storage areas, shipboard machinery spaces, aircraft cockpits, military vehicles, and token booths in subways (see Figure 11.6).

SYSTEM COMPONENTS

Pre-engineered Halon 1301 total-flooding systems consist of components designed to be installed according to pretested limitations as approved or Listed by a testing laboratory. They may use special nozzles, spaced strategically throughout the protected enclosure, to distribute agent evenly at the prescribed concentration commensurate with the hazard. Some of these hazards include flammable liquids and gases and electrically energized equipment. Halon 1301 should not be used on combustible metals (sodium, potassium, magnesium, or titanium), metal hydrides (aluminum hydride, lithium hydride, or zirconium hydride), chemical mixtures that are capable of rapid oxidation (cellulose nitrate or gunpowder), chemicals capable of undergoing rapid oxidation in the absence of air, or certain organic peroxides.[6]

The amount of Halon 1301 in the total-flooding system must be sufficient to protect the largest single hazard in the enclosure or group of hazards being protected simultaneously. Halon 1301 has a minimum design concentration for machinery spaces, turbine enclosures, or pump rooms containing diesel fuel, gasoline, or crude oil and other similar petroleum products of 6% of the gross volume of the space. When other flammable or

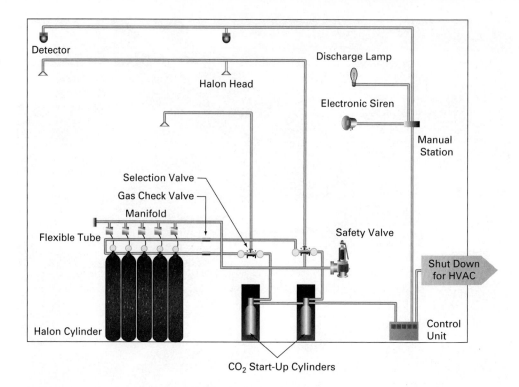

FIGURE 11.7 Halon 1301 total-flooding system

combustible materials are present in the protected enclosure, higher concentrations of the agent may be required. Where required by the authority having jurisdiction (AHJ), reserve quantities of this extinguishing agent must also be available. The Halon 1301 supply will be stored in containers designed to hold this gaseous extinguishing agent in liquefied form at ambient temperatures. Nitrogen or carbon dioxide is generally used to increase the pressure inside the containers typically to 360 psi (2,482 kPa) or 580 psi (3,999 kPa).[7]

NFPA 70: National Electrical Code and *NFPA 72: National Fire Alarm and Signaling Code* installation, testing, and maintenance requirements are used for detection, actuation, alarm, and control systems. Detection and actuation are automatic, although manual-only actuation may be permissible within an AHJ (see Figure 11.7).

ACTIVATION PROCESS

Halon 1301 total-flooding systems should ideally discharge their agent when the fire is in its Incipient phase. In general, this is accomplished automatically by detection devices that control the release of the fire extinguishing agent. Smoke (or other products of combustion) detectors are often used. Backup electrical power (24-hour standby energy source) is also required for detection, signaling, control units, and actuation components.

The nozzles for Halon 1301 systems can be located in the ceiling or installed in the floor depending on the hazard. Upon detection, there is approximately a one-minute time delay between devices triggering the agent release. More than one detector must usually activate. An alarm will sound, warning occupants of impending discharge and allowing for escape from the enclosure. Pull station (manual) activation will cause halon to dump automatically. Accidental/malicious activation can be stopped by using abort switches, which are normally required for these systems.

Upon discharge, vapors will quickly expand, filling the space with a cold fog. This condition is a direct result of condensation from the moisture in the air. This cloud will normally dissipate in a few minutes. Discharge can create air turbulence that can unseat ceiling tiles. The reduced visibility, chilling of the air, and high noise level upon activation can be frightening to the unaware.

EMERGENCY PROCEDURES FOR EMPLOYEES

When a Halon 1301 system activates, occupants should be trained to leave the area of discharge immediately. If feasible, an attempt should be made to close doors to the enclosure once it is assured that all occupants have escaped. This action should only be performed by designated personnel as denoted in a safety plan. Re-entering the area of discharge can be dangerous. It should not be allowed until the room is clear of all residual halon gas.

If required, the onsite fire brigade must be provided appropriate personal protective equipment (PPE) and positive-pressure, self-contained breathing apparatus (SCBA). Suitable safeguards should be installed to both ensure prompt evacuation and prevent entry into the enclosure. Safety procedures should include awareness training for all personnel as well as warning signs.

CLEAN AGENT SUBSTITUTE FOR HALON 1301

Halotron II is a clean total-flooding agent based on a blend of hydrofluorocarbons (HFCs) and carbon dioxide. It has zero ODP. Halotron® II replaces Halon 1301 with great effectiveness in certain applications. This agent is mainly used in Scandinavia, but it is also approved for use in the United States, although it is subject to limitations.

Decommissioning

Decommissioning procedures have been formulated to ensure that halon resources are not accidentally discharged into the atmosphere during the recovery process (see Figure 11.8). These procedures are also critical in maintaining a safe working environment for personnel involved in the decommissioning process. Halon cylinders are under pressure and must be handled with caution. Improperly moved cylinders may cause agent release in an uncontrollable manner. The cylinder can also act like a projectile, leading to serious injury or death to anyone in the area. Dangerous pressure release can occur if the cylinder valve is damaged or can be caused by inadvertent activation of the discharge mechanism.

FIGURE 11.8 Need for halon decommissioning is evident as portable halon fire extinguishers are shown abandoned in Greenland. *Source:* Ashley Cooper Pics/Alamy

Documented incidents of injury or death have been reported in the United States due to implementation of improper procedures while performing agent recovery operations. The predominant causes of accidental discharge of halon systems include:

- Accidental automatic firing at the control unit
- Accidental manual activation
- Accidental operation of the cylinder valve
- Damage to the discharge head/neck
- High cylinder temperature above the working pressure

Ideally, systems should be decommissioned by those who installed and serviced them; however, this is not always possible. Reports submitted to the Fire Suppression Systems Association (FSSA) frequently attribute the cause of the accidents to improper handling of pressurized cylinders by untrained and unqualified personnel. In addition, actuating devices had not been removed from the valves and protection caps were not installed prior to removal of the cylinders from service.[8]

CHAPTER REVIEW

Summary

Manmade chemicals, especially chlorine and bromine compounds, have been verified as destructive to the Earth's ozone layer. New production of halon therefore ceased in the United States in 1994. Recycling the existing supply of halon, however, allows this agent to be used for "essential applications" in the military on ships, aircraft, and land vehicles as well as for critical commercial purposes. Halon is an effective fire extinguishing agent, even at low concentrations. The ability of halon to suppress fire without production of residue that can damage valuable assets and its nonelectrical conductivity, are two key benefits of this agent.

Review Questions

1. What two halons are still in general use today in the United States?
2. Why was the production of the halons ceased in the United States on January 1, 1994, under the Clean Air Act?
3. Scientists have developed a relative scale called the ozone-depletion potential. What common refrigerant was assigned the arbitrary value of 1 as a reference point?
4. What military organization developed the halon numbering system?
5. Name the replacement extinguishing agent for Halon 1211 that is a US EPA approved clean agent.
6. Pre-engineered Halon 1301 total-flooding systems should not be used to protect what type of materials?
7. Who should ideally be tasked with the decommissioning of halon systems?

Endnotes

1. National Aeronautics and Space Administration (NASA), "Ozone Hole Watch," last updated July 20, 2013, http://ozonewatch.gsfc.nasa.gov/facts/hole.html.
2. National Fire Protection Association (NFPA), *NFPA 12A: Standard on Halon 1301 Fire Extinguishing Systems* (Quincy, MA: NFPA, 2009).
3. M. Matrat, M. F. Laurence, Y. Iwatsubo, C. Hubert, N. Joly, K. Legrand-Cattan, J. P. L'Huillier, C. Villemain, and J. C. Pairon, "Reactive Airways Dysfunction Syndrome Caused by Bromochlorodifluoromethane from Fire Extinguishers," *Occupational and Environmental Medicine* 61 (2004): 712–714, doi: 10.1136/oem.2003.009837.
4. Eric Peterson, *Standards and Codes of Practice to Eliminate Dependency on Halons—Handbook of Good Practices in the Halon Sector* (Hertfordshire, UK: UNEP/Earthprint, 2001).
5. National Fire Protection Association (NFPA), *NFPA 12A: Standard on Halon 1301 Fire Extinguishing Systems* (Quincy, MA: NFPA, 2009).
6. Daniel F. Sheehan, *An Investigation into the Effectiveness of Halon 1301 (Bromotrifluoromethane CBTF3) as an Extinguishing Agent for Shipboard Machinery Space Fires* (New London, CT: US Coast Guard, Office of Research & Development, 1972).
7. National Fire Protection Association (NFPA), *NFPA 12A: Standard on Halon 1301 Fire Extinguishing Systems* (Quincy, MA: NFPA, 2009).
8. H3R Clean Agents, Inc., "Decommissioning a Halon System," *FSSA Safety Alert*, October 28, 1993, www.h3rcleanagents.com/support_faq_7.htm.

Additional References

Gary Taylor "Halongenated Agents and Systems," in *Fire Protection Handbook*, 17th ed., ed. Arthur E. Cote (Quincy, MA: NFPA, 1991), 241–256.

National Fire Protection Association (NFPA), *NFPA 2001: Standard for Clean Agent Fire Extinguishing Systems* (Quincy, MA: NFPA, 2012).

CHAPTER 12

Clean Agent Fire Suppression Systems

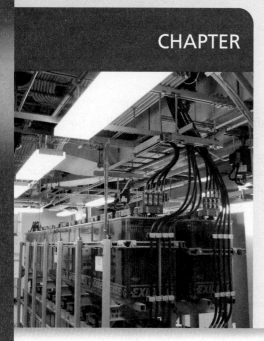

Source: Ronald R. Spadafora

KEY TERMS

Covalent bonds, *p. 209*
LOAEL, *p. 210*

NOAEL, *p. 210*

"Drop-in" replacement, *p. 214*

OBJECTIVES

After reading this chapter, the reader should be able to:

- Describe halon replacement criteria
- List and explain the Significant New Alternatives Policy (SNAP) program items
- Describe and explain halocarbon and powdered aerosol agents
- Explain halocarbon total-flooding design applications
- Understand powdered aerosol design applications

Introduction

Clean agents extinguish fires as a gas giving them the ability to permeate into hard-to-reach and obstructed areas. Obscured fires are rapidly extinguished. Unlike water, these fire-suppression agents are nonconductive and noncorrosive. They are therefore safe to use on electrified equipment. They also leave no residue to clean up. Clean agents are used in total-flooding applications in which an entire space is protected (see Figure 12.1). Once the fire is extinguished, the clean agent is ventilated from the room along with the products of combustion. Operations are brought back to full service in a very short time. Although the National Fire Protection Association (NFPA) recommends that occupants always exit the

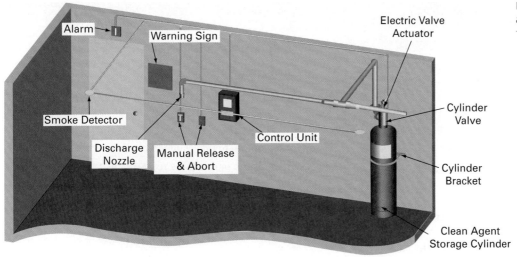

FIGURE 12.1 Clean agent system layout

area of hazard in the event of fire, extensive toxicity studies have shown that most clean agents are people friendly and can be used for the protection of occupied spaces.

Fires have a tremendous negative impact on business. Industry studies show that 43% of businesses closed by a substantial fire never reopen. Another 29% fail within three years.[1] Consider the effect of the loss of a priceless work of art in a museum or a one-of-a-kind piece of equipment in a factory. These lost items are irreplaceable.

Overview

There is a critical need for clean extinguishing agents. Clean agent fire-suppression characteristic criteria include low toxicity, environmentally friendly, fast acting and effective, noncorrosive, electrically nonconductive, and leaves no residue. In total-flooding applications, Halon 1301 has been replaced by halocarbon compounds*(fluorinated ketones and hydrofluorocarbons [HFCs]). These replacement agents contain no ozone-depleting chlorine or bromine. Halocarbon extinguishing agent is effective on Class A, B, and C fires.

Although powdered aerosols are not considered clean agents, they are acceptable substitutes for Halon 1301 in total flooding systems. Powdered aerosol fire extinguishing agent technology originated in the 1980s in what was then the Soviet Union. Since then, it has been introduced globally. Fixed powdered aerosol systems are used primarily in total-flooding applications. Their composition is a varied mixture that may contain dry chemical extinguishing agent, gelled halocarbons, water, inert gas, carbon dioxide, and oxygen. These mixtures are stored inside modular units shaped as box-like generators with discharge outlets or applicators.

Significant New Alternatives Policy

In 1994, the US Environmental Protection Agency (EPA) established the Significant New Alternatives Policy (SNAP) program pursuant to section 612 of the Clean Air Act. This section required the EPA to evaluate the overall effects on human health and the environment of all substitutes for ozone-depleting substances (ODS). The data collected from these tests help in selecting the agents. Rules promulgated under SNAP make it unlawful

Covalent bond ■ A chemical joining that involves the sharing of electron pairs between atoms. It creates a stable balance of attractive and repulsive forces.

---

*Halocarbon compounds are chemicals in which one or more carbon atoms are linked by **covalent bonds** with one or more halogen atoms (fluorine, chlorine, bromine, or iodine).

TABLE 12.1 Acceptable Substitutes for Halon 1301 in Total-Flooding Systems

SUBSTITUTE	COMMENTS
Powdered aerosol A	For use in unoccupied areas only
Powdered aerosol C	For use in unoccupied areas only
Powdered aerosol D	*
Powdered aerosol E	For use only in normally unoccupied areas *
Carbon dioxide	System design must adhere to OSHA 1910.162(b)(5) and NFPA Standard 12
Water	None
Water mist systems using potable or natural sea water	None
HFC-227ea (FM-200®)	**
IG-100	**
IG-55	**
IG-541(INERGEN®)	**
N2 Towers® System	**
C6-perfluoroketone (Novec™ 1230)	**
Haltron® II	Acceptable in areas that are not normally unoccupied only

*Use of this agent should be in accordance with the safety guidelines in the latest edition of *NFPA 2010: Standard for Aerosol Extinguishing Systems*.[4]

**Use of this agent should be in accordance with the safety guidelines in the latest edition of *NFPA 2001: Standard for Clean Agent Fire Extinguishing Systems*.[5]

LOAEL (Lowest Observable Adverse Effect Level) ■ Lowest concentration or amount of a substance, found by experiment or observation, that causes an adverse alteration of morphology, functional capacity, growth, development, or lifespan of a target organism distinguishable from normal (control) organisms of the same species and strain under defined conditions of exposure.

NOAEL (No Observable Adverse Effect Level) ■ The level of exposure of an organism, found by experiment or observation, at which there is no biologically or statistically significant increase in the frequency or severity of any adverse effects in the exposed population when compared to its appropriate control.

to replace an ODS with an agent that may present adverse effects to human health and the environment.

The EPA provides toxicity guidance for the use of new clean extinguishing agents through the use of **LOAEL** (lowest observable adverse effect level) and **NOAEL** (no observable adverse effect level) values. In assessing the toxicity of halocarbon alternatives, the EPA is primarily concerned with human exposure during a fire. Occupied hazard areas can be safely protected by agents up to an agent's LOAEL concentration, provided the area can be exited within a short period of time. The SNAP program reviews substitutes in eight industrial sectors, including foam blowing, refrigeration and air conditioning, solvents, and aerosol propellants. It also reviews fire suppression and explosion protection.[2] Since 1994, SNAP has reviewed over 300 substitutes and approved over 90%, most with some use condition to minimize environmental or health risks. Lists of acceptable and unacceptable substitutes are updated several times each year.[3] Table 12.1 presents a short, noninclusive summation of acceptable substitutes for Halon 1301 in total-flooding systems.

Industry Standards

Standards pertaining to clean agents have also been developed by the fire protection industry. *NFPA 2001: Standard on Clean Agent Fire Extinguishing Systems* serves as the primary document covering these agents. NFPA 2001 addresses total-flooding systems as well as local-application systems, in which a concentration of agent is discharged directly over the burning material in close proximity to the fire source. For local-application systems, concentrations of clean agent in the vicinity of the discharge may exceed the maximum permitted exposure limits determined through the SNAP program in order to achieve the agent concentration necessary for fire extinguishment. Occupant exposure to agent discharge from local-application systems can vary greatly and therefore is a complicated assessment for the design professional.

Agents proposed for inclusion in NFPA 2001 must first be evaluated in a manner equivalent to the process used by the SNAP program. NFPA 2001 is a technical standard that examines the design, installation, testing, inspection, operation, and maintenance of clean agent fire suppression systems. It also denotes components for these systems. Features include agent supply, distribution, detection, actuation, and control systems. Minimum requirements are included for all clean agents that are approved by the SNAP program.

Halocarbons

Halocarbons are synthetic organic substances that contain a carbon–halogen (chlorine, fluorine, bromine, or iodine) chemical bond, either individually or in some combination. Halocarbons remove heat when discharged onto the surface of combustible materials. The heat is absorbed from the surface of the burning material, thereby lowering its temperature below the autoignition point.

Halocarbons are stored in cylinders as a liquid and pressurization is achieved with nitrogen. Because halocarbon concentrations in the air for total-flooding extinguishment applications are relatively low, they take up many times less storage space than systems based on carbon dioxide and inert gases. When discharged, the liquid flows through a network of piping into the protected area, where it vaporizes. Halocarbons can be safely used where people are present. Common occupancies using halocarbon agents for fire suppression are businesses, government buildings, universities, hospitals, and museums.

NOVEC 1230

Novec™ 1230 is produced by the 3M™ Company. It is a high molecular weight material compared to the first-generation halocarbon clean agents. It is a liquid at room temperature, but it gasifies immediately after being discharged. Chemically, it is a fluorinated ketone* with a chemical formula of $C_6F_{12}O$. It is a waterless, fluid fire suppressant that principally extinguishes fire by removing heat from the fire. Novec 1230 has the widest margin of safety when used in total-flooding systems in occupied spaces (see Figure 12.2).

Upon discharge from its nozzle/applicator, Novec 1230 creates a gaseous mixture with air that has a high heat capacity (see Figure 12.3). At an accurate system design concentration, Novec 1230 absorbs sufficient heat to upset the balance of the fire tetrahedron. This cools the fire to the point at which it can no longer sustain itself. A low design concentration of 4% to 6% by volume for total-flooding applications in combination with a high NOAEL of 10% gives a safety margin of up to 100%.[6]

FM-200®

The halocarbon FM-200, manufactured in the United States by the Great Lakes Chemical Corporation, is a liquefied gas that is stored under pressure in containers (see Figure 12.4). It extinguishes fire primarily by physically cooling the flame at a molecular level. FM-200 is a compound of carbon, fluorine, and hydrogen (CF_3CHFCF_3). It is a colorless, odorless, and electrically nonconductive extinguishing agent. Upon automatic

---

*Fluorinated ketones (fluoroketones) are clear, colorless, odorless liquids with low to moderate boiling points that are compressed and pressurized with nitrogen and stored in tanks.

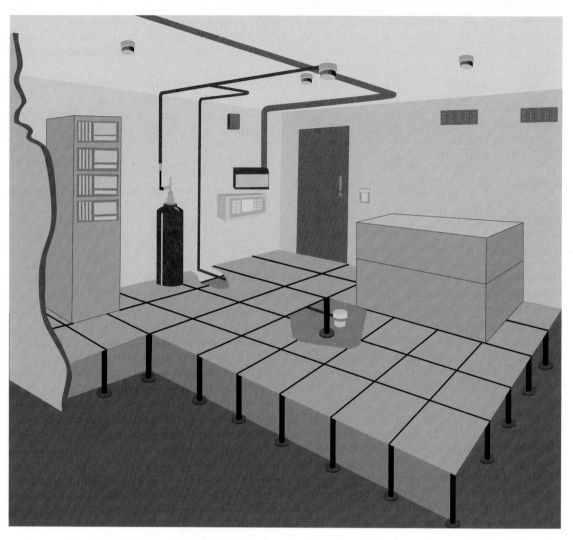

FIGURE 12.2 Novec 1230 total-flooding extinguishing system

or manual discharge (completed in 10 seconds or less) it quenches flame to the extent to which the combustion reaction cannot sustain itself. Activation is tied into an alarm system that prewarns occupants to leave the protected area (see Figure 12.5). Most metals, such as steel, aluminum, and brass, as well as plastics, rubber, and electronic components are not damaged by exposure to FM-200.

FM-200 is considered nontoxic to humans in concentrations necessary to extinguish most fires. Safety precautions should still be taken when applying and handling this clean agent. The major disadvantage of FM-200 is that it decomposes to form halogen acids when exposed to open flames. This discharge creates a health hazard to occupants from the agent itself as well as from the products of decomposition that result when the agent is exposed to the fire.[7]

ADDITIONAL HALOCARBON AGENTS

DuPont™ manufactures several of the most popular halocarbon fire extinguishing agents. They include FE-25™, FE-13™, and FE-36™.

FIGURE 12.3 Novec 1230 nozzle (left); INERGEN nozzle (right). *Source:* Ronald R. Spadafora

FIGURE 12.4 FM-200 storage containers for a total-flooding system protecting electrical equipment. *Source:* Ronald R. Spadafora

FIGURE 12.5 FM-200 manual activation pull station and alarm unit for a total-flooding system.
Source: Ronald R. Spadafora

FE-25

FE-25 is a liquefied gas stored in sealed cylinders that mirrors the fire extinguishing capabilities of Halon 1301 for total-flooding system applications. It is generally used in concentrations ranging from 8% to 12%. FE-25 is considered a **"drop-in" replacement** and retrofit system for Halon 1301 extinguishing systems by fire protection engineers, because its flow characteristics and required vapor pressure are similar. It generally requires only minimal adjustments to existing piping. This agent is discharged through nozzles attached to a piping network, activated by a fire alarm control unit (FACU) that includes smoke detection.

"Drop-in" replacement ■ An easy retrofit for an existing halon extinguishing system with minimal modifications.

FE-13

FE-13 is a gaseous extinguishing agent also used as a replacement for Halon 1301 and is ideal for cold-temperature areas (such as offshore oil platforms) due to its high boiling point and high vapor pressure. It can be used in temperatures as low as −40°F (−40°C) and its high-pressure discharge means a quick, even distribution within the hazard. With a maximum nozzle height of 25 ft. (8 m), it also is suitable in spaces with high ceilings. However, the high pressure of an FE-13 system limits the amount that can be stored within the piping. FE-13 is stored at room temperature as a liquid in high-pressure cylinders and uses its own storage pressure for discharge and distribution.

FE-36

FE-36 is a gaseous fire suppressant used in total-flooding fire extinguishing equipment as a substitute agent for Halon 1301. It can also replace Halon 1301 in explosion suppression and explosion inerting applications. As a replacement for Halon 1211 in portable fire extinguishers, it discharges as a stream of gas and liquid droplets, with a range up to 16 ft. (5 m). FE-36 suppresses fire through heat absorption and a chemical interaction. It has a very low toxicity level, is noncorrosive, electrically nonconductive, and leaves no residue.

TABLE 12.2 Safety Comparison of Clean Agents

AGENT	USE CONCENTRATION	NOAEL	SAFETY MARGIN
Novec 1230	4%–6%	10%	67%–150%
FM-200	4.5%–8.7%	9%	3%–20%
FE-25	8%–11.5%	7.5%	0%

Source: "Safety Comparison of Clean Agents" from Willis Technical Advisory Bulletin-Halon Alternatives. Published by Willis Group Holdings, © 2005.

A comparison of the three most popular halocarbon clean agents in Table 12.2 reveals that Novec 1230 is the safest for use in occupied spaces.[8]

Halocarbon Total-Flooding System Design

Halocarbon design includes determining the agent quantity, piping layout, and pressure drop through the piping and accessories as well as fixing the location and number of discharge nozzles for uniform distribution of the agent throughout the enclosure. The fire protection design engineer must also consider pressure drop through the system, because this will affect the filling density in the halocarbon containers or cylinders and their number. The agent quantity required for total flooding is determined independently based on the design concentration of the agent necessary for the type of fire to be extinguished, the amount of time the agent is required to remain in the protected enclosure, and the reserve quantity needed to circumvent leakage of agent. Sudden discharge of an agent within a short period of time will raise the pressure inside the space because of the agent's expansion. The rapid rise in pressure must be evaluated. Ceilings and walls must be examined and reinforced if necessary to withstand the impact of pressure change.

System components consist of agent containers or cylinders, supports (racks), piping, discharge nozzles, detectors, alarm appliances, and a fire alarm control unit (FACU). These systems generally have an electric actuator (solenoid) attached to one or more agent-storage containers. A signal from the FACU allows the electric actuator to operate, which in turn initiates rapid operation of the discharge valve(s) and release of the agent.

Protection under Raised Floors

Raised-floor configurations can be found in enclosures (telecommunications facilities, for example) protected by total-flooding halocarbon clean agent systems. Electrical cables or ventilation components designed to cool equipment may be installed in these low-level voids. If this space is not protected with its own clean agent system, discharged halocarbons in the upper level of the enclosure will leak underneath the floor and dilute the design concentration necessary for extinguishment of the fire in the main enclosure. In addition, a fire occurring in the void space under the raised floor may not be extinguished by a discharge of halocarbon clean agent above, because the agent migrating into the space below will be unlikely to be of a concentration needed for total suppression.

In addition, if a halocarbon clean agent system only protects the void below the raised floor, a fire in the area above might activate the clean agent system below. During and after agent discharge, some of the agent from the space under the raised floor can leak into the enclosure above. The discharged clean agent could be exposed to the fire. Thermal decomposition of the halocarbon will occur. This could have a detrimental effect for sensitive equipment.

A similar scenario could occur in which a halocarbon extinguishing agent protects the underfloor area and another type of fire protection system protects the area above. If the system above does not extinguish the fire in the main area, the halocarbon clean agent system below the raised floor could activate. A concentration of released halocarbon agent from the underfloor space could migrate to the enclosure above and come in contact with the fire, creating the formation of harmful decomposition products.[9]

Deep-Seated Smoldering Fires

Two types of fires can occur when solid materials burn: flaming combustion and smoldering combustion. They often occur concurrently, although one type can precede the other. Lower concentrations of clean extinguishing agent are required to suppress flaming combustion due to its location and exposure to system discharge. In the absence of deep-seated smoldering combustion, it is also less likely to reignite.

In contrast, smoldering combustion—especially if it is deep-seated—will not be as rapidly extinguished. Smoldering fires create lower heat release rates (HRR) and thereby slower rates of heat loss within the burning material. The fuel remains above its autoignition temperature, allowing it to react with oxygen to sustain combustion. Smoldering deep-seated fires can cause serious damage to valuable building contents and equipment. Higher clean agent concentrations in the protected space for a longer period of time is generally prescribed for this type of hazard protection. This design will provide the agent enough time to penetrate into the fuel.[10]

Powdered Aerosols

Powdered aerosol is a chemical extinguishing agent consisting of micronized solid particles that are created in a chemical reaction and released in aerosol form (see Figure 12.6). The contents of the powdered aerosol can vary according to the type of agent being used. These systems pulverize dry chemical ingredients into aerosol particulate, thereby enhancing its fire extinguishment capability by reducing its mass while increasing its surface area. Powdered aerosol can also be mixed with other fire extinguishing agents and released from its container under pressure for fixed total flooding and local applications as well as portable fire extinguisher use. When dispersed, it provides a 20-minute or longer fire extinguishing or inert mixture in air. This technology, however, is generally designed for normally unoccupied spaces, because powdered aerosol particulate can raise human blood pH levels and is a respiratory and eye irritant.

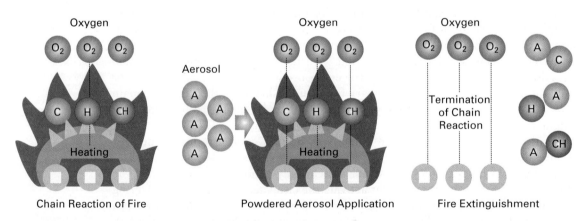

FIGURE 12.6 How powdered aerosol works

Powdered aerosol systems involve simple installation and negligible maintenance. Unlike the halocarbons, these systems do not require pressurized cylinders or piping. They provide highly efficient fire extinguishment of Class A, Class B, and Class C fires. There are two types of powdered aerosol systems: dispersed and condensed.

DISPERSED SYSTEMS

In a dispersed system, the powder forming the aerosol is stored in a pressurized cylinder containing a gas (halocarbon or inert gas). When the system is activated, the aerosol is introduced through a delivery system similar to that used for gaseous agents.

CONDENSED SYSTEMS

In a condensed system, the aerosols are produced pyrotechnically by using a solid compound in the generator. The aerosol particles are released in the exhaust of the burning compound along with the degradation products of the pyrotechnic compound (nitrogen, oxygen, carbon monoxide, carbon dioxide, and water). Potassium salts are generally used to produce the aerosol particles (see Figure 12.7).

Powdered aerosol generators are designed for placement around the enclosure being protected. An electrical impulse from a separate detection system or a self-contained detection component built into the generator itself provides the catalyst for activation. This energy penetrates into the powdered aerosol mixture, dispersing it very rapidly (0.1–1 second) over long distances onto the hazard. The small particle size (1–2 micrometre [μm])* of the agent and enhanced surface area allow it to effectively inhibit the combustion process.

FIGURE 12.7 Condensed powdered aerosol extinguishing system that uses generators to produce aerosols pyrotechnically from a solid compound. *Source:* AFG Flame Guard USA

---

*1 μm = 0.000039 in.

Powdered Aerosol System Design

Powdered aerosol technology has introduced into the fire protection field new ideas for the total-flooding and local-application market. The phase out of halon combined with the limitations of halocarbon substitutes (weight and volume requirements) has created a real need for powdered aerosol extinguishing products that are economical and cost-effective.

PROPELLED EXTINGUISHING AGENT TECHNOLOGY TOTAL-FLOODING SYSTEMS

These systems incorporate several generators of various shapes and sizes that hold powdered aerosol extinguishing agent. The generators are connected in a loop configuration to a FACU. Propelled extinguishing agent technology (PEAT) systems are simple to install and require minimal maintenance. They are not dependent upon an agent-delivery system (high-pressure cylinders, valves, gauges, piping, connectors, and nozzles). This type of system can be used in unheated enclosures and is reliable at temperatures down to −40°F (−40°C).

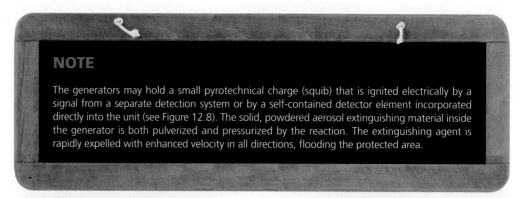

NOTE

The generators may hold a small pyrotechnical charge (squib) that is ignited electrically by a signal from a separate detection system or by a self-contained detector element incorporated directly into the unit (see Figure 12.8). The solid, powdered aerosol extinguishing material inside the generator is both pulverized and pressurized by the reaction. The extinguishing agent is rapidly expelled with enhanced velocity in all directions, flooding the protected area.

FIGURE 12.8 Powdered aerosol suppression system protecting a backup battery bank room.
Source: Ronald R. Spadafora

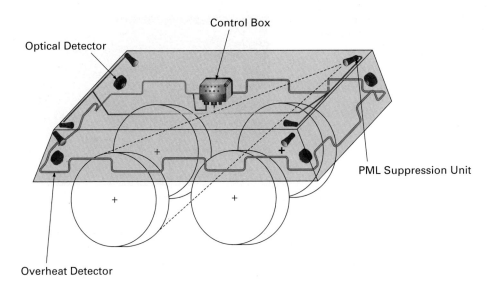

FIGURE 12.9 Propelled extinguishing agent technologies (PEAT) combine powdered aerosol and propelled launchers to provide fire protection for modern battlefield vehicles.

PROPELLED LAUNCHER TOTAL-FLOODING AND LOCAL-APPLICATION SYSTEMS

Propelled launchers (PL; see Figure 12.9) may be used instead of generators for both total flooding and fixed local applications in which the agent is delivered through a tubular barrel directly onto the fire.

Propelled launchers are activated in the same ways as generators. They are used to protect paint spray booths, pump stations, boiler rooms, transformers, and engine rooms. Propelled mini-launchers (PML) with smaller amounts of powdered aerosol are also used for unique fire protection needs within small, confined spaces. PMLs are installed for the protection of military vehicles (crew cabs, engine compartments, chassis, and tire areas).

PNEUMATIC IMPULSE TOTAL-FLOODING AND LOCAL-APPLICATION SYSTEMS

Another kind of PEAT system uses pneumatic high pressure instead of pyrotechnical charges and squibs to provide impulses to activate containers that are filled with powdered aerosol extinguishing agent. In total-flooding applications, this system requires air cylinders and piping. A compressed air pump may also be needed for large area protection.

HEAT-ACTUATED TOTAL-FLOODING SYSTEMS

Heat-actuated systems consist of generators, canisters, or drums that hold a solid, powdered aerosol extinguishing agent. When the container is heated by the fire to approximately 900°F (482°C), it converts its contents to a highly fluidized aerosol and is ready to activate automatically. Heat-actuated containers may also have a fire-conducting cord that can be ignited by direct flame contact. The ignited cord will provide the energy to pulverize/pressurize the agent. Discharge is *not* dependent upon a fire detection system or an external electrical power supply. Heat-actuated systems are used for total-flooding purposes only. In an emergency, the containers can be manually carried and/or rolled into position inside the fire area by trained personnel from a safe distance.

Firefighter Operations

Powdered aerosol portable, handheld fire canisters and discs (see Figure 12.10) are new firefighting weapons in the firefighter's tool box. These devices are designed for

FIGURE 12.10 Powdered aerosol portable discus-shaped device. *Source:* AFG Flame Guard USA

use by firefighters during extinguishment operations. The canisters and discs provide an effective aerosol fire-suppression agent.

Firefighters arriving at the scene of an incipient structural fire prior to a positive water source being obtained can pull the pin (canister) or ripcord (disc) and toss the device into a room on fire. Within seconds, it activates. These devices are also heat-actuated, making them fail-safe. They emit a nontoxic, potassium-based aerosol cloud of powder for several minutes that suppresses the fire and cools the room without reducing the oxygen level by interfering with the chemical reaction required for a fire to intensify and spread. This new extinguishing tool can buy time for further rescue operations and for firefighters to use hose streams. This application can also eliminate many of the contributing factors of a flashover.

Portable powdered aerosol is designed to rapidly control and/or extinguish Class A, B, and C fires inside structures with a volume up to 2,000 ft.^{3} (57 m^3). Additional uses include small, tough-to-access spaces (attics, cellars, or subcellars, for example). Within a short time period (10 seconds), it releases a white cloud of powdered aerosol in all directions, filling the enclosure.

ROCKFORD, ILLINOIS, FIRE

At approximately 0830 hours, heavy black smoke was pushing out from a basement of a house in Rockford, Illinois. A District Chief pulled the pin of a powdered aerosol handheld fire extinguisher and tossed the device through a basement window. The flames were reaching to the upper levels of the structure when the device activated and appeared to slow down the fire for a few minutes. This extinguishing action bought firefighters enough time to force entry and get a better position; fire companies on the scene were in the process of stretching hose lines into the building. Within 22 minutes, the fire was brought under control, and nobody was injured. It was the second time that morning that firefighters had been dispatched to the house, which was under renovation and not occupied. The Rockford Fire Department bought a substantial number of these extinguishing devices to equip 13 of its vehicles that do not carry water (command vehicles and ambulances).[11]

A benefit of portable handheld powdered aerosol fire suppressant device use by first responders is the lowering of the temperature of a fire room, making it safer for firefighters to enter. This agent is capable of bringing preflashover room temperatures down considerably. It can also be beneficial in reducing the amount of water firefighters need to extinguish the fire. This correlates to less water damage for the building and lowers the potential for mold development. In addition, the less water used, the better preserved a possible crime scene will be.

National Fallen Firefighters Foundation: 16 Firefighter Life Safety Initiatives Program

The eighth initiative of the NFFF has a focus on technology and how the identification of technological fixes may contribute to the bigger picture of reducing line-of-duty deaths. This initiative asks the fire service to use existing technology and be aware of emerging technologies.[12]

Getting the best benefits from any new tool requires proper training and putting the tool in perspective in addition to common sense. The use of handheld powdered aerosol suppressants should be explored by the fire service to evaluate what applications are deemed useful. Fire departments must review their resources and operational procedures prior to purchasing this new suppressant technology. Will the use of portable powdered aerosol devices provide a safer and more efficient work environment? Review what fire departments who already have this technology are saying about its benefits and drawbacks. When considering a new technology, the NFFF warns that you should anticipate resistance and think about ways to counter arguments.

Portable Fire Extinguishers

Clean agents are also used in portable fire extinguisher applications, in which the agent discharges from a handheld or wheeled fire extinguisher. They are available from numerous manufacturers in various sizes and Underwriters Laboratories Inc. (UL) ratings. The UL Listed units carry a 2-B:C, 5-B:C, 1-A:10-B:C, or 2-A:10-B:C rating. Portable fire extinguishers are a suitable option to protect high-value assets such as computer rooms, telecommunications facilities, process control rooms, museums, archives, marine facilities, banks, laboratories, and airplanes.

CHAPTER REVIEW

Summary

Clean agents are being introduced throughout the world to replace the environmentally unfriendly halons. The agents reviewed in this chapter are only a few from the EPA's SNAP program list. The ideal clean agent alternative has still not been manufactured. The fire protection engineer must assess many factors in order to determine the appropriate clean agent to be used. Firefighters must become familiar with this new technology during training drills, site visits, and inspections to better understand its capabilities and potential hazards. This information and knowledge will enhance operational strategy at fires and emergencies. The use of powdered aerosol handheld portable extinguisher devices during fire operations should also be explored by the fire service.

Review Questions

1. Name the two main types of halocarbon clean agents that have replaced Halon 1301 in total-flooding systems.
2. In 1994, the US EPA established what program pursuant to section 612 of the Clean Air Act that required the EPA to evaluate the overall effects on human health and the environment of all substitutes for ozone-depleting substances?
3. What is the major way in which halocarbon clean agents extinguish fire when discharged onto the surface of combustible materials?
4. In total-flooding systems, how long does it take for the halocarbon FM-200 to be completely discharged into the enclosure?
5. What halocarbon clean agent is considered by fire protection engineers as a "drop-in" replacement and retrofit system for Halon 1301 extinguishing systems due to the fact that its flow characteristics and required vapor pressure are similar?
6. How do firefighters use powdered aerosol to suppress fires?
7. How can raised floor configurations, found in spaces that are protected by total-flooding halocarbon clean agent systems, negatively impact the effectiveness of discharge for a fire in the upper level of the enclosure?
8. Why is FE-36 an ideal replacement for Halon 1211 around susceptible (high-risk) populations (for example, in surgical operating rooms)?
9. Why are total-flooding propelled extinguishing agent technology systems simple to install and maintain?

Endnotes

1. BFPE International, "Suppression Systems," www.bfpe.com/article1_why_clean_agents.htm.
2. US Environmental Protection Agency, "FACT SHEET: Proposed Rule—Update to the List of Acceptable Substitutes in Fire Protection," last modified August 19, 2010, www.epa.gov/ozone/snap/fire/SNAP_Fire_Factsheet.html.
3. US Environmental Protection Agency, "Substitutes for Halon 1301 as a Total Flooding Agent," last modified August 15, 2012, www.epa.gov/ozone/snap/fire/lists/flood.html.
4. National Fire Protection Association (NFPA), *NFPA 2010: Standard for Aerosol Extinguishing Systems* (Quincy, MA: NFPA, 2010).
5. National Fire Protection Association (NFPA), *NFPA 2001: Standard for Clean Agent Fire Extinguishing Systems* (Quincy, MA: NFPA, 2012).
6. Wilhelmsen Technical Solutions, "Extinguishing Fire Using Novec 1230 fluid," www.wilhelmsen.com/services/maritime/companies/wts/safetysolutions/firesuppressionsystems/u1230afes/Pages/ExtinguishingfireusingNovec1230fluid.aspx#.
7. Fenwal Protection Systems, "FM-200® Fire Suppression Agent," www.fenwalfire.com/utcfs/ws-390/Assets/f-93-1900_print.pdf.
8. Willis Property Risk Control, "Halon Alternatives," *Willis Technical Advisory Bulletin*, February 2005.

9. Milosh Puchovsky, "Clean Agent Fire Suppression Systems," *Consulting-Specifying Engineer*, May 6, 2011, www.csemag.com/single-article/clean-agent-fire-suppression-systems/c917ebc25d02cca3500e210f3dbecbb0.html.
10. Ibid.
11. Jeff Kolkey, "Rockford Firefighters Deploy New Weapon in Suspicious House Fires," rrstar.com, June 19, 2012, www.rrstar.com/news/x681119678/Firefighters-respond-to-2-fires-at-same-Rockford-house.
12. National Fallen Firefighters Foundation, "Initiative 8: Utilize Available Technology Wherever It Can Produce Higher Levels of Health and Safety," *White Paper Series on the 16 Firefighters Life Safety Initiatives* (Emmitsburg, MD: National Fallen Firefighters Foundation, 2007).

CHAPTER 13

Water Mist Fire Suppression Systems

Source: Ronald R. Spadafora

KEY TERMS

Application specific, *p. 225*

Deionized, *p. 225*

Hypoxia atmosphere, *p. 235*

OBJECTIVES

After reading this chapter, the reader should be able to:

- Understand the differences between water mist and sprinkler systems
- Explain the extinguishing properties of water mist
- List the pressure classifications of water mist systems
- Describe the four types of water mist systems
- Discuss water mist system applications

Introduction

A water mist system is a water-based fire protection system. It has nozzles that are able to distribute its stored water onto a variety of hazards. Water mist fire-suppression systems date back almost 100 years, with the first systems (steam) used to protect lumber-drying kilns. Subsequently, water mist systems were developed for use on ships.

Overview

Water mist research has focused on the fire protection problem of fire suppression aboard ships and spacecraft. The International Maritime Organization (IMO), for example, requires water suppression systems (sprinkler systems) on ships. The weight of the water, however, and the amount of storage area required makes many water-based systems impractical. Water mist is considered a viable alternative, because its delivery system is much lighter than sprinkler and water spray systems. Properly designed water mist systems can be effective on both solid fuel (Class A) and liquid fuel (Class B) fires. Droplets at 400 micrometres (µm)* or larger are used for Class A combustibles, whereas smaller water droplets are designed for extinguishment of Class B fires. Unlike sprinkler systems, however, water mist systems are **application specific** and each particular hazard or occupancy requires its own design. Water mist systems that use **deionized** water are designed to safely protect Class C hazards (energized electrical equipment).

In 1993, researchers and engineers, along with water mist manufacturers, insurance company representatives and industrial users, assembled to develop a new standard for the design and installation of these systems. In 1996, the first National Fire Protection Association (NFPA) standard on water mist systems, *NFPA 750: Standard for the Installation of Water Mist Fire Protection Systems*, was generated. Water mist technology is an evolving fire protection technology. It has suppression characteristics that make it a logical choice for applications in which low water requirements for the fire protection system are beneficial.

Application specific ■ Pertains to the requirement for assessment of the hazard and volume of space that needs to be protected on a case-by-case basis prior to installation.

Deionized ■ Water that has had its ions (iron, sodium, and calcium) removed. Water containing ions is significantly more electrically conductive than water that does not contain ions.

Extinguishing Properties

Water mist is an effective, nongaseous fire extinguishing replacement agent for the halons. The minute water droplets generated are a major feature of the technology. Small drops of water, with a greater surface area to mass ratio, are more effective in extinguishing fire than large droplets. Pressure is another important feature of these systems. Pressure design for water mist systems can produce fine water mist droplets at the minimum design operating pressure of the water mist nozzle.[1]

Unlike gaseous fire extinguishing agents in total-flooding applications, water mist systems have the ability to successfully fight fires within an enclosed room that is not tightly sealed. The discharge of water mist is normally continuous, and loss of extinguishing agent via leaks is not crucial to performance. Water mist in enclosures also has the ability to limit combustible gas temperature.

Droplet Size

NFPA 750 has divided the droplets produced by a water mist system into three classifications. These classes distinguish between "coarser" and "finer" droplet sizes. The optimum size of droplets for fire suppression is strongly dependent on many factors, including the properties of the combustibles, the degree of obstruction in the compartment, and the size of the fire. The droplet size distribution that is most effective in extinguishing one fire scenario will not necessarily be the best for other scenarios. Furthermore, any changes in spray velocity (momentum) and enclosure effects will change the optimum droplet size for fire suppression.

It is very difficult for small droplets to penetrate into the seat of the fire. The fire plume hinders these droplets from reaching the fuel surface. Water droplets with low momentum are easily carried away from the fire by air currents. Large droplets can penetrate the fire plume easily to provide direct impingement, which will wet and cool the combustibles. As the size of the water droplet increases, however, the capability of water mist extinguishing systems in suppressing obstructed/shielded fires is reduced. Large droplets with high velocities also can cause liquid fuels to be splashed, with the potential result of increasing the size of the fire.

Upon discharge of a water mist system, clouds of water droplets are formed both in the upper area of an enclosure and close to the floor. The steam generated by these systems only displaces the oxygen in the close vicinity of the fire, allowing occupants to safely be exposed and be capable of exiting the area (see Figure 13.1).

---

*1 µm = 0.000039 in.

FIGURE 13.1 Water mist extinguishing properties

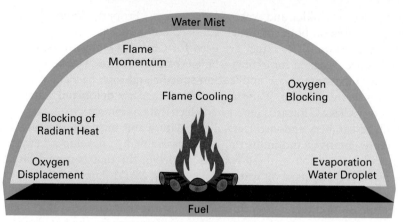

The evaporation of a water droplet, turning it into steam, involves a change in phase for the water. The steps involved are:

- Heating of the droplet
- Boiling of the droplet
- Evaporation

Through testing, a prediction can be made by the fire protection engineer regarding the heat absorbed by a droplet (given the initial diameter) as well as the heat absorbed by the steam produced by the evaporation of the droplet. Testing results also confirm that smaller droplets evaporate faster than larger droplets. In addition, droplets closely spaced evaporate more slowly than droplets positioned wider apart. Efficient evaporation is therefore created by the design professional through the implementation of a nozzle delivering water at relatively high pressure to produce small water droplets over a wide area.[2]

Fires can release tremendous amounts of heat, which is readily used by water mist systems to convert the water droplets to water vapor. The vapor displaces oxygen and acts like an inert extinguishing gas. Water mist droplets are also effective in scattering radiated heat. In the fire service, hose lines spraying wide fog patterns are used by firefighters to shield them from heat as they approach an intense fire. This principal is similar to water mist systems. The mist is able to protect occupants in a similar fashion.[3] In addition, the fine water mist droplets help to scrub the air during a fire. Smoke particles attach themselves to the droplets and are carried down to the ground, thereby enhancing visibility.[4]

 SCANDINAVIAN STAR FERRY DISASTER

Water mist applications further developed following the fire on the passenger ferry *Scandinavian Star* in April 1990, which killed 158 people (156 passengers and two crew members).

Early in the morning of April 7, 1990, a fire broke out on this ferry when en route between Oslo in Norway and Fredrikshavn in Denmark. There were 99 crew members and 383 passengers on board. Most passengers came from Norway. This tragedy galvanized the water mist industry. In 1993, it conducted a large series of ship cabin and corridor fire tests. These tests were independently witnessed and led to improved IMO fire safety requirements on passenger ships and the development of installation guidelines and fire test procedures for alternative sprinkler systems.[5]

Fire tests further concluded that a significant general property of water mist systems is their low water consumption compared with conventional sprinkler systems. Standard sprinkler fire protection systems use five to 10 times more water to control and extinguish fires than water mist. The benefits of low water consumption are that the water supply, the pump size, and the diameter of the piping can be substantially smaller for water mist systems than for sprinkler systems. This is a valuable benefit when space and weight requirements are restrictive. In some cases, an adequate water supply may be small enough to enable a water mist system to be a self-contained, standalone unit that is not dependent on a connection to an external water supply. Besides being used for the protection of engine rooms and machinery spaces, water mist systems should also be installed in accommodation, public space, and service areas.[6]

Water Mist Advantages over Conventional Sprinkler Systems

Conventional sprinkler systems rely on an enormous amount of water in order to perform successfully. The use of water mist fire suppression, compared to sprinkler systems, has revealed many more advantages, as shown in Table 13-1.

Water Mist Disadvantages over Conventional Sprinkler Systems

When compared to conventional sprinkler systems, water mist systems have been developed comparatively recently. They therefore do not have the same comprehensive amount of performance and testing data on fire hazards. Water mist systems are also very "project specific," which makes it difficult to produce a broad set of guidelines for their use. Additional disadvantages when compared to sprinkler systems are listed in Table 13-2.

TABLE 13-1	Water Mist Advantages Compared to Conventional Sprinkler Systems

- Control of flammable liquid fires
- Electrically nonconductive
- Enhanced visibility
- Improved aesthetics
- Less clean-up time
- Less smoke damage
- Less water damage
- Low space and weight requirements for extinguishing agent
- Low water supply requirements
- Reduced downtime and business interruption
- Reduced water flow rates
- Uses fewer materials in construction
- Washing of toxic and corrosive gases

TABLE 13-2	Water Mist Disadvantages Compared to Conventional Sprinkler Systems

- High engineering skill level required to install system
- Do not perform as well in large, open areas
- More expensive
- Requires more design time and resources
- High level of maintenance and testing requirements
- More nozzles are required for a given floor area making them more intrusive
- Individual components cannot be used on an ad-hoc basis

Pressure Classifications

The fire protection system designer selects a water mist extinguishing system when it is decided that the creation of heat-absorbing vapor is the primary performance objective. Water mist system design takes into consideration both the volume and the probable fire dynamics of the hazard to determine the number of nozzles and nozzle flow rates. When compared to the water demand required for sprinkler systems, the amount of water used for a water mist system is significantly less.

Pressure is a strong contributor to the size of a water mist droplet. At the nozzle, pressure allows the device to project water mist droplets for extended distances. In general, the higher the pressure at the nozzle, the farther the nozzle is required to be from the axis of the fire plume centerline. Three classifications for water mist system pressurization have been established by NFPA 750.

LOW PRESSURE

Low-pressure system piping is subjected to pressures of 175 psi (1,207 kPa) or less. This is approximately the same pressure range used for standard sprinkler systems. Low-pressure water mist systems are designed to retain all the performance and reliability of traditional water-based systems and also take advantage of the efficiency of small water droplet technology. They eliminate the requirement for filters in the water supply system. Low operating pressure also provides the option of using either copper or plated carbon steel piping as alternatives to stainless steel. Typical flow rates are 2 to 3 gpm (7–11 L/min.) from each nozzle in the system. The result is a reduction in costs compared to higher-pressure systems, due in part to ease of installation, reduced pipe diameters, smaller pumps and water tanks, less system weight, and minimal water discharge.

Low-pressure systems are well suited for the protection of rooms and spaces (residential, hotels, or offices, for example) that have light to moderate, Class A fire loads. Open-nozzle design is ideal for the protection of hazards involving Class B combustibles such as flammable liquids and gas turbines. Low-pressure systems using deionized water protect Class C hazards and are commonly tied into automatic shutdown of the power supply to the electrical equipment being protected by a fire-detection system.

MEDIUM PRESSURE

Medium-pressure system piping is subjected to pressures between 175 psi (1,207 kPa) and 500 psi (3,447 kPa).

Medium-pressure water mist systems are generally used for local applications. They are good, low-cost alternatives to the more efficient high-pressure systems. They protect diesel engines, generators, and boilers. Recently, however, the water mist industry has polarized its efforts within the high- and low-pressure markets and withdrawn from medium-pressure technology.[7]

HIGH PRESSURE

High-pressure system piping is designed for pressures of 500 psi (3,447 kPa) or greater. These systems use less water compared to low-pressure water mist systems or traditional sprinkler systems (up to 90%) to control and suppress fires. They are therefore more environmentally sustainable than these other two types of systems. High-pressure water mist systems deliver water from cylinders at a high velocity with the aid of inert gas (commonly nitrogen) via specially designed nozzles (see Figure 13.2). Their mist discharge has significantly greater surface area than low-pressure water mist nozzles and traditional sprinkler heads. High-pressure water mist systems, in general, also require fewer nozzles than low-pressure water mist systems. They typically use high-purity water due to the higher operating

FIGURE 13.2 High-pressure water mist cylinder rack. High-pressure water mist systems use less water compared to low-pressure water mist systems and traditional sprinkler systems. The cylinder on the far left contains nitrogen. *Source:* Ronald R. Spadafora

pressures and lower tolerance of water containing particulate matter, which could block the very small nozzle orifices.

High-pressure water mist systems also help to minimize the impact on natural resources. These systems use small, lightweight, bendable stainless steel tubes, giving this type of system a distinct advantage in retrofit settings, where buildings are being updated to meet new fire codes (see Figure 13.3). High-pressure water mist systems also take up less volume than traditional sprinklers, because their tubing is relatively easy to install and maneuver around corners and obstructions. The tube network is small enough to be hidden from view with crown molding to preserve the building aesthetics. The stainless steel tubes do not readily corrode or rust as does the black iron piping used for sprinkler systems. Their useful life can reach 100 years, versus less than 50 years for traditional sprinkler systems. The build-up of impurities and bacteria associated with the contents of black iron pipe is also avoided.

Following high-pressure water mist discharge, clean-up is faster, safer, and easier compared to traditional sprinkler systems. The cost of business interruption and downtime is minimized, allowing companies to recover more quickly after system activation.[8]

System Components

The various components that make up a water mist system are discussed ahead.

WATER SUPPLY

A pumped supply with strainers to prevent clogging of nozzles or a rack of cylinders containing water pressurized by stored gas in high-pressure cylinders are usually used.

FIGURE 13.3 High-pressure stainless steel water mist system piping (tubes). *Source:* Ronald R. Spadafora

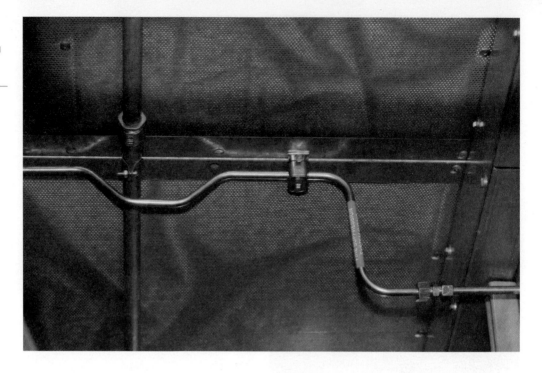

PUMPS/CYLINDERS

Centrifugal fire pumps are used in low-pressure water mist systems. High-pressure systems require positive displacement pumps. These pumps will be driven by electricity, diesel fuel, or gas motors. Fire pumps for water mist extinguishing systems are designed to exceed flow rate and pressure demands by a minimum of 10%. Pump installations must have a metal plate that provides the rated capacity and pressure of each pump.

In lieu of pumps, gas cylinders may be used. Self-contained systems consist of cylinders filled with compressed nitrogen or air and cylinders filled with deionized water. An air compressor keeps the discharge piping under pressure. When the system is in standby, water is in the cylinders and not in the pipes. This design avoids leakage in the piping network. These systems are ideal for the protection of Class C hazards, such as electrical transformers.

DISTRIBUTION PIPING

Piping material choice is contingent upon pressure design for the system. Low-pressure systems may use steel threaded pipe and fittings. Stainless steel tubing with compression fittings is commonly used for high-pressure systems. Filters and strainers are provided at each water supply connection or system riser.

NOZZLES

Water mist nozzles are defined in NFPA 750 as special-purpose devices containing one or more orifices designed to produce and deliver a water spray or as hybrid (water and gas) types. Single-orifice, low-pressure, thermally activated (closed) nozzles use quick-response frangible bulbs that have operating temperature ratings from 135°F to 650°F (57°C to 343°C). Frangible bulb nozzles are color coded based on their temperature classification. Other types of nozzles do not use frangible bulbs. They are opened by valves that can be set to function manually or automatically by an electrical, hydraulic, or pneumatic signal. Open, high-velocity nozzles contain multiple orifices and are designed to produce a unique, highly pressurized, micronized water spray pattern. A high-capacity filter shields against impurities in the water. Nozzles have total-flooding, local, and zone applications.

Low-Pressure Water Mist Nozzles

The design of low-pressure system nozzles still produces small water droplets. The large opening reduces susceptibility to clogging. Low-pressure systems with heat-sensitive nozzles ensure that only the nozzles within the immediate vicinity of the fire operate. These systems are suitable for the following applications:

Gas turbines
Semiconductor production equipment
- Wet benches
- Spin dryers
- Solvent tools
- Flammable liquid baths

Machinery Spaces, Auxiliary Turbine Hazards
- Oil pumps
- Oil tanks
- Fuel filters
- Generators
- Gearboxes
- Drive shafts
- Lubrication skids
- Compressors and diesel engine rooms

High-Pressure Water Mist Nozzles

These nozzles are manufactured from stainless steel due to the required high pressures of operation. They can produce 8,000 droplets of water from the amount of water required to produce a single droplet from a conventional sprinkler head (see Figure 13.4). The large number of small water droplets results in a vaporization rate of 400 to 1 compared with a conventional sprinkler head. A 10-minute water mist discharge will usually result in a residual water depth of less than .25 in. (6 mm) on floor surfaces.[9]

High-pressure water mist extinguishing systems have many applications in the following occupancies: telecommunications, power generation, manufacturing, industrial, transportation facilities, institutional, and residential.

Fire Detection and Alarm

Detection systems used to actuate water mist systems are automatic. Very early (air-aspirating) smoke detection systems are commonly installed in conjunction with water

FIGURE 13.4 High-pressure water mist nozzle protecting a backup rooftop generator in a local-application system. *Source:* Ronald R. Spadafora

mist systems. Detectors trigger alarm appliances to activate locally and may also initiate remote central monitoring station notification.

Actuation and Service Disconnect

An emergency release device activated by a single manual operation is required for all water mist systems. This device must be readily accessible, and its intended purpose must be clearly recognizable. A service disconnect key switch is yet another component.

Types of Systems

Although water mist systems differ significantly from standard sprinkler systems, their types are the same. The following subsections discuss the four types of water mist systems.

WET PIPE SYSTEM

Wet pipe systems use nozzles that actuate individually. Automatic, heat-sensitive nozzles are fitted into distribution piping that is permanently charged with water. These systems are used to protect areas in which temperatures are above 39°F (4°C) or are unlikely to fall below freezing. Wet pipe systems are generally used in accommodation and similar areas in which the fire hazard is solid combustible materials.

DRY PIPE SYSTEM

Dry pipe water mist systems also use automatic, heat-sensing nozzles fitted into pipework. However, the distribution piping, instead of being filled with water, is permanently charged with pressurized air or inert gas (nitrogen, for example). Water is kept from entering the system by a closed valve on the pump side. During a fire scenario, the heat-sensing nozzles in the fire area open, and pressure within the distribution piping begins to drop. This event allows the pump-side valve to open. Water is then released into the distribution piping, from which it is discharged through the open nozzle. Dry pipe systems are typically used in spaces subject to freezing.

DELUGE SYSTEM

A deluge system has open nozzles along its distribution piping and water flow controlled by closed valves. This type of system is designed to bring a large number of open nozzles into action simultaneously in the event of a fire, during which the valve opens. All nozzles are activated automatically through detection system initiation or manually via a mechanical override switch. Subsequently, a fire alarm control unit (FACU) operates to release the system's water supply through all the nozzles. Many current applications of water mist technology, both total flooding and local application, use this kind of system in spaces in which fuel fires can occur.

PREACTION SYSTEM

This system type is essentially the same as a dry pipe system, except that it is connected to fire detection. Preaction water mist systems consist of nozzles with individual actuating devices connected to piping filled with air. The system is designed to prevent water flow into pipework until both a reliable detector and a nozzle activate. Preaction systems are used to protect areas in which the risk of false discharge or leakage must be kept to an absolute minimum.

Water Supply Methods

Water mist can be delivered through system distribution piping to discharge nozzles using either a single- or twin-fluid method. The two methods are as follows.

FIGURE 13.5 Twin-fluid, low-pressure water mist system. *Source:* Ronald R. Spadafora

SINGLE FLUID

A single-fluid system creates water mist by delivering water through the nozzle under pressure from either pressurized cylinders or high-pressure pump system. Another type of single-fluid system stores unpressurized water in a stainless steel tank. A pump is installed to pressurize the water and flow it to the nozzle through distribution piping. These systems create high-pressure that can be conveyed long distances. Single-fluid systems feature a simple structure requiring low maintenance and reliable performance.

TWIN FLUID

A twin-fluid system generates water mist by mixing inert gas (nitrogen) or compressed air fed from separate piping to the point of discharge (see Figure 13.5). The discharge pressures for water and gas are separately controlled. The gas/air and water are mechanically mixed at the water mist nozzle, which delivers the extinguishing agent at low pressure. The nozzles consist of a water inlet as well as a gas inlet. When water is mixed with the gas at the nozzle, it degenerates into droplets.

Twin-fluid systems have good reliability and are less likely to clog. The main disadvantage of a twin-fluid system compared to a single-fluid system is cost. Twin-fluid systems require two supply lines instead of one as well as the storage of a sufficient quantity of compressed gas. In addition, changes in the ventilation conditions in the space being protected can have a greater influence on the fire-suppression performance of twin-fluid water mist systems than on single-fluid/high-pressure water mist systems because of less spray momentum at the nozzle.[10]

Applications

Several types of application systems are discussed next, including total flooding, local application, zoned application, and handheld extinguishers.

TOTAL-FLOODING APPLICATION SYSTEMS

Total-flooding systems are installed to provide complete protection of an enclosure or space. This is achieved by the simultaneous operation of all nozzles by manual or automatic means. Total-flooding water mist systems were initially designed for enclosures limited to less than 10,000 ft.^{3} (283 m^3). Recent technology advancements, however, allow protection of volumes up to 60,000 ft.^{3} (1,699 m^3). They are used to protect ordinary combustibles, flammable/combustible liquids, electronics and telecommunications equipment, machinery rooms, gas turbine enclosures, semiconductor equipment, and generator/transformer enclosures.

LOCAL-APPLICATION SYSTEMS

Local-application systems have been developed for the protection of individual areas or equipment. These systems do not require any enclosure and typically are not limited to specific dimensions. They are designed and installed to provide complete distribution of mist directly on or around the hazard to be protected. They can be actuated by automatic nozzles or by an independent detection system.

ZONED-APPLICATION SYSTEMS

Zoned-application systems are a subset of total-flooding systems. They protect a predetermined segment of the enclosure by the activation of a selected group of nozzles. Zoned application systems are designed and installed to provide complete water mist protection throughout a portion of the space. This objective is achieved by simultaneous operation of nozzles through manual or automatic means via automatic nozzles or activation by an independent fire detection system.

PORTABLE HANDHELD FIRE EXTINGUISHERS

Water mist portable extinguishers (see Figure 13.6) were introduced in the United States in the mid-1990s. At first, they were modified, stored-pressure water extinguishers that included a spray nozzle that allowed them to meet the requirement of American National

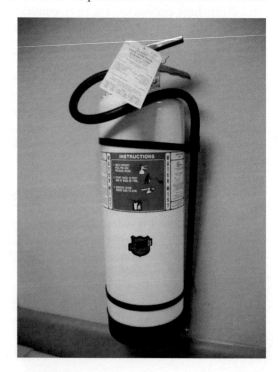

FIGURE 13.6 Water mist portable fire extinguisher.
Source: Ronald R. Spadafora

Standards Institute/Underwriters Laboratories (*ANSI/UL 711: Rating and Fire Testing of Fire Extinguishers*[11]) to earn a 2A, 1-B:C rating.

Today, portable water mist fire extinguishers have a specially designed conical nozzle that greatly enhances the cooling and soaking characteristics of the agent. The concept of an operator being protected from electrical shock while applying water to an electrically energized appliance is equivalent to a handheld fog nozzle being used by firefighters. The special nozzle of the water mist extinguisher generates water droplets in a spray form that is not a solid stream; therefore the electric current path is broken between the droplets, preventing current flow-back to the operator of the extinguisher.[12] Water mist portable fire extinguishers use deionized water rather than plain water, allowing them to easily comply with Class C conductivity test requirements under ANSI/UL 711. Water mist portable fire extinguisher applications include hospitals, healthcare facilities, MRI rooms, telecommunications areas, and electronic equipment–manufacturing occupancies.

WATER MIST WITH FOAM ADDITIVES

The extinguishing properties of foam, especially its ability to spread over flammable liquid pool fires, can improve the effectiveness of water mist for certain applications. Water–foam mixtures can minimize the reignition potential of the hazard. Foam can be considered in applications for which there are obstructions between the water mist nozzles and the fire. The water-foam mist system must have been fully fire tested in combination with this foam additive. The corrosive properties of foam concentrate on protected equipment must also be considered.

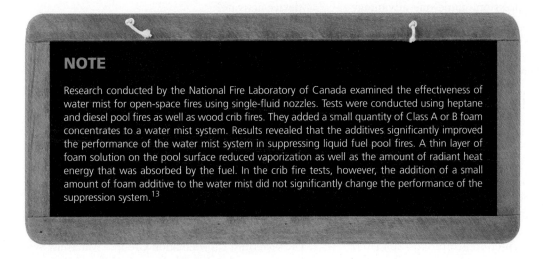

NOTE

Research conducted by the National Fire Laboratory of Canada examined the effectiveness of water mist for open-space fires using single-fluid nozzles. Tests were conducted using heptane and diesel pool fires as well as wood crib fires. They added a small quantity of Class A or B foam concentrates to a water mist system. Results revealed that the additives significantly improved the performance of the water mist system in suppressing liquid fuel pool fires. A thin layer of foam solution on the pool surface reduced vaporization as well as the amount of radiant heat energy that was absorbed by the fuel. In the crib fire tests, however, the addition of a small amount of foam additive to the water mist did not significantly change the performance of the suppression system.[13]

OCCUPANT SAFETY

Water mist using deionized or potable water is benign and does not present a toxicological or physiological hazard to human beings. These systems are therefore safe for use in occupied spaces. The use of additives or blends in the systems, however, should be assessed on a case-by-case basis. **Hypoxia atmospheres** can be produced, however, with twin-fluid systems and also in unusual cases in which steam generation is significant. It is important that normally occupied areas protected by a water mist system have safety plans in place denoting evacuation in the event of the activation of a fire alarm.

Hypoxia atmosphere ■ Reduced oxygen content in the air.

CHAPTER REVIEW

Summary

Water mist is an emerging fire protection technology. It has properties that make it an attractive choice for select applications, in particular those in which low water storage and consumption is beneficial. In locations in which water supplies are limited, water mist systems are an ideal alternative to traditional sprinkler systems. Using water mist systems to protect occupants in residential buildings and private dwellings, where most fire deaths in the United States occur, is still in the experimental stage.

Water mist protection systems should only be designed and installed for applications for which testing results have demonstrated fast and effective results. The characteristics of the specific application should be consistent with this testing.

Review Questions

1. Properly designed water mist systems can be effective on what classification(s) of fire?
2. Why is deionized water used in water mist systems?
3. Why can water mist droplets absorb heat energy more readily than conventional sprinkler heads?
4. Regarding fire suppression, name one advantage large droplets have over small droplets.
5. Name the four types of water mist systems.
6. Name the type of water mist system that is designed to bring a large number of open nozzles into action simultaneously in the event of a fire when a valve opens.
7. What type of water mist system is designed to prevent water flow into pipework until both a reliable detection device and a nozzle heat-activation device activate?
8. What mechanisms are used in single-fluid systems to generate pressurized water through their nozzles?
9. How do twin-fluid systems generate water mist?

Endnotes

1. National Fire Protection Association (NFPA), *NFPA 750: Standard on Water Mist Fire Protection Systems* (Quincy, MA: NFPA, 2010).
2. Ohio State University, "The Theory of Evaporation Enabling the Design of the Turbomister," www.turbomisters.com/wp-content/uploads/2011/07/design.pdf.
3. Jack R. Mawhinney, "Water Mist Fire Suppression Systems," in *Operation of Fire Protection Systems*, ed. Arthur E. Cote (Quincy, MA: NFPA, 2003), 415–464.
4. Maarit Tuomisaari, "Smoke Scrubbing in a Computer Room," *Proceedings of the Halon Options Technical Working Conference, Albuqerque, NM* (1999): 308–319.
5. Socialstyrelsen "The Fire on the Passenger Liner *Scandinavian Star* April 7, 1990—KAMEDO-report 60," www.socialstyrelsen.se/publikationer1993/thefireonthepassengerlinerscandinavianstarapril7-1990kamedo-report60.
6. Jukka Vaari, "Water Mist for Offshore Applications," *Business Briefing: Exploration & Production: The Oil & Gas Review* (2005): 36–38.
7. UK Watermist Co-ordination Group, "Watermist Systems—Compliance with Current Fire Safety Guidance," *Technical Guidance Note* no. 1, February 2012, www.fia.uk.com/download.cfm?docid=7b7df0b4-6729-4e23-b949f1d59b05d5b1.
8. Larry W. Owen, "Using High Pressure Water Mist Fire Protection Systems for Offshore Oil Drilling and Producing Facilities," *Business Briefing: Exploration & Production: The Oil & Gas Review* (2004): 56–58.
9. Ibid.
10. Zhigang Liu and Andrew K. Kim, "A Review of Water Mist Fire Suppression Systems—Fundamental

Studies," *Journal of Fire Protection Engineering* 10, no. 3 (2000): 32–50.
11. American National Standards Institute/Underwriters Laboratories, *ANSI/UL 711: Rating and Fire Testing of Fire Extinguishers* (New York, NY: ANSI, 2004).
12. ZhigangLiu and Andrew K. Kim, "A Review of Water Mist Fire Suppression Systems—Fundamental Studies," *Journal of Fire Protection Engineering* 10, no. 3 (2000): 32–50.
13. Adnrew K. Kim, Bogdan Z. Dlugogorski, and Jack R. Mawhinney, "The Effect of Foam Additives on the Fire Suppression Efficiency of Water Mist," *Proceedings of the Halon Options Technical Working Conference, Albuquerque, NM* (1994): 347–358.

Additional References

Lance D. Harry, "Next Generation Water Mist Fire Protection Systems: Safe, Effective, and Environmentally Sustainable Solutions" (press release, February 1, 2012), www.marioff.com/about-marioff/news-and-press-releases/next-generation-water-mist-fire-protection-systems-safe.

United States Fire Administration (National Fire Academy), "Portable Extinguishers: Water Mist Extinguishers," no. FP-2012-46REV *Coffee Break Training—Fire Protection Series*, November 13, 2012, www.usfa.fema.gov/downloads/pdf/coffee-break/cb_fp_2012_46.pdf.

CHAPTER 14

Fire Alarm and Detection Systems

Source: Ronald R. Spadafora

KEY TERMS

Automatic fire detectors, *p. 239*
Microprocessors, *p. 240*
Light-Emitting Diodes, *p. 240*
Stratification, *p. 248*
Thermal lag, *p. 249*
Eutectic, *p. 250*
Thermocouple, *p. 251*
Thermistor, *p. 251*

OBJECTIVES

After reading this chapter, the reader should be able to:

- Describe basic fire alarm components
- Understand the features of the fire alarm control unit
- Analyze conventional and addressable fire alarm systems
- Explain the study of fire detector initiating devices
- List and explain signaling and supervisory technology
- List and explain notification appliances
- Discuss the differences among fire alarm connection stations

Introduction

Fire alarms are used in many premises as part of a fire protection system. Fire codes over the years have expanded the requirements for fire alarm systems. Buildings and occupancies that commonly are required to install such systems include, but are not limited to, the following: office buildings, hotels, motels, schools, shelters, multiple dwellings, hospitals, marinas, commercial malls, theaters, dance halls, stadiums, arenas, industrial complexes, and manufacturing plants. The primary purpose of fire alarm systems is to warn building personnel and occupants of a possible dangerous condition. In some cases, they will also transmit signals that indicate a potential hazard to the local fire department. Often, this is accomplished via an approved central monitoring station company.

There are many different types of automatic initiating devices. Some require movement of water through piping, loss of air pressure from a system, or the operation of extinguishing systems in order to trigger an alarm. Many automatic initiating devices, however, are **automatic fire detectors** that provide early awareness of danger. Classification of these detectors is according to the element of fire they are programmed to monitor. Elements include aerosol (smoke), energy (thermal and light), and gas (carbon monoxide and natural gas). The classification of fire and the nature of the hazard being protected also are taken into consideration.[1]

Automatic fire detector ■ A device that automatically initiates an alarm signal once it detects the presence of the fire signature element (flame, smoke, heat, gas, and light) it was designed to detect.

Overview

A fire alarm system consists of components and circuits arranged to monitor and annunciate the status of alarm and supervisory signal-initiating devices. It also will begin an appropriate response. A fire alarm system typically is classified as automatic and/or manually activated. These systems must be inspected, maintained, and tested on a regular basis. Records must be kept by building personnel responsible for such tasks. Defective equipment should be replaced immediately by authorized service technicians.[2]

Fire Alarm System Basics

Fire alarm systems are classified according to the function they are designed to perform (see Figure 14.1). Basic components include the following:

- Fire alarm control unit
- A primary, or main, power supply
- A secondary, or standby, power supply
- One or more initiating/signaling/supervisory circuits to which fire alarm boxes, fire detectors, and supervisory devices are connected
- One or more fire alarm notification appliance circuits connected to audible/visual appliances
- Connection to a central station, proprietary supervisory station, or remote supervisory station[3]

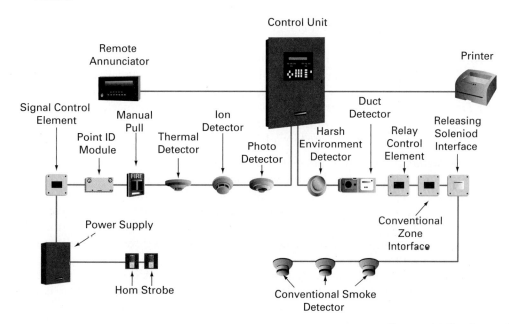

FIGURE 14.1 FACU operational functions

FIGURE 14.2 Fire alarm control unit. *Source:* Ronald R. Spadafora

Fire Alarm Control Unit

A fire alarm control unit (FACU) monitors inputs and controls output through various types of circuits (see Figure 14.2). The FACU processes all unusual conditions (alarm, supervisory, and trouble) and initiates an appropriate action. It is commonly housed inside a cabinet and is considered the "brain" of the alarm system. The FACU may consist of small **microprocessors**, electrical terminals, batteries, **light-emitting diodes** (LEDs), relays, and fuses, or it may be a sophisticated computerized terminal that employs touchscreen technology. The location of the FACU is subject to approval by the authority having jurisdiction (AHJ), typically following the guidelines set forth by *NFPA 72: National Fire Alarm and Signaling Code*.[4]

The FACU is commonly installed in an open lobby or entry area, within a closed or secure electrical or communications equipment room, or at the building's fire command station (FCS). The FACU should be secure from malicious or inadvertent tampering.[5] Typical conditions monitored by the FACU would be backup power battery status, HVAC failure, ground fault, open or short circuit on a wire, phone line outage, or internal component failure.

All of the equipment that makes up the FACU is Listed for use in fire protection service. The Listing standards are therefore more stringent than standards solely for electrical safety use. Listed FACUs are published in the current Underwriters Laboratories (UL) *Fire Protection Equipment Directory* and are tested for performance against *ANSI/UL 864: Standard for Control Units and Accessories for Fire Alarm Systems*.

Three types of signaling initiated by a FACU are as follows.

- *Alarm Signal:* An indicator initiated by a fire alarm initiating device such as a manual pull station, automatic fire detector, water flow device, or other mechanism of which activation is suggestive of the presence of an element of the fire. When a fire signal is generated, the FACU activates the building's audible and visual appliances connected to the fire alarm. It also may actuate and control certain building functions (fire dampers, electromagnetic door holders, and fire protection systems) as well as send a signal to a central monitoring station.
- *Supervisory Signal:* Indicates that a system (fire suppression) or equipment that is being monitored by the fire alarm system has been compromised or is in an

Microprocessors ■ Microprocessors incorporate the functions of a computer's central processing unit (CPU) on a single integrated circuit (IC).

Light-emitting diodes (LEDs) ■ Semiconductor devices that emit infrared or visible light when charged with an electric current.

abnormal state. A supervisory signal will audibly/visually arrive at the FACU or annunciator panel to indicate that the condition needs to be investigated and corrected.
- *Trouble Signal:* Is initiated by a system or device indicative of a fault in a monitored circuit, system, or component of the fire alarm system. A trouble signal will audibly/visually be denoted at the FACU.

Annunciator Panel

An annunciator panel is a visual display component of a fire alarm system that provides the identification and location of initiating devices. It can denote zone, floor, and type of device activated. It will also provide trouble signal indication that results from a fault in a monitored circuit, system, or component of the fire alarm system. Moreover, annunciator panels can indicate the status of elevators, HVAC fans and dampers, fire pumps, and voice alarm speakers. The annunciator panel is generally installed in a fire control room or adjacent to the main entrance to the building, where first responders can view the valuable information presented upon entering.

Conventional Fire Alarm System

Conventional fire alarm systems have one or more circuits routed through the protected space or building into a FACU. Initiating devices are located along a circuit to address the needs of the fire protection design. Selection and placement of these devices is dependent upon a variety of factors, including the need for automatic or manual initiation, the anticipated type of fire, and the desired speed of response.

To ensure that conventional systems are functioning properly, they monitor the condition of each circuit by sending a small amount of current through the wires. Should a fault occur, this current cannot flow and a trouble signal is registered at the FACU. During a fire incident in which a device has initiated an alarm, it will be indicated at the FACU along a circuit or zone (area). It will not, however, denote the specific device or location within this zone.

Emergency responders may need to search the entire zone serviced by the wiring to pinpoint where the initiating device is located, and service technicians must survey the entire circuit to identify the problem. Other drawbacks of conventional alarm systems include the need to operationally test each device in the fire alarm system to verify working condition. Advantages of conventional systems are that they are relatively simple to install for small- to medium-size buildings, and servicing does not require a large amount of specialized skill or training.[6]

Addressable Fire Alarm System

An addressable fire alarm system provides the status of the initiating devices that comprise the system. The FACU features detailed information about each device's location or "address." The location of an operating addressable device is commonly denoted through an annunciation panel or printout and is located according to building, floor, or fire zone.

Analog Addressable (Intelligent) Fire Alarm System

Analog addressable fire alarm systems represent the current state of the art in fire alarm technology. The location of the device that initiated the alarm can be identified precisely. Detector devices on these systems are individually electronically coded with a programmed address (number) on the wiring loop. The FACU shares information with each device by using its unique address number. Under normal conditions, the FACU monitors each device and evaluates replies to ascertain whether each device is functioning properly. This technology allows the FACU to make "intelligent" decisions concerning whether to act on the signal being received by a detector and go into alarm or not. The FACU may decide to send out a warning or prealarm to building personnel, providing the opportunity to investigate and replace a defective detector rather than transmit a false alarm.

Analog addressable alarm systems may be less expensive to install than conventional systems. They typically require less wiring and connections. Sensitivity testing costs are reduced for systems with smoke detectors, because they do not have to be checked manually as you would have to do with a conventional system.

Primary and Secondary Power Supplies

Fire alarm systems have both primary and a secondary power to ensure minimal interruption of operation. A brief review of these two types of electrical supply follows.

PRIMARY (MAIN) POWER SUPPLY

The main power supply for a fire alarm system is provided with a dedicated circuit from a local utility. In nonresidential applications, a branch circuit is dedicated to the fire alarm system. In remote areas, utility-generated power may not be available. Under these conditions, electrical power is commonly provided by a generator.

SECONDARY (BACKUP) POWER SUPPLY

Secondary power supply for a fire alarm system automatically supplies electrical energy to the system within 10 to 30 seconds whenever the primary supply is down or cannot provide the minimum voltage required for proper system operation. Secondary power will come from sealed, lead-acid storage batteries alone or from storage batteries in conjunction with dedicated, fuel-driven generators.

Initiating Devices

Many types of initiating devices and detectors are installed to perform a myriad of fire protection actions. Pull station alarm boxes are manual devices, usually wall mounted. In its simplest form, the user activates the alarm by pulling the handle down, which completes a circuit sending an alarm signal to the FACU.

Detectors are automatic devices that are activated by smoke, heat, flame, gas, or pressure. The term "spot" denotes that the detector's monitoring element is concentrated in the immediate area where it is installed and will generally not activate due to conditions occurring at a distance. Examples include certain smoke and heat detectors. "Line" types of detectors are devices in which detection is continuous along a path. Examples include projected beam smoke detectors and heat-sensitive cable.

Manual (Pull Station) Alarm Boxes

Manual alarm boxes are hand-operated devices used to initiate an alarm signal. A building occupant who notices a fire or emergency must activate the alarm with a demonstrative physical action. Fire alarm systems that are manually activated use fire alarm pull stations. These initiating stations are required to be located near the exits throughout the protected area so that they are conspicuous, unobstructed, and accessible. Approved

FIGURE 14.3 Double-action manual pull station.
Source: Ronald R. Spadafora

plastic covers are permitted to protect fire alarm manual pull stations and provide deterrence from false alarms. There are two types of manual fire alarms: single-action and double-action stations.[7]

Single-action stations require only one step to activate the alarm. Pulling down on a lever is a common way in which the alarm can be activated. Single-action alarm stations are often installed indoors. The cover on these alarm station serves as the lever. When the cover is pulled down, a switch inside the station closes, sending the alarm signal to the FACU.

Double-action stations require two steps to activate the alarm. The occupant must first break a glass, open a door, or lift a cover. The user then gains access to a lever or handle, which must then be operated (the second step) to initiate the alarm. Double-action stations tend to reduce the number of malicious false alarms, because they require more of a commitment from the operator to transmit (see Figure 14.3).

Smoke Detectors

A *smoke detector* is a device that identifies the unburned particles of combustion, which is typically an indicator of fire. Many high-occupancy buildings of large dimensions have their smoke detectors powered by an FACU. Most smoke detectors work either by optical detection (photoelectric) or by physical process (ionization). Many manufacturers are now producing smoke detectors that use both detection methods. Smoke detectors have variable sensitivity to smoke generated by fires. Photoelectric smoke detectors sense smoldering fires better than flaming fires, because the smoke from smoldering fires is typically composed of large particles. In contrast, ionization smoke detectors are better at detecting fast, flaming fires than slow, smoldering fires, because these fires generate much smaller (microscopic) particles.[8] Because fires develop in different ways and are often unpredictable in their growth, however, neither type of detector is always better than the other.

FIGURE 14.4 Spot-type smoke detector powered by a FACU. *Source:* Ronald R. Spadafora

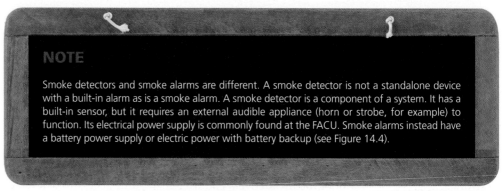

NOTE

Smoke detectors and smoke alarms are different. A smoke detector is not a standalone device with a built-in alarm as is a smoke alarm. A smoke detector is a component of a system. It has a built-in sensor, but it requires an external audible appliance (horn or strobe, for example) to function. Its electrical power supply is commonly found at the FACU. Smoke alarms instead have a battery power supply or electric power with battery backup (see Figure 14.4).

SPOT PHOTOELECTRIC LIGHT-SCATTERING SMOKE DETECTORS

These types of smoke detectors are the most common. In a photoelectric light-scattering detector, there is a T-shaped chamber with a light-emitting diode (LED) that directs a beam of light straight across the top, horizontal bar of the T. A photocell (light sensor), positioned at the bottom of the vertical base of the T, generates a current when it is exposed to light. Under smoke-free conditions, the LED beam crosses the top of the T in an uninterrupted straight line. No light strikes the photocell positioned at a right angle below the beam, and thus no current is generated. When smoke is present, however, the LED light is scattered by smoke particles, and some of the light is directed down the vertical base of the T, striking the photocell. When sufficient light hits the cell, the current generated triggers the alarm (see Figure 14.5).

SPOT PHOTOELECTRIC LIGHT-OBSCURATION SMOKE DETECTORS

These detectors position the photocell directly in the beam of the LED or laser light source. During fire conditions, smoke interferes with the light source transmission, and therefore the light received by the photocell diminishes. At a predetermined level, a signal is sent out to the FACU (see Figure 14.6).

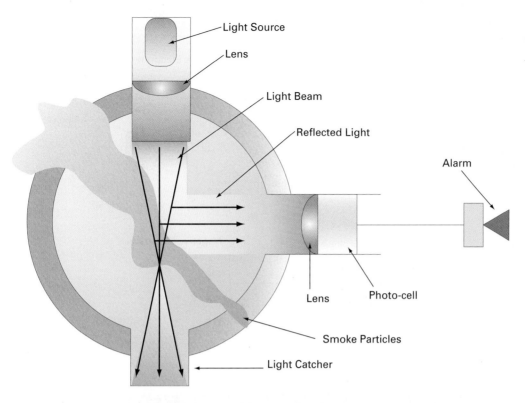

FIGURE 14.5 Typical photoelectric light-scattering chamber with smoke

SPOT IONIZATION SMOKE DETECTORS

An ionization smoke detector uses a radioisotope to produce ionizing radiation in the form of alpha particles into an ionization chamber (open to the air) and a sealed reference chamber. These ionized air molecules allow the passage of small electric current between charged electrodes. Under fire conditions, smoke particles pass into the chamber. The alpha particles attach themselves to the smoke, making them less able to carry the current. An electronic circuit detects the current drop and initiates the alarm process. Ionization detectors are more sensitive to active flaming fires than are photoelectric detectors (see Figure 14.7).

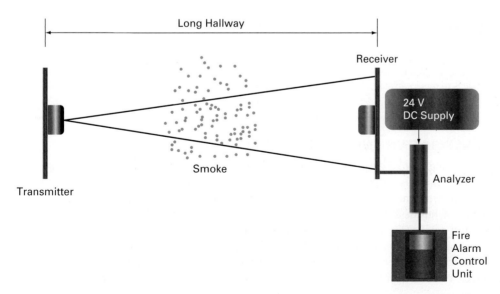

FIGURE 14.6 Typical photoelectric light-obscuration chamber with smoke

Chapter 14 Fire Alarm and Detection Systems

FIGURE 14.7 Ionization smoke detectors use an ionization chamber and a source of ionizing radiation to detect smoke. When smoke enters the ionization chamber, it attaches to the ions and neutralizes them, disrupting the current. The smoke detector senses the drop in current between the plates and sets off the horn.

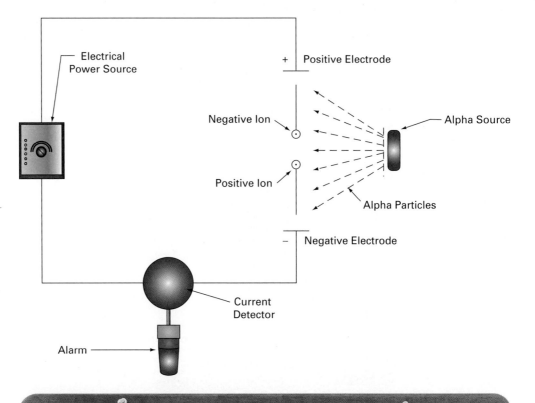

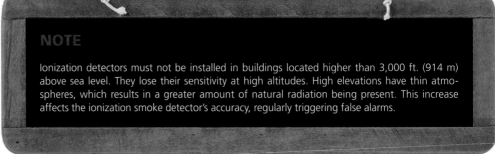

NOTE

Ionization detectors must not be installed in buildings located higher than 3,000 ft. (914 m) above sea level. They lose their sensitivity at high altitudes. High elevations have thin atmospheres, which results in a greater amount of natural radiation being present. This increase affects the ionization smoke detector's accuracy, regularly triggering false alarms.

 NATIONAL FIRE PROTECTION ASSOCIATION (NFPA) RECOMMENDATION

The National Fire Protection Association (NFPA) recommends both detector technologies for optimal protection, because it is impossible to predict the type of fire that can occur. One way to achieve this goal is to replace existing detectors with dual-sensor devices, which combine photoelectric and ionization technologies in a single unit.

AIR-ASPIRATING SMOKE DETECTION SYSTEMS

Air-aspirating smoke detector systems (see Figure 14.8) provide very early warning smoke detection solutions with continuous air sampling. This allows building personnel the time needed to investigate an alarm and initiate an appropriate response to prevent injury, property damage, or business disruption. Air-aspirating smoke detection systems have multiple alarm levels and a wide range of sensitivity that does not degrade or change over time. Minimal levels of smoke can be detected before a fire has time to escalate. Often standard, ceiling-mounted spot-detectors will not recognize the danger until it is too late since they passively rely on the products of combustion to rise up to them. High ceilings can create greater delays.

246 **Chapter 14** Fire Alarm and Detection Systems

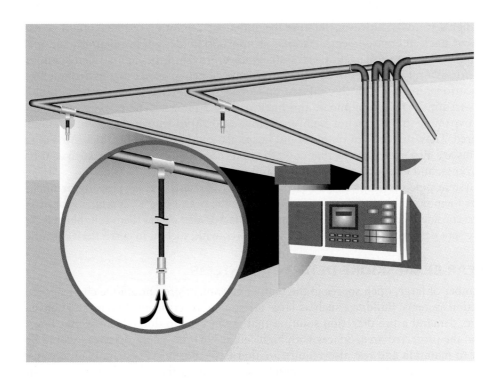

FIGURE 14.8 Air-aspirating systems are high sensitivity smoke detectors. *Source:* Air Sampling—High Sensitivity Smoke Detector (HSSD) from VESDA http://xtralis.com/. Used by permission of Xtralis Pty Ltd.

An air-aspirating system actively draws air into its detector through a pipe network by use of an air pump. Air samples first pass through a filter that removes dust and dirt before entering the detector chamber. From the filter, the air sample goes through a calibrated, light-scattering detection chamber. When smoke is present, it is instantly identified by the highly sensitive receiver system. The signal is then processed and presented via a hard-copy graphic display and/or alarm threshold indicators on a control unit (see Figure 14.9). Air-aspirating systems can also relay this information to a FACU.

FIGURE 14.9 Air-aspirating smoke detector system control unit. *Source:* Ronald R. Spadafora

Chapter 14 Fire Alarm and Detection Systems

High sensitivity, however, does not mean that air-aspirating systems cannot be used in dusty or unclean environments. As long as appropriate design, installation, and maintenance processes are followed, these systems have proven to be highly reliable and effective. Air-aspirating systems can accommodate a broad range of environments and applications, including both confined and open spaces, telecommunications buildings, waste-treatment facilities, mines, and nuclear power plants.[9] These systems are suitable for environments in which a highly sensitive and rapid smoke detection capability is required. They are ideal for clean rooms, hospitals, and healthcare facilities. Air-aspirating systems are also used in areas that contain goods easily damaged by fire, such as tobacco, electronics, and highly flammable liquid and gases.

System pipe networks can be easily hidden in an aesthetically pleasing manner. This makes them desirable for occupancies (office buildings and hotels) in which spot detectors can be considered aesthetically displeasing. This factor also makes them suitable in locations where spot detector systems can be easily tampered with, such as in correctional facilities.[10]

LINEAR-BEAM SMOKE DETECTOR SYSTEMS

A number of large, open spaces in the built environment present unique challenges to fire detection systems. Buildings such as large atria, airport terminals, hotels, and convention centers demand a fire-detection solution that is sensitive to diluted smoke but nonintrusive in the protected area. Spaces with high ceilings in which spot detectors would be difficult to maintain and test also require an alternative detection system. High ceilings can also cause **stratification** of smoke, whereby warmer air at the ceiling level prevents smoke from the fire plume, that has been cooled by entrained air, from rising.

A linear-beam smoke detection system measures light attenuation (obscuration) between a transmitter and a receiver. Within the transmitter, one or two wavelengths of light, infrared (IR) and ultraviolet (UV), flash periodically. The light is outside the human visible range. This light is focused into a compact beam by a lens. An adjustable mechanism inside the transmitter allows the beam to be accurately directed towards the receiver at the other end of the coverage area. The receiver also has an alignment mechanism and a lens that focuses the beam onto its light sensor. Transmitter and receiver can be installed parallel to each other or at an angle (the transmitter positioned on the designated floor and the receiver placed at a higher or lower elevation).

The electrical output from the light sensor is calculated so that the signal reduction due to smoke obscuration between the transmitter and receiver can be identified. The transmitter and receiver generally are wired together so that the light flash is in sequence with the receiver. Alternative installations may have both the transmitter and the receiver housed together in a single enclosure. In this design, the light beam is directed and aligned onto a remote reflector that returns the light back towards the receiver.

Linear beam smoke detectors can be difficult to align and prone to false alarms. Components normally require a very precise alignment upon installation. Some designs use software-controlled, motor-driven mechanisms to adjust alignment automatically and maintain an accurate signal path. Building movement caused by temperature changes will also affect alignment. Equipment booms, banners, dust in the air, insects crawling upon linear beam components, or birds flying into the light beam's path can trigger false readings and subsequent unfounded alarms.[11]

VIDEO-IMAGING SMOKE DETECTION SYSTEMS

Video-imaging detection systems are based upon computer analysis of video images. Motion patterns of smoke or flames are identified, alerting the system operator in a very short period of time. These detection systems therefore enable fast response to potential incipient fires. This action correlates to enhanced occupant safety, faster extinguishment, and a reduction in property damage.

Stratification ■ Predisposition of smoke in the fire plume to stop rising up to the ceiling space as it entrains cool air and takes on the temperature of surrounding air.

A video-imaging detection system is a volume sensor. It monitors a large area and has a high probability of successfully detecting smoke and flames. Video imaging is a good option for occupancies in which smoke spreads in an unusual manner. Examples include road and rail tunnels, mines, excavations, and areas in which forced ventilation is being introduced. Video imaging also has been used effectively inside aircraft hangars, shopping malls, factories, and warehouses in which air stratification can occur. Large open areas where there might be no heat or smoke in one defined space is yet another potential application of this detection system. A warehouse, for example, would have video detection installed to protect loading docks, packing areas, battery and forklift maintenance facilities, and high-hazard areas (aerosol rooms). In addition, increased security can reduce the risk of arson through the use of video-imaging detection technology.

Heat Detectors

A *heat detector* (see Figure 14.10) is a device designed to respond when the convected thermal energy of a fire increases the temperature of a heat-sensitive element. The thermal mass and conductivity of the element regulate the rate of heat flow. In order to predict response of a heat detector, two factors must be known: set-point temperature and response time index (RTI). Set-point temperature is the temperature the detector is designed to operate. The RTI is a measure of how fast the detector will respond based upon heat transfer to its sensing element per unit of time. It is evaluated during Listing tests.

Heat detectors have three main classifications of operation: rate of rise, fixed temperature, and rate compensation. Each type of heat detector has its advantages. For example, a rate-of-rise heat detector installed above a large, closed oven may cause a false alarm every time the oven door is opened due to the sudden influx of heat upon the detector. In this instance, a fixed-temperature heat detector should be installed. In contrast, a rapidly moving fire in an area with an abundance of flammable materials that is protected with a fixed-temperature heat detector can exceed the alarm threshold of the device due to **thermal lag**. For this scenario, a rate-of-rise detector may be preferred. Rate-compensation heat detectors, however, are designed to eliminate the thermal lag dilemma associated with a fixed-temperature detector as well as the problem of false alarms and the risk of missing slow heat release combustion that are inherent with rate-of-rise detectors.

Thermal lag ■ Describes a material's thermal mass with respect to time. A material with high thermal mass (high heat capacity and low conductivity) will have a large thermal lag. The surrounding air temperature must be considerably higher than the heat detector rating in order to raise the heat-detector element to the operating temperature. Due to thermal lag, fires generating high heat rapidly can exceed the alarm threshold of fixed-temperature heat detectors causing a delayed alarm.

FIGURE 14.10 Spot-type Heat detector with dust cap.
Source: Ronald R. Spadafora

Eutectic ■ The proportion of constituents in an alloy or other mixture that yields the lowest possible complete melting point.

SPOT FIXED-TEMPERATURE HEAT DETECTORS

Fixed-temperature heat detectors are the simplest type of heat detectors. They are designed to alarm when the sensing element reaches a predetermined temperature. Spot fixed-temperature heat detectors include fusible-element and bimetal varieties.

The *fusible-element* type of heat detector has a fusible link made of an **eutectic** alloy (a mixture of two or more metals) that melts at a predetermined temperature, releasing a spring under tension that makes an electrical circuit and actuates an alarm signal.

The *bimetal* type of heat detector comprises a bimetal strip that forms one part of an electrical circuit. Each metal has a different coefficient of expansion. One end of the strip is fixed, and the other end is free to move. During a fire, a rise in temperature will cause the free end to expand and deflect toward the metal with the lower coefficient of expansion. The deflective action is designed to close an electrical contact and initiate an alarm. A variation on this design utilizes two bimetalic strips (see Figure 14.11).

SPOT RATE-OF-RISE HEAT DETECTORS

These detectors initiate an alarm if the temperature inside its chamber rises faster than a predetermined design rate (normally 12°F [−11°C] to 15°F [−9°C]). Inside the chamber of the detector is a relief orifice to regulate the rate of temperature rise by modulating pressure. Under ambient conditions, slow temperature buildup will not initiate an alarm,

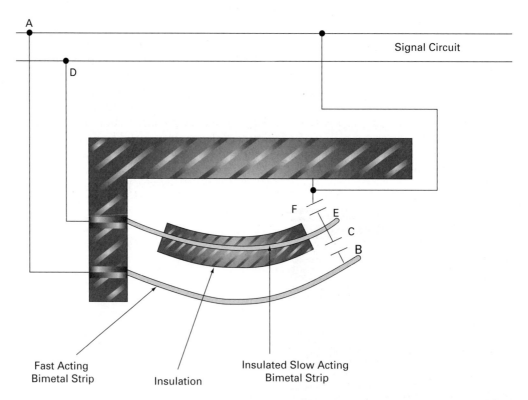

FIGURE 14.11 Rate-of-rise heat detector operation using two bimetallic strips. One bimetallic strip has been insulated to make it less sensitive to heat. On an appreciable temperature rise, the contact 'B' on the faster response bimetallic strip closes on contact 'C' of the slow acting bimetallic strip. This causes an alarm signal to be produced by an alarm circuit connected between points 'A' and 'D'. In the case of a very slow rate of rise in temperature, however, the difference in movements between contacts 'C' and 'B' will be such that a high temperature will be reached before an alarm sounds. Therefore, to ensure that the alarm signal is initiated when the designed activation temperature is reached, a second contact 'F' is provided on the slow acting bimetallic strip. At a designated temperature, contact 'E' closes to contact 'F' and an alarm signal is initiated.

because pressure increases are gradual and within the design parameters of the relief orifice. During a fire, however, heat transfer into the air chamber causes temperature to rapidly rise and air to expand. Pressure is now beyond the capacity of the relief orifice. A diaphragm, also inside the chamber, will bend as air expands. This action will close electrical contacts and initiate an alarm.

Heat detectors are also designed to use **thermocouples** (two dissimilar metals joined at their ends) to detect the rate of temperature rise. These *thermocouple rate-of-rise heat detectors* use the voltage differences that are proportional to the temperature difference at the metal junction to initiate an alarm.

Another type of rate-of-rise detector employs pneumatic tubing (diaphragm) filled with air and a relief vent. During normal circumstances, heated air will expand with the excess volume exhausted through the vent port prior to pressure buildup. Under fire conditions, however, air expanding at a rate that exceeds the relief capacity of the vent will build pressure and cause the diaphragm to move sufficiently to create an electrical circuit and initiate an alarm. *Pneumatic* rate-of-rise heat detectors are prone to a significant number of false alarms as a result of their operating mechanism.

Thermocouple ▪ A device consisting of two dissimilar metals, joined at their ends, that produce voltage when heated.

SPOT RATE-COMPENSATION HEAT DETECTORS

Rate-compensation heat detectors are designed to initiate an alarm when the temperature of the surrounding air reaches a predetermined level, regardless of the rate of temperature rise. The detector has temperature-sensitive struts made of brass alloy with electrical contact points attached within an aluminum shell casing. The coefficient of expansion of the struts is different than the casing. A slow increase in air temperature will cause the struts to heat up and resist expansion; no alarm will be initiated. If the temperature rises slowly to a high temperature, however, both casing and struts will be heated and the detector will perform like a fixed-temperature device. If the temperature rises rapidly, the struts do not have time to heat up, but the outer shell casing expands, causing the contact points to close and sound the alarm.

These detectors are designed to eliminate the thermal lag associated with a fixed-temperature detector, the problem of false alarms, and the risk of missing slow heat release combustion that are inherent with rate-of-rise detectors. They are considered the most reliable spot heat detectors (see Figure 14.12).

COAXIAL CONDUCTOR LINE HEAT DETECTORS

These detectors use a semiconductor material inside a stainless steel capillary tube. The tube contains a coaxial center conductor separated from the tube wall by a temperature-sensitive material known as a **thermistor**. During ambient conditions, a small amount of

Thermistor ▪ A temperature-sensing element composed of semiconductor material that exhibits a large change in resistance proportional to a small change in temperature.

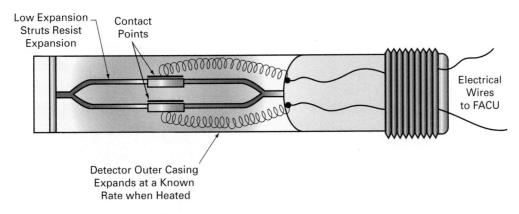

FIGURE 14.12 Rate-compensated heat detector. Robert M. Gagnon, *Design of Special Hazard and Fire Alarm Systems*, 2nd ed. (Clifton Park, NY: Delmar Cengage Learning, 2008), 369.

electrical current that is below the alarm threshold flows through the line circuit. Under fire conditions, however, as the temperature rises the electrical resistance of the temperature-sensitive material decreases, and more current flows to initiate an alarm. The looped wiring is connected to special controls that can pinpoint the spot of the temperature change. Once the coaxial conductor and thermistor cool to their normal temperature, the system will reset.[12] This detector can be designed to operate as a fixed-temperature device, a rate-of-rise device, or both.

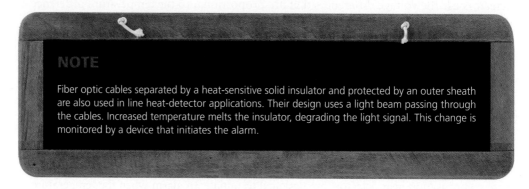

NOTE

Fiber optic cables separated by a heat-sensitive solid insulator and protected by an outer sheath are also used in line heat-detector applications. Their design uses a light beam passing through the cables. Increased temperature melts the insulator, degrading the light signal. This change is monitored by a device that initiates the alarm.

PAIRED-WIRE LINE HEAT DETECTOR

The paired-wire line detector is a fixed-temperature design. It consists of a parallel pair of wires wrapped in a braided-sheath exterior cable. The two wires are separated by thermally sensitive insulating material that is manufactured to melt at a specific temperature. The wires are arranged so that when the insulation melts due to fire they will come into contact with one another, and the short circuit created causes an alarm. Like the coaxial conductor, the paired-wire line device can determine the location of the alarm with great accuracy. The paired-wire system measures electrical resistance on the wires to identify the alarm point. Unlike coaxial conductor line heat detectors, however, paired-wire systems are not self-restoring. After the wire has fused, it must be replaced. The electrical continuity of the loop also must be tested to verify that it is within the manufacturer's recommended limits.[13]

Paired-wire line heat detectors and coaxial conductor line heat detectors are used to monitor hazards in facilities that have cable trays, conveyor belts, and bulk flammable liquid tanks and valves (see Figure 14.13).

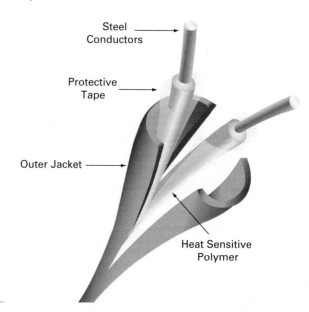

FIGURE 14.13 Paired-wire linear heat detector

Flame (Radiant) Detectors

Flame detectors are the fastest-responding type of detector; they initiate an alarm based on visible and invisible radiant energy (electromagnetic radiation) that is emitted from a fire (see Figure 14.14). Flame detectors are typically installed in occupancies in which fire hazards are high. These include fuel loading platforms, industrial processing facilities, hyperbaric chambers, and places where rapid flaming fires or explosions can occur. These detectors must be installed in an area devoid of obstructions, because they are line-of-sight devices. Spacing is determined based on fire protection objectives and should include the minimum size of flaming fire or spark that must be detected and the response time.[14] There are three general types of flame detectors: infrared (IR), ultraviolet (UV), and combination IR/UV.

IR flame detectors use a lens and filter system that screens out unwanted wavelengths. Radiant energy is directed onto a photovoltaic cell that is sensitive to IR. Solar radiation, however, can interfere with the activation mechanism of this device, leading to false alarms. IR detectors can respond to reflected levels of radiant energy off a wall and therefore may not have to be installed within line of sight of the hazard under certain conditions.[15] IR detectors are typically used to protect large open areas where flammable liquids are used and stored, posing the potential for a sudden, rapid-flaming fire. They are also commonly installed in waste-handling facilities, color-printing industries, and paper-manufacturing plants.

UV flame detectors are designed to measure radiant energy in the UV wavelengths. These detectors are solid-state devices that typically employ a vacuum photodiode Geiger-Muller gas-filled tube for detection of UV radiation produced by fire. They are effective on most fires, including hydrocarbons and metals. UV detectors also have explosion-suppression system applications. Close proximity of the detector to the hazard as well as careful installation are essential for proper performance. Potential false alarm sources include lightning, arc welding, X-rays used in testing of metals, and radioactive materials.[16] Uses include engine rooms in ships, factories affected by drafts/wind, and warehouses.

FIGURE 14.14 Infrared flame detector. *Source:* Ronald R. Spadafora

Combination IR/UV flame detectors sense radiant energy from both radiant spectrums to initiate an alarm. They usually require both IR and UV sensor response to activate, resulting in a better track record for false alarms. Combination IR/UV detectors are used if activation time is of lesser concern than nuisance alarms. They are found inside aircraft hangers, generator rooms (diesel and gas turbines), and paint-processing facilities.

Spark detectors are flame detectors that use a photodiode to sense radiant energy from sparks and embers. Radiant energy from sparks/embers is primarily infrared. They are installed on ducts and conveyors to monitor potential ignition sources and have very fast response times. Spark-detector activation when used in dust collectors and ventilation duct systems can open water spray nozzles for extinguishment purposes. Spark detectors also have applications in explosion-suppression systems.[17]

Spark detectors have a wide range of industrial fire safety applications in dust-collection systems for production machines, dryers, mills, sanders, ovens, grinders, pelletizers, buffers, and furnaces (which are all sources of sparks that can cause fires and explosions).

Gas-Sensing Detectors

Gas-detection systems provide alarm output signals to alert people and initiate corrective action. Gas detectors use a sampling system in which an air sample is extracted and directed to a sealed sensor for analysis. Sampling system components include a vacuum pump, sensors, flow meters, filters, and flow-control elements. *Point* gas detectors/sensors monitor a specified area within a facility and are strategically located for early detection of gas. They can be fitted with either combustible or toxic gas sensors. These detectors monitor a specific area or point within the facility and must be strategically located. They also require calibration for the gas type to be detected (see Figure 14.15).

FIGURE 14.15 Point gas detector/sensor (right) and rate-of-rise heat detector (left) at an LNG facility. *Source:* Ronald R. Spadafora

Line-of-sight gas detectors/sensors monitor the presence of combustible hydrocarbon gases within a beam of infrared light over a larger area than point detectors cover. The light beam is projected between modules. To ensure that the gas/vapor hazard passes through the light beam, the modules must be properly located and aligned. Like point detectors, line-of-sight detectors must be calibrated for the gas type to be detected. For detection of combustible gases, the most common choices are catalytic and infrared sensors. Catalytic sensors detect a wide range of combustible vapors, including hydrocarbon-based gases, hydrogen, and acetylene and offer good accuracy and fast response times. IR sensors can detect only hydrocarbon-based gases. IR sensors do not detect the presence of substances such as hydrogen, carbon disulfide, or acetylene.[18] Toxic gas detectors use electrochemical sensors to identify a wide range of different toxic gases in a variety of applications. They analyze the concentration of a target gas by oxidizing or reducing it at an electrode and measuring the resulting current.

Carbon monoxide (CO) detectors initiate an alarm based on an accumulation of carbon monoxide over time. Detectors may be based on an electrochemical reaction that produces current to produce an alarm, or a semiconductor sensor that changes its electrical resistance in the presence of carbon monoxide.

Carbon monoxide is a colorless, tasteless, and odorless compound produced by incomplete combustion of carbon-containing materials. It is virtually undetectable and is often referred to as the "silent killer." It is a highly toxic gas that attaches to the oxygen carrying red blood cells in the blood stream with an affinity 200 times stronger than oxygen. This action produces inadequate amounts of oxygen traveling through the human body.

Elevated levels of carbon monoxide can be dangerous to humans depending on the amount present and length of exposure. Smaller concentrations can be harmful over longer periods of time whereas increasing concentrations require diminishing exposure times to be harmful.

Table 14.1 provides an overview of the human health effects of carbon monoxide inhalation.[19]

Pressure-Sensing Detectors

Pressure-sensing detectors are components found in an explosion-suppression system. Prior to a detonation or deflagration, a pressure wave ripples out from the epicenter of the flame front. Explosion-suppression systems use pressure-sensing detectors to identify and act on the developing explosion. A properly operating pressure sensor will detect a sharp

TABLE 14.1 Carbon Monoxide Health Effects

CO IN AIR PPM	EXPOSURE	TOXIC SYMPTOMS
9	short term	*ASHRAE recommended maximum allowable concentration in a living area
35	8 hours	**OSHA maximum allowable exposure in the workplace
200	2–3 hours	Slight headache, fatigue, nausea, dizziness
800	45 minutes	Dizziness, convulsions, unconscious within 2 hours, death after 2-3 hours
1,600	20 minutes	Dizziness, nausea, death within 1-hour
6,400	1–2 minutes	Dizziness, nausea, death within 25-30 minutes
12,800	1–3 minutes	Death within 1–3 minutes

*ASHRAE was formed as the American Society of Heating, Refrigerating, and Air-Conditioning Engineers through the merger in 1959 of American Society of Heating and Air-Conditioning Engineers (ASHAE), founded in 1894, and The American Society of Refrigerating Engineers (ASRE), founded in 1904.

**OSHA is the Occupational Health & Safety Administration under the jurisdiction of the U.S. Department of Labor.

increase in pressure in one millisecond. The resulting signal from the pressure-sensing device sends a warning signal to a FACU, which in turn sends a release signal to the container holding the fire-suppressant agent.

In order to avoid false alarms and unnecessary activation of suppression agent into a protected vessel or connecting ductwork due to fluctuations in system pressure, these detectors are designed to analyze changes in pressure and determine if it fits the parameters associated with an explosion prior to sending an activation signal. Depending upon system design, the FACU can wait for a confirming signal or require three sensors to activate or "vote." A voting system increases system reliability.

Notification Appliances

Notification appliances alert the occupants of a building of a fire. There are two ways to do this. The first method is the general alarm. It activates all audio/visual appliances throughout the building when a fire is detected. For certain occupancies, this may be the only feature available. The second way notification appliances work is called the selective method. This type of system activates the audio/visual appliances only on the floor on which the alert signal activated and the floor immediately above. Audible and visual notification appliances are additionally used to indicate trouble signals in the fire alarm system. Fire alarm system notification appliances include horns, strobes, speakers, and gongs (see Figure 14.16).

There are two operating modes for fire alarm notification appliances: public and private. In the public operating mode, audible and visual signaling is intended for building occupants. In the private operating mode, however, these signals are intended for building personnel responsible for monitoring fire protection systems and occupant safety. When notified, building staff are trained in initiating emergency action and evacuation procedures. In general, public-mode signaling has more stringent requirements than private-mode signaling, because it is designed for building occupants who are not designated or trained to take on a safety role during an emergency.[20]

Fire Alarm Connection Stations

Automatically notifying building personnel and the fire department, as early as possible, is extremely important to effectively reduce losses due to a fire. Early detection helps to give in-house fire safety role players and firefighters time to respond and then to control and suppress the fire. The most common type of fire alarm connecting stations include central stations, proprietary supervisory stations, and remote supervisory stations.

FIGURE 14.16 Combination horn/strobe fire alarm.
Source: Ronald R. Spadafora

CENTRAL STATION

A central station supervises and monitors a protected property through the use of a transmitter device at the premises. The transmitter receives alarm signals and retransmits them to the central station. Depending upon the nature of the signal, the central station will arrange for any investigations or repairs needed or will relay an alarm to the local fire communications center or fire department. Central stations are normally privately owned and separate from the municipal fire department.

PROPRIETARY SUPERVISORY STATION

A proprietary supervisory station monitors property that is privately owned. The station is located at the premises being supervised. Specially trained personnel work at the station, normally on a 24/7 basis. The station verifies fire alarm signals received and notifies fire brigade members or the local fire department to respond. Inspection, testing, and maintenance of fire alarm equipment at the station and throughout the protected area is the responsibility of the building's owner.

REMOTE SUPERVISORY STATION

The section in NFPA 72 pertaining to remote supervisory station fire alarm systems does not contain any requirements for the physical supervising of the station itself. Alarm signals from distant protected properties that are received at the station are relayed to a public fire service communications center, a fire station, or a governmental agency with the responsibility to take prescribed action.

CHAPTER REVIEW

Summary

The fire alarm design professional establishes goals that use NFPA 72 as a guide to attain at least the minimal levels of protection mandated by the AHJ. Equipment specifically manufactured for these purposes are chosen and installed. Alarm connections to a central station, proprietary supervisory station, or a remote supervisory station allow for notification of maintenance personnel and first responders either directly or indirectly.

Review Questions

1. Name a drawback of conventional fire alarm systems regarding locating and servicing activated and defective initiating devices.
2. What are the operational parameters for a fire alarm system's secondary power supply?
3. What advantage do double-action fire alarm manual pull stations have over single-action stations?
4. What is the main difference concerning the operational capabilities of photoelectric smoke detectors and ionization smoke detectors?
5. Describe the functionality of a spot photoelectric light-scattering smoke detector.
6. What do air-aspirating smoke detector systems use to draw air into a detection control unit from the area being protected?
7. Linear beam smoke detectors can be difficult to align and prone to false alarms. Provide some examples of how unfounded alarms can be caused.
8. What type(s) of heat detectors are suitable for exterior exposures or interior spaces in which the ambient temperature changes during the course of the day?
9. What is the fastest-responding type of detector?
10. Spark detectors are flame detectors that sense radiant energy from burning embers. Where are these detectors commonly installed?
11. What type of vapors can gas-sensing IR devices detect?
12. Why are the requirements for public-mode signaling more stringent than those for private-mode signaling?

Endnotes

1. Maurice A. Jones, Jr., *Fire Protection Systems* (Clifton Park, NY: Delmar Cengage Learning, 2009).
2. Fire Department of the City of New York (FDNY), "Study Material for the Examination for the Certificate of Fitness for Fire Alarm Systems Inspection, Testing and Service Technician," S-98 (New York, NY: Fire Department of the City of New York, 2012).
3. Wayne D. Moore, "Fire Alarm Systems," in *Operation of Fire Protection Systems—A Special Edition of the Fire Protection Handbook*, ed. Arthur E. Cote (Quincy, MA: NFPA, 2003), 5–16.
4. National Fire Protection Association (NFPA), *NFPA 72: National Fire Alarm and Signaling Code* (Quincy, MA: NFPA, 2013).
5. United States Fire Administration (National Fire Academy), "Fire Alarms & Detection: Fire Alarm Control Unit Overview," *Coffee Break Training—Fire Protection Series*, www.usfa.fema.gov/downloads/pdf/coffee-break/cb_fp_2012_20.pdf.
6. A. Bhatia, "Overview of Fire Alarm and Detection Systems," *Course no. E04-021*, www.cedengineering.com/upload/Overview%20of%20Fire%20Alarm%20&%20Detection%20Systems.pdf.
7. Fire Department of the City of New York (FDNY), "Study Material for the Certificate of Fitness for Supervision of Fire Alarm Systems and Other Related Systems," S-95 (New York, NY: Fire Department of the City of New York, 2012).

8. Minnesota Rural Electric Association Loss Control Services, "Fire and Property Protection Placement and Spacing of Smoke Detectors," www.mrea.org/membersonly/.../Placement_of_Smoke_Detectors.doc.
9. Xtralis, "Very Early Warning Aspirating Smoke Detection in Nuclear Power Facilities," www.ansul.com/en/us/DocMedia/18434.pdf.
10. Xtralis, "VESDA by Xtralis Very Early Warning Aspirating Smoke Detection," http://xtralis.com/p.cfm?s=22&p=244.
11. Ron Knox, "Open-Area Smoke Imaging Detection (OSID)," *FPRF SUPDET 2010 Full Paper*, www.nfpa.org/~/media/Files/Research/Research%20Foundation/foundation%20proceedings/openareasmokeimagingdetectionosidknox.pdf. (Weblink used by permission of National Fire Protection Association)
12. United States Fire Administration (National Fire Academy), "Fire Alarms & Detection: Line-Type Heat Detectors," *Coffee Break Training—Fire Protection Series*, no. FP-2012-28, July 10, 2012, www.usfa.fema.gov/downloads/pdf/coffee-break/cb_fp_2012_28.pdf.
13. Ibid.
14. James C. Roberts, "Automatic Fire Detectors," in *Operation of Fire Protection Systems—A Special Edition of the Fire Protection Handbook*, ed. Arthur E. Cote (Quincy, MA: NFPA, 2003), 17–33.
15. Lawrence J. Marchetti, "Introduction to Fire Protection Systems," www.pdhcenter.com/courses/m110a/Module5.pdf.
16. James C. Roberts, "Automatic Fire Detectors," in *Operation of Fire Protection Systems—A Special Edition of the Fire Protection Handbook*, ed. Arthur E. Cote (Quincy, MA: NFPA, 2003), 17–33.
17. Ulrich Krause, ed. *Fires in Silos: Hazards, Prevention, and Fire Fighting* (Weinheim, Germany: Wiley-VCH, 2009).
18. Dave Opheim, "Selecting and Placing Gas Detectors for Maximum Application Protection," http://cmastage.det-tronics.com/Documents/ar-0800-gasopheim.pdf.
19. The Engineering ToolBox, "Carbon Monoxide—Health Effects," www.engineeringtoolbox.com/carbon-monoxide-d_893.html.
20. Robert P. Schifiliti, "Notification Appliances," in *Operation of Fire Protection Systems—A Special Edition of the Fire Protection Handbook*, ed. Arthur E. Cote (Quincy, MA: NFPA, 2003), 35–41.

Additional Reference

American National Standards Institute/Underwriters Laboratories, Inc., *ANSI/UL 864: Standard for Control Units and Accessories for Fire Alarm Systems*, 9th ed. (New York, NY: ANSI, 2008).

CHAPTER **15**

Smoke Control Systems

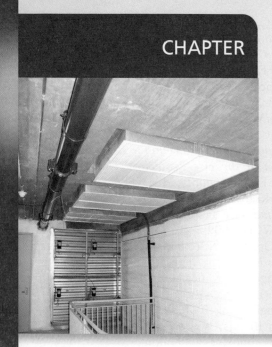

Source: Ronald R. Spadafora

KEY TERMS

Cleanroom, p. 260
Lethal concentration in air (LC$_{50}$), p. 262
Static pressure, p. 265
Rapid intervention teams, p. 271
Stack effect, p. 271
Neutral pressure plane, p. 273

OBJECTIVES

After reading this chapter, the reader should be able to:

- Understand the dangers of smoke and toxic gases
- Explain the design features of nondedicated smoke control systems
- Explain the design features of dedicated smoke control systems
- Identify the causes and ramifications of stack effect and reverse stack effect
- Describe passive smoke control protection components and devices

Introduction

Smoke control (management) systems are employed as life safety systems. They are often dictated by authority having jurisdiction (AHJ) code requirements for certain building types, sizes, and configurations. They are designed to protect occupant means of egress, building content, machinery, and the area surrounding a fire from contamination. They can also provide a tenable environment for firefighters to stage and begin operating in their effort to extinguish the fire. It is not unusual for smoke control requirements to be driven by insurance requirements rather than code due to the capital liability represented by loss of production and/or damage to equipment that could result from even a small fire. Many **cleanroom** facilities, for example, have extensive smoke management systems because adjacent processes could be severely compromised by relatively small amounts of smoke generated in a fire.

Cleanroom ■ An environment, typically used in scientific research, with a controlled level of contamination.

260

Overview

A common approach to limiting and controlling smoke movement involves pressurizing the areas on either side of the fire area and then exhausting the smoke inside the fire enclosure. This concept creates a "pressure sandwich." Although fresh air (oxygen) may be introduced to the fire area, increasing the smoke generated, most smoke control systems are aimed at protecting the occupants.

Smoke exhaustion, also known as purging, can be a part of a smoke control system. It should be understood, however, that simply evacuating large amounts of air from the fire area cannot by itself control smoke movement. This is because of the large quantities of smoke that are produced during a fire. Exhausting smoke also may not provide required airflow at open doorways and pressure differentials across barriers. Smoke exhaustion may dilute the smoke, but it cannot ensure that the air in the fire area will be breathable.

NFPA Smoke Control and Management Strategies

The strategy recommended in *NFPA 92: Standard for Smoke Control Systems* consists of creating differential pressures across zone boundaries to contain smoke to its area of origin.[1] This has proven effective at limiting the spread of smoke. Two techniques used when following these guidelines are mentioned below.

- Lower air pressure in the smoke-filled zone by controlling the airflow into it and turning *on* the exhaust fans from the area. This is a *negative* air pressure technique that pulls smoke out of the zone and vents it to the outside of the structure.
- Pressurize the areas and floors impinging on the fire by forcing air into the surrounding space and turning *off* all exhaust fans and exhaust dampers. This *positive* air pressure technique allows airflow into the burning zone via natural building openings (holes and cracks) and keeps the smoke from spreading.

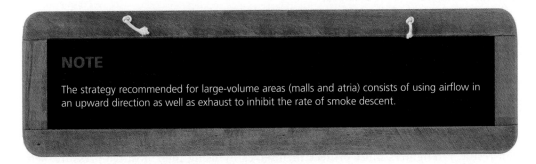

NOTE
The strategy recommended for large-volume areas (malls and atria) consists of using airflow in an upward direction as well as exhaust to inhibit the rate of smoke descent.

Smoke and Smoke Inhalation

Smoke consists of airborne solid, liquid, and gaseous particles emitted when materials undergo combustion. Smoke inhalation is the primary cause of death in victims of structural fires. It kills by a combination of thermal damage, poisoning, and pulmonary irritation.

An estimated 50% to 80% of fire deaths are the result of smoke inhalation rather than burns.[2] Combustion can use up the oxygen in a room near the fire. This can result in death by simple asphyxiation (lack of oxygen). Chemicals produced in a fire can cause injury to skin and mucous membranes, interfering with the normal lining of the respiratory tract. Swelling and respiratory distress can also result. Chemical irritants found in smoke include hydrogen chloride, sulfur dioxide, and ammonia. Toxic compounds produced in a fire can also interfere with the body's capability to deliver and use oxygen at the cellular level, causing cells to die. Carbon monoxide (CO) and hydrogen cyanide (HCN) are examples of such compounds.

Toxic Gases

One of the major lethal factors in fires is toxic gas. The predominant toxic gas is CO, which is generated from the combustion of wood, wood products, and other cellulosic materials. With the increased use of plastics, however, HCN has been thrust to the forefront as a major concern.

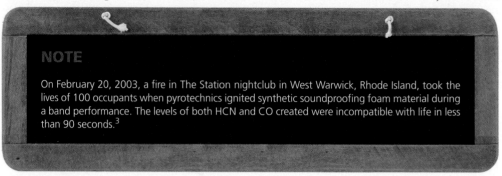

NOTE

On February 20, 2003, a fire in The Station nightclub in West Warwick, Rhode Island, took the lives of 100 occupants when pyrotechnics ignited synthetic soundproofing foam material during a band performance. The levels of both HCN and CO created were incompatible with life in less than 90 seconds.[3]

HYDROGEN CYANIDE

Hydrogen Cyanide (HCN) is emitted from plastics when heated. Some synthetic (man-made) materials that emit HCN include nylon, polyurethane, polystyrene, and melamine. Today, plastics are commonly used in products found in everyday life (foam insulation, furnishings, upholstery, appliances, draperies, carpets, and clothing, for example). HCN is much more toxic than CO.

Victims of a fire inhaling HCN associated with smoke often experience cognitive dysfunction and drowsiness that can impair the ability to escape. Exposure to low concentrations may result in confusion, anxiety, perspiration, headache, drowsiness, and rapid breathing. Exposures to higher amounts can cause tremors, irregular heartbeat (which can be delayed two to three weeks after exposure), coma, respiratory depression, respiratory arrest, and cardiovascular collapse.[4]

HCN causes rapid death by metabolic asphyxiation. The **lethal concentration in air** (**LC_{50}**, or the concentration estimated to kill 50% of the test population) depends on the duration of the exposure. Table 15.1 details the LC_{50} of HCN in air, estimated for humans.

Many firefighters can probably recall incidents during which they experienced dizziness, weakness, and rapid heart rate (just to name a few symptoms) and did not realize that they may have been exposed to cyanide. Research is increasingly pointing toward HCN as a toxic gas that is as much of a threat as CO to first responders.

Firefighters must fully understand the dangers and consequences of inhaling these toxic gases during fire operations. Both short-term and long-term negative health effects can result. Firefighters must protect themselves by using their positive-pressure, self-contained breathing apparatus (SCBA) throughout a fire until toxic gas levels have been assessed with atmospheric air monitoring.

Lethal concentration in air (LC_{50}) ■ The potential toxicity of chemical substances is commonly presented as their LC_{50}. It is the concentration of a substance that is lethal to 50% of the organisms in a toxicity test. LC_{50} can be determined for any exposure time, but the standard exposure period is 96 hours. Other durations are 24, 48, and 72 hours. As a general rule, the longer the exposure, the lower the LC_{50}.

TABLE 15.1	Hydrogen Cyanide—LC_{50} in Air for Humans
LC_{50} PPM	**EXPOSURE DURATION**
3,404 ppm	1 minute
270 ppm	6–8 minutes
181 ppm	10 minutes
135 ppm	30 minutes

Note: The average fatal concentration of hydrogen cyanide has been estimated at 546 ppm for 10 minutes.[5]

Source: "Hydrogen Cyanide–Lethal Concentration in Air for Humans (LC50 Concentration Estimated to Kill 50% of the Test Population)" by Hathaway, G.J., Proctor, N. H, Hughes, J.P., and Fischman, M.L. from *Proctor and Hughes' Chemical Hazards of the Workplace*. Published by John Wiley & Sons, © 1991.

PROVIDENCE, RHODE ISLAND, FIRES

Three fires occurring in Providence, RI, reviewed during a two-day period on March 23–24, 2006, demonstrate the danger of cyanide poisoning to firefighters. A fast-food restaurant fire on March 23 resulted in a firefighter complaining of headache, dizziness, difficulty breathing, cough, and random incoherent speech. When transported to a hospital trauma center, the firefighter had high levels of blood cyanide (57 micrograms [µg]/dl).* Four additional firefighters were later diagnosed with levels of blood cyanide above 20 µg/dl. During the next tour of duty (most of the firefighters who had worked the fast-food fire had been relieved), two additional fires occurred. The first fire, in an apartment building, resulted in no reported injuries. The second fire was inside a house. At this incident, a firefighter collapsed at the scene due to a heart attack. The firefighter was immediately resuscitated and transported to a hospital. Lab testing revealed that the firefighter had a blood cyanide level of 66 µg/dl. After consulting with doctors at the hospital, fire officials instructed all firefighters who had responded to any of the three fires and experienced symptoms of cyanide poisoning to go to the hospital for further examination. Twenty-seven firefighters had their cyanide levels tested. Eight firefighters tested above 20 µg/dl.[6]

*Normal levels of whole blood are believed to be between 0 to 20 micrograms per deciliter (µg/dl). Cyanide levels as low as 50 µg/dl in the blood have proven to be toxic. Blood cyanide levels of 250–300 µg/dl are fatal.[7]

Nondedicated Smoke Control Systems

Smoke control equipment can be either nondedicated or dedicated. Nondedicated systems use heating, ventilation, and air-conditioning (HVAC) system components to create differential pressure zones for smoke control. Zone differential pressurization is attained using 100% outside supply air without any return or exhaust in the outlying zones. The fire area's air supply, however, is cut off, and full exhaust of this zone to the outside air is implemented. The result of this action prevents smoke from the fire zone from spreading to other parts of the building.

SINGLE-ZONE HVAC SYSTEMS

Single-zone HVAC systems serve one floor or small area from a mechanical station located adjacent to the space. In this type of system, air is cooled directly in the exchange of heat from the refrigerant. When used for smoke control (see Figure 15.1), each individual system serves one fire zone; therefore, in multifloor buildings several single-zone systems must be employed.

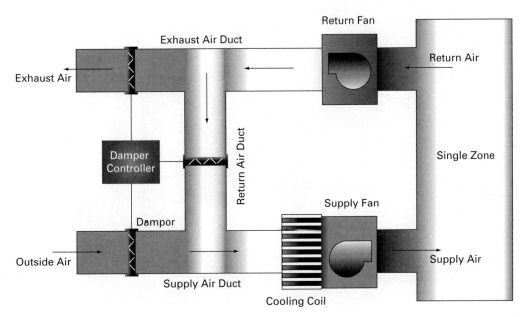

FIGURE 15.1 Single-zone HVAC system with direct outside air supply and direct exhaust air

CENTRAL HVAC SYSTEMS

Central HVAC systems provide conditioned air to multiple floors in high-rise buildings and therefore are used for smoke control on more than one floor. Typically, chilled water is the cooling medium. These systems use extensive ductwork to distribute air.

MGM GRAND HOTEL AND CASINO FIRE

The MGM Grand Hotel and Casino fire (see Figure 15.2) occurred in the early morning hours of November 21, 1980, in Las Vegas, Nevada. The origin of the fire was determined to be in one of its five casino restaurants: The Deli, located on the second floor. This fire killed 85 people and injured approximately 650 others. Fire spread across the casino, which had no installed fire sprinklers. Smoke then spread into the hotel tower section of the MGM. At the time of the fire, approximately 5,000 people were in the building, a 23-story luxury resort with more than 2,000 hotel rooms. The Clark County Fire Department was the initial first responder agency to arrive. Military helicopters were a major part of the rescue effort that removed 1,000 people from the roof of the building.

Most of the fatalities were located on the upper floors. These deaths were the result of smoke inhalation. Openings in vertical shafts (elevators and stairwells) and seismic joints allowed toxic smoke to spread to the upper levels of the structure. This fire demonstrated how buoyant gases can travel rapidly through a building in which protection of floor penetrations does not exist or is not maintained. Faulty smoke dampers within the ventilation duct network of the central HVAC system also allowed smoke to circulate, accelerating the spread of the poisonous gases. The HVAC system did not shut down automatically during the fire. Deaths occurred in the stairwells, where the only re-entry doors were at the roof and on the ground floor level. Victims also died from smoke inhalation while they were sleeping.

This fire led to the general publicizing of the fact that during a building fire smoke inhalation is a more serious threat than flames. Fire death statistics revealed that 75 people succumbed from smoke inhalation and carbon monoxide poisoning, four from smoke inhalation alone, three people died from burns and smoke inhalation, one person died from burns alone, and one person died from massive skull trauma caused by jumping from a high window.[8]

FIGURE 15.2 Fire at the MGM Grand Hotel and Casino in Las Vegas, Nevada. *Source:* Keystone/Stringer/Hulton Archive/Getty Images

Dedicated Smoke Control Systems

Dedicated equipment is designed to be used only for smoke control. It features **static pressure** controls and safety devices and is sized to accommodate the required airflows. Dedicated systems are not impacted by HVAC design constraints involving energy, operating costs, noise level, or the need for comfort. Examples of dedicated smoke control systems include stairwell pressurization, elevator shaft pressurization, smoke shaft exhaust, and atrium smoke control.

Static pressure ■ When pertaining to airflow, static pressure is used to evaluate the amount of impact various ventilation system components have within a given system. For mechanical ventilation systems, fans create positive static pressure to move air, whereas other components create negative static pressure that causes resistance to air moving through a system.

STAIR PRESSURIZATION SYSTEMS

Stairwells are frequently the primary escape routes for the occupants of a building in a fire emergency. In addition, the stairwells are normally the primary access routes for the firefighting teams operating at the fire. It is critical that stairs remain clear of smoke for as long as is necessary to safely evacuate the building and bring the fire under control. Many high-rise buildings employ a stairwell pressurization system to keep stairs clear of smoke while operating at a low enough pressure that an average person can open the doors to enter or exit the stairwell on their way to safety.

The main objective of a stairwell pressurization system is to maintain a livable environment in the stairwell. This is accomplished by keeping the stairwell static pressure above that of the smoke zone area. A static pressure that is too low in relation to the smoke zone will not provide enough resistance against the tendency of smoke to enter into the stairwell.

A stairwell pressurization fan (see Figure 15.3) is designed to supply sufficient airflow to overcome the air losses of the stairwell. Leakage can occur through clearances around floor exit doors and the normal porosity of masonry. Fans must have the capacity to maintain the

FIGURE 15.3 Stairwell pressurization fan.
Source: Ronald R. Spadafora

FIGURE 15.4 Stairwell pressure/smoke-relief dampers. *Source:* Ronald R. Spadafora

stairwell pressure with a nominal number of floor exit doors open and the airflow out a pressure relief damper at the top of the shaft (see Figure 15.4). The relief damper should be remotely operated to prevent stairwell overpressurization and vent smoke.

Fans (propeller, centrifugal, or axial type) can be positioned at the bottom, top, or at both ends of the stairwell. Fans mounted at the bottom of the stairwell are less likely to draw smoke into the stairwell than top-mounted fans. Bottom-mounted fans, however, can funnel airflow out of an open ground-level exit door rather than pressurizing the stairwell. Fan air intakes must supply 100% outside air to the fan. Numerous injection points, set up with multiple fans or a ducted air-distribution system employing modulating dampers, are normally required for adequate static pressure when buildings are more than eight stories. The dampers should be controlled based upon the differential pressure between the building interior and the stairwell. Ducted air distribution systems must have a smoke detector that will stop the fans if smoke is detected.[9]

There are two types of stairwell pressurization systems: compensated and uncompensated. A compensated system can adjust the airflow to maintain a specific static pressure level. This system uses multiple injection points that respond to changes in static pressure levels caused by building occupants opening and closing stairwell doors (see Figure 15.5). Noncompensated systems do not have static pressure control provisions.

ELEVATOR SHAFT PRESSURIZATION SYSTEMS

Elevator shaft pressurization systems are designed for several purposes. They prevent smoke from traveling upward through the shaft to various floors above. This can provide occupants with the opportunity to use the elevators for evacuation. Elevators employed for this purpose must be under the approval and supervision of building safety personnel according to preplanning guidelines. Elevators used for egress can be especially helpful in occupancies in which a large number of nonambulatory persons live and work. This benefit will keep stairwells unobstructed for fleeing ambulatory occupants. Another purpose of elevator shaft pressurization design is that elevators which service the fire floor may be used safely by first responders operating at a fire. It allows them to transport manpower, tools, and equipment to the upper floors of high-rise buildings.

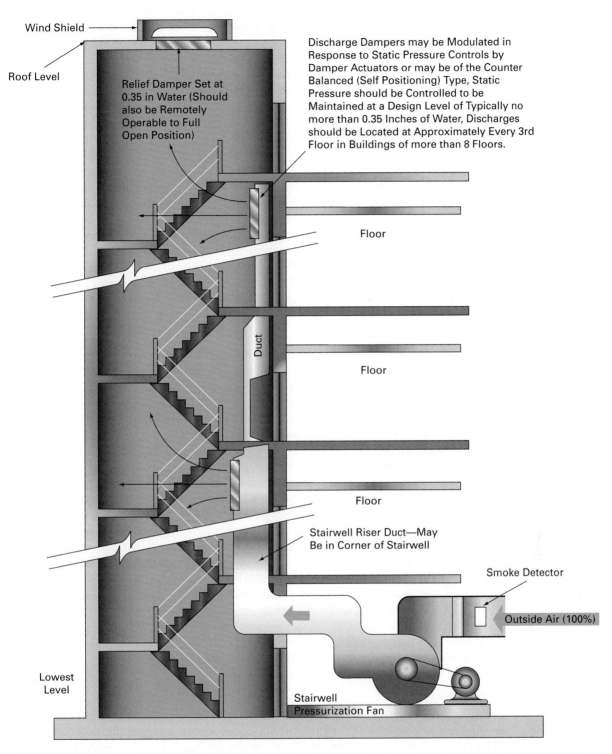

FIGURE 15.5 Compensated stairwell pressurization system with multiple injection. Siemens Building Technologies, Inc., *Smoke Control System Application Guide*, 125-1816, rev. 6, August 2000, p.18.

Although small amounts of smoke may be tolerated in the elevator if it is being used by firefighters, smoke in an elevator used for evacuation purposes is not acceptable. Therefore, elevators designated for evacuation must have smoke-protection barriers at each elevator floor lobby that prevent smoke in the hallways from entering. A successful elevator pressurization system is difficult to attain and is complicated by many variables. Unlike stairwell doors, elevator doors commonly have significant clearance that allows substantial amounts of air to leak out of the shaft. Several elevator doors per floor can compound this leakage problem.

Moreover, the "piston effect" created by the elevator moving up and down the shaft creates problems for efforts to maintain control over the static pressure within the elevator shaft. This has an impact in controlling smoke in the elevator car and in the areas adjacent to the shaft. Elevator cars moving within the shaft create higher pressures ahead of the car and lower pressures behind it. The decreased pressure behind the moving elevator car can draw smoke into a normally pressurized shaft. This problem is most pronounced in single-car shafts.[10] One or more fans are used to control the elevator pressurization system. In general, the design calls for the pressurization fans to discharge directly into the elevator shaft near the main floor. Airflow is directed upward toward a pressure-relief damper at the top of the shaft. Components and the control system for an elevator pressurization system should be on the building's emergency power circuit if one is provided.

SMOKE SHAFT SYSTEMS

A smoke ventilation shaft is a vertical enclosure of noncombustible construction that extends from the bottom to the top of a building and has openings at the top to the outside air along with openings to floor spaces at each story. It provides a mechanical way to exhaust smoke from a selected floor with dampers or automatic opening devices. They can be placed into operation automatically or controlled remotely. Smoke shaft systems are used to protect means of egress from smoke entry and to reduce smoke levels in corridors and lobbies. The smoke shaft is designed to keep smoke from spreading laterally and throughout the building. The venting action of the smoke shaft creates a suction pressure that induces the flow of air from adjacent areas.[11]

Dampers are kept closed at each building zone connection to the smoke shaft. Upon detection of smoke, the fire zone smoke shaft damper fully opens and the smoke shaft fan is activated. Only the damper on the fire floor opens to provide smoke release.

A smoke shaft exhaust fan typically has cubic feet per minute (CFM) capacity equal to the volume of the largest floor area served by the smoke shaft. Smoke shafts provide a natural path for the upward movement of smoke. Used in conjunction with pressurization of adjacent nonsmoke zones, they assist in controlling smoke spread by reducing the static pressure on the fire floor as well as providing greater exhaust capabilities. A caution about smoke shafts, however, concerns their tightness. Openings (dampers, for example) in the shaft do leak. If they are located on every floor, it is unlikely that the smoke shaft will be effective for buildings higher than 650 ft. (200 m).[12]

ATRIUM SYSTEMS

An atrium (plural: atria) is a large, multi-story open space with a glazed roof. It is commonly built just beyond the main entrance doors within a larger, multistory building. Building codes describe the atrium as an opening connecting two or more stories other than enclosed stairways, elevators, hoistways, escalators or other equipment that is closed at the top and not defined as a mall.

The objective of the atrium smoke control system is twofold: to maintain the bottom of the smoke layer at a specific height above the highest walkway open to the atrium (generally 6 to 10 ft. [1.8 to 3 m]) and to do so for a specified period of time (typically 20 minutes). Activation of an atrium smoke control system for a fire involves turning *on* all exhaust and supply fans (they are normally *off*) and opening all smoke dampers (they are normally closed). Active smoke control systems (sprinkler heads and supply/exhaust fans) are activated by smoke/heat detectors and sprinkler water flow devices.

A large, unenclosed shaft that extends upward through multiple floors is contrary to fire safety, which asserts that compartmentalization is required to limit the spread of fire and smoke from the point of origin. Atria present the fire service with some unique design challenges. Without the benefit of a smoke-management system, fire originating at the base of an atrium will produce smoke and toxic gases that rise up to the ceiling level. From this area, with nowhere to travel vertically, smoke moves horizontally and then downward as the volume of smoke increases. In time, smoke will enter into the occupied floors that are on and connected to the atrium, threatening occupants of the building. Occupants using escape routes within the atrium space during fires may not reach an exit.

Fire protection systems and firefighting concerns are two important aspects of atria design. Designers often build their smoke control systems based on an evaluation of actual anticipated fuel loads and heat release rate (HRR) within the atrium. Specific objectives must include providing a viable means of egress for occupants and facilitating occupant rescue. Smoke temperature and toxicity of the materials expected to put in atria must be examined to ascertain if there is sufficient time in the design for safe egress of occupants.

Airflow and Exhaust

Atrium smoke control components aim to prevent smoke from filling the atrium from a fire originating in it or on an adjoining floor. They rely on airflow and exhaust rather than differential pressurization to maintain the smoke layer at its apex above occupied floor/balcony areas (see Figure 15.6). The height of a smoke layer can be controlled naturally or mechanically. Natural venting at the top of the atrium space, which uses automatically

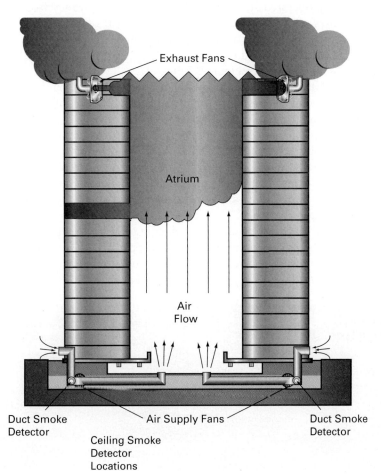

FIGURE 15.6 Atrium smoke control system. *Siemens Building Technologies, Inc., Smoke Control System Application Guide, 125-1816, rev. 6, August 2000, p. 25.*

FIGURE 15.7 Atrium exhaust fans at ceiling level.
Source: Ronald R. Spadafora

opening or manually operated roof hatches, relies on the buoyancy force caused by the elevated temperatures of the layer of hot gases. Mechanical venting is more efficient with regard to the venting rate and less affected by wind than natural venting. Fans along the ceiling are the primary means used to mechanically exhaust smoke and other products of combustion from atria spaces (see Figure 15.7).

Supply air inlets that provide makeup replacement air for air that has been exhausted are commonly installed at the first floor or lowest level of the atrium. Atria over a certain height may be mandated to have more supply fans for additional outside supply air. These fans will push air in an upward direction toward the ceiling exhaust system.

Incoming air should not enter the atrium through exhaust ventilators. This design reduces the flow of smoke leaving and thereby drastically reduces the ventilator's efficiency. In addition, as the hot exhaust air mixes with cooler incoming air coming through the ventilator the smoke layer is disturbed and cooled, lowering its buoyancy. Therefore, separate inlet ventilation should be provided. The air inlet must be installed at low level and remote from the smoke layer. This can be achieved by the fire protection engineer through air inlet ventilators in other zones or doors or windows that open automatically. To determine the appropriate size of supply and exhaust fans, they use the equations in NFPA 92B, *Standard for Smoke Management Systems in Malls, Atria, and Large Spaces*.[13]

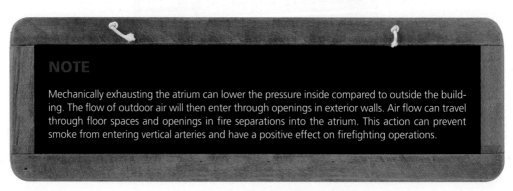

> **NOTE**
>
> Mechanically exhausting the atrium can lower the pressure inside compared to outside the building. The flow of outdoor air will then enter through openings in exterior walls. Air flow can travel through floor spaces and openings in fire separations into the atrium. This action can prevent smoke from entering vertical arteries and have a positive effect on firefighting operations.

Visibility

The reduction in visibility is a major hazard in atrium fires that needs to be addressed in any smoke-management design. Occupants evacuating through smoke may become confused and disoriented. In general, if there is enough visibility through the smoke for people to see the emergency exits, then toxic products are unlikely to prevent them from escaping. To achieve adequate visibility, building occupants should be physically separated from the smoke, or the smoke concentration should be limited.[14]

Firefighting Planning, Strategy, and Tactics for a Fire in an Atrium

Atria design error can occur when smoke control systems are installed based on only one fire scenario. In reality, multiple fire situations should be anticipated. For example, if an atrium is used temporarily as an exhibition space or showroom venue, this adds a substantial fuel load that was not calculated into the design. Chief Officers must be vigilant in their recommendations and guidance to building owners, stakeholders, managers, and agents regarding such ancillary use. Additional fuel loading should be kept to a minimum to ensure that atrium smoke control and other fire protection systems will not be overwhelmed and defeated.[15]

Planning is important to enhance situational awareness. Chief Officers should ascertain how many street entrances lead into the base of the atrium. Can fire apparatus access these streets? How many hose lengths are required to stretch from the engine to reach atria base areas in lieu of using standpipe outlets? Also, predetermine the maximum floor levels that apparatus ladders can reach on all sides of the atrium. Exterior ladder placement will be essential to successfully completing firefighting operations in a structure containing an atrium.

During fire operations, the chief officer in command of the incident should request a building engineer to respond to the command post. This person's expertise regarding the atrium's smoke control equipment can be extremely useful. Properly activating fans, vents, hatches, and exhausts will allow firefighters to complete vital objectives successfully. Floor plans can also aid in locating atrium living spaces and their relationship to means of egress. Firefighters should note which staircases, elevators (passenger and freight), and passageways are interconnected to the atrium. Formulate strategy and tactics based on the location of the fire and where fire and smoke is most likely to travel into occupied areas. Public announcements over building loudspeakers can also prove valuable. Conveying to occupants what actions they should perform will help to protect them as well as facilitate firefighting operations.

Chief Officers must also develop tactics to prevent units from operating conflicting hose lines. This dangerous situation can easily occur, because atria often are designed with several entrances. Engine Officers must communicate with each other regarding apparatus placement, point of entry, hose line stretches, and hose stream direction to enhance safety. Engine companies should be teamed up when stretching hose lines. Anticipated long hand stretches dictate this policy. In general, large-diameter hose lines should be stretched for fires of unknown size. Stairs that service the atrium must be determined and assigned proper designations (attack or evacuation). Ladder companies must also team-up to accomplish their goals. Thermal-imaging cameras and search ropes are two primary tools these firefighters should carry. Multiple **rapid intervention teams** (RITs) may be required based upon the atrium's height and dimensions. When warranted by fire conditions, consider positioning a RIT inside the building in an uncontaminated area, to monitor firefighters working in a dangerous area on the upper floors.

Principles of Smoke Control

Smoke is also driven by the **stack effect**, which is created by differences between the internal and external temperature of the building. A review of both the positive and negative stack effect phenomena follows.

Rapid Intervention Teams ■ Firefighting companies designated to stand by and be available as a rescue team at a fire or emergency scene for operating firefighters who may become injured, lost or trapped or who find themselves in immediate danger.

Stack effect ■ The movement of air into and out of a building. The stack effect, also known as the "chimney effect," is driven by buoyancy, which occurs due to a difference in indoor-to-outdoor air density that results from temperature and moisture differences. The result is either a positive (positive stack effect) or negative (negative stack effect) buoyancy force.

POSITIVE STACK EFFECT

When the weather is cold, air conditioning for the building will warm interior air for human comfort. This heated air will generally rise into vertical openings (elevator shafts, utility shafts, and staircases). Under these conditions, for a fire on the lower half of a high-rise building that is below the Neutral Pressure Plane (NPP), smoke will be drawn into the hallways and up these voids. This is known as a positive stack effect (see Figure 15.8). The lower the fire floor and greater the difference between the temperature inside the building and the temperature outside the building, the greater the pull of smoke will be.

POSITIVE-PRESSURE VENTILATION FOR FIREFIGHTING PURPOSES

During the past 10 years, the National Institute of Standards and Technology (NIST) has conducted numerous experiments that examine the effectiveness of positive-pressure ventilation (PPV) fans for firefighting. PPV is used as a tactic to facilitate firefighter hose-line advance onto a fire by enhancing visibility and driving back smoke and heat. It is also used to ventilate a building subsequent to extinguishment.

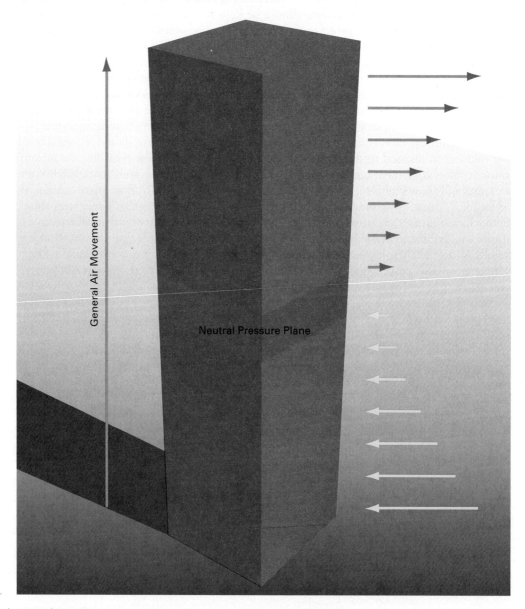

FIGURE 15.8 The positive stack effect will draw smoke into the hallways on the lower sections of high-rise buildings and up vertical shafts. The lower the fire floor and the greater the difference between the inside and outside air temperatures, the greater the pull will be. This will occur during winter and will often be much stronger than a negative stack effect

NIST POSITIVE-PRESSURE VENTILATION EXPERIMENTS

In 2008, fire experiments conducted by the National Institute of Standards and Technology (NIST) in a two-story, large area, vacant, noncombustible high school in Toledo, Ohio, sought to examine the ability of positive-pressure ventilation fans to limit smoke spread or to remove smoke from areas in an educational occupancy. NIST investigated ways for the fire service to get the products of combustion away from occupants rather than the more difficult effort of removing the occupants from the hazard. Test fires during the experiments generated vast amounts of smoke and hot gases. Instrumentation measuring temperature and pressure assessed fire and firefighting conditions as well as how PPV tactics would improve or worsen the chances of survival for occupants.

Six experiments were conducted in the 340,080 ft.^{3} (9,630 m^3) gymnasium using four different fan configurations and three different types of fans. These experiments analyzed fire growth and smoke removal pertaining to firefighting activities. Portable fans were used in three experiments, whereas the three other experiments used mounted fans. The PPV fans were all positioned outside the gymnasium lobby and were started after natural ventilation began. The portable fans were eventually moved inside to the doors between the lobby and the gymnasium. Two ventilation points were used at the rear of the gymnasium as well as an opening in the roof.

During the test fires, NIST engineers determined when the smoke layer had dropped to the point at which tenability at the floor level would be compromised. At this point, the structure was ventilated naturally via rear gymnasium doors/roof openings or ventilated with PPV fans. NIST researchers recorded the ventilation effects inside the gym relating to different fan configurations prior to the fire being extinguished. After fire suppression, the smoke was ventilated outside the gymnasium.

Findings in this limited number of experiments showed that PPV created a reduction in temperatures that provided occupants with a more tenable environment and enhanced firefighter effectiveness and safety. PPV also limited smoke spread, offering occupants better visibility and thereby improving their chances of reaching an exit.[16]

NEGATIVE STACK EFFECT

During the summer months, building air-conditioning will cool interior air. For a fire on the upper floors of a high-rise building, smoke being drawn into the hallways will flow downward into vertical openings. This is known as a negative or reverse stack effect. The higher the fire floor is in the building, the stronger will be the negative stack effect (see Figure 15.9).

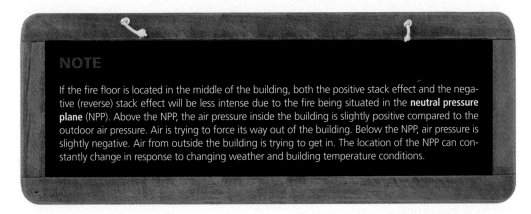

NOTE

If the fire floor is located in the middle of the building, both the positive stack effect and the negative (reverse) stack effect will be less intense due to the fire being situated in the **neutral pressure plane** (NPP). Above the NPP, the air pressure inside the building is slightly positive compared to the outdoor air pressure. Air is trying to force its way out of the building. Below the NPP, air pressure is slightly negative. Air from outside the building is trying to get in. The location of the NPP can constantly change in response to changing weather and building temperature conditions.

Neutral pressure plane ■ The area or bank of floors inside a building at which no pressure difference between the interior and the exterior exists.

Passive Protection

Passive protection is the design of floor, ceiling, roof, and wall assemblies that are constructed to limit the transfer of flame, heat, or smoke from one area of a structure to another. It helps to shield occupants from the products of combustion during fire emergencies via horizontal (door passageways) and vertical (elevator shafts) openings.

FIGURE 15.9 The negative (reverse) stack effect will draw smoke into hallways at the top floors of high-rise buildings and down vertical shafts due to the cooling effect of air conditioning during warm or hot days

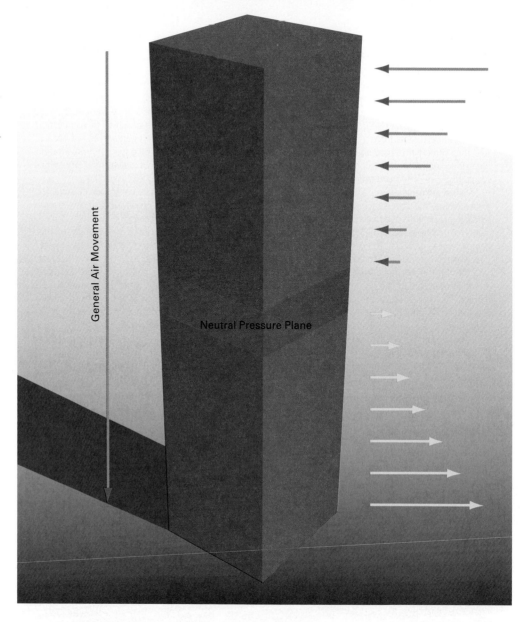

Passive smoke control components (see Figure 15.10) may include smoke and fire dampers, smoke curtains, and smoke-proof enclosures. Fire separations and stair vestibules are often required to prevent smoke from encroaching on public egress passageways.

SMOKE DAMPERS

Smoke dampers are passive fire protection products used in HVAC ductwork. They may also be installed in physical smoke barriers to prevent smoke spread from the area of fire origin. Smoke dampers are activated by detectors, the fire alarm system, or interlocked with a fire-suppression system in accord with the *International Building Code* (IBC) or AHJ code requirements. They close by an electric or a mechanical (spring) actuator and may be designed to automatically open (reset), or they may have to be opened manually.

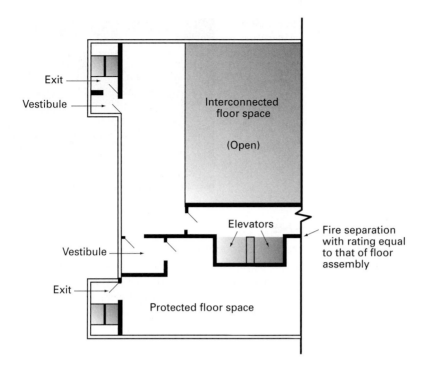

FIGURE 15.10 Passive smoke control components incorporated to protect an atrium's means of egress as well as occupied floor areas. *Source:* "Passive Smoke Control Components" from 2009 International Building Codes Commentary (ICC). Copyright 2010. http://publiccodes.cyberregs.com/icod/ibc/2009f2cc/icod_ibc_2009f2cc_9_sec009.htm. Used by Permission of www.iccsafe.org, International Codes Council (ICC).

FIRE DAMPERS

Fire dampers prevent the spread of fire inside ductwork through fire-resistance-rated walls and floors. Under fire conditions, the fire damper closes (normally activated by a thermal element [link] that melts at high ambient temperatures). Actuation allows springs to close the damper blades. Fire dampers can also close following receipt of an electrical signal from a fire alarm system using detectors. If the installation of a fire damper interferes with the operation of a required smoke control system in accordance with the IBC or AHJ code requirements, approved alternative protection must be used.[17]

COMBINATION SMOKE/FIRE DAMPERS

Combination smoke/fire dampers are also manufactured. Combination dampers provide the smoke-sealing capabilities of a standard smoke damper with the high temperature–resistance capability of a fire damper. They close upon the detection of smoke or heat. Their actuation must also be in accord with the IBC or AHJ code requirements.[18]

SMOKE CURTAINS

Smoke curtains (see Figure 15.11) are designed as vertical or horizontal barriers to contain and/or to channel away smoke, heat, and superheated gases. They activate upon receipt of a signal from the fire alarm control unit (FACU). Smoke curtains can be designed to form smoke layer reservoirs to minimize lateral smoke spread. They can also drop down from a ceiling to a preset position above the floor below, automatically, after a predetermined period of time, to assist with protecting means of egress.

A smoke curtain is comprised of a fire-resistant fabric wrapped around a roller, which is driven by an internal, electrically powered, tubular motor. The complete assembly is housed within a casing that is mounted either directly on or in the void above the ceiling. An effective smoke curtain should deploy automatically in a safe and controlled manner in the event of an alarm signal or a prolonged power failure. When combined with smoke ventilation, a smoke curtain will allow a far greater escape time for occupants.

FIGURE 15.11 Smoke curtain with flame-resistant coating designed to restrict smoke from spreading from the stage to the audience. Also used to contain smoke prior to smoke control exhaust system activation. *Source:* BorjaxS

SMOKE-PROOF ENCLOSURE

Smoke-proof enclosures are fire protected exit stairwells designed to prevent smoke and the products of combustion created during a fire from entering so that building occupants may exit safely. Rather than entering the stairwell directly from the building interior, occupants enter (via a doorway) a smoke-proof enclosure by means of an open-air balcony or open vestibule. From this area, fleeing occupants then proceed through another doorway to the exit stairwell itself. The balcony or vestibule is ventilated either by mechanical exhaust fans or naturally (smoke shaft).

CHAPTER REVIEW

Summary

Smoke and the products of combustion can be lethal to building occupants as well as first responders. Smoke is driven through openings by the expanding gases and can also travel upwards or downwards by positive and negative (reverse) stack effect as a result of temperature differences found between the interior and exterior of the building. Smoke control principles generally involve managing smoke by use of differences in air pressure to minimize its spread while exhausting it from the building. Smoke control systems for atria, however, rely on airflow and exhaust rather than differential pressurization. PPV is used by firefighters as a tactic to facilitate hose-line advancement by enhancing visibility and pushing away smoke and heat. Passive systems use dampers, curtains, and enclosures to contain smoke and keep it from spreading.

Review Questions

1. Why is it that simply exhausting smoke cannot by itself control smoke movement?
2. What is the primary cause of death in victims of structural fires?
3. At the MGM Grand Hotel and Casino fire, what HVAC system defect(s) allowed smoke to circulate to the upper floors of the building, resulting in fatalities?
4. A stair pressurization system is an example of what kind of smoke control system?
5. Elevators used for egress during fires can be especially helpful in what type occupancies?
6. What do smoke control systems for atria rely on to maintain the smoke layer at its apex above occupied floor/balcony areas?
7. What fire safety principle relating to limiting the spread of smoke is compromised in atria design?
8. What are the major objectives of an atrium smoke control system?
9. Define the stack effect.
10. During what season of the year will positive stack effect create the most pull on smoke in an upward direction?
11. In what section of a high-rise building will both the positive stack effect and the negative stack effect be less intense?
12. Explain passive smoke control protection.
13. Describe the passive smoke control component known as the smoke curtain.
14. How do smoke-proof enclosures prevent smoke and the products of combustion from entering so that building occupants may exit safely?

Endnotes

1. National Fire Protection Association (NFPA), *NFPA 92: Standard for Smoke Control Systems* (Quincy, MA: NFPA, 2012).
2. eMedicineHealth, "Smoke Inhalation," www.emedicinehealth.com/smoke_inhalation/article_em.htm#smoke_inhalation_overview.
3. William Grosshandler, Nelson Bryner, Daniel Madrzykowski, and Kenneth Kuntz, *Report of the Technical Investigation of The Station Nightclub Fire: Appendices, NIST CSTAR 2: Vol. II* (Washington, DC: US Government Printing Office, 2005).
4. G. J. Hathaway, N. H. Proctor, J. P. Hughes, and M. L. Fischman, *Proctor and Hughes' Chemical Hazards of the Workplace, 3rd ed.* (New York, NY: Van Nostrand Reinhold, 1991).
5. US Department of Health and Human Services, Public Health Service, Agency for Toxic Substances and Disease Registry, "Toxilogical Profile for Cyanide," www.atsdr.cdc.gov/toxprofiles/tp8.pdf.
6. J. Curtis Varone, Thomas N. Warren, Kevin Jutas, Joseph Molis, and Joseph Dorsey, *Report of the Investigation Committee into the Cyanide Poisonings of Providence Firefighters*, www.firefightercancersupport.org/wp-content/uploads/2013/06/cyanide_poisonings_of_providence_firefighters.pdf.
7. Ibid.

8. National Fire Protection Association (NFPA), *Investigation Report on the MGM Grand Hotel Fire, Revised Edition* (Quincy, MA: NFPA, 1982).
9. Siemens Building Technologies, Inc., *Smoke Control System Application Guide*, 125-1816, rev. 6, August, 2000.
10. Ibid.
11. George T. Tamura and C. Y. Young, "Basis for the Design of Smoke Shafts," *Fire Technology* 9, no. 3 (1973): 209–222.
12. Leslie E. Robertson and Takeo Naku, eds., *Tall Building Criteria and Loading* (Reston, VA: ASCE Publications, 1980).
13. National Fire Protection Association (NFPA), *NFPA 92B: Standard For Smoke Management Systems in Malls, Atria, and Large Spaces* (Quincy, MA: NFPA, 2009).
14. Vincent Dunn, *Command and Control of Fires and Emergencies* (Saddle Brook, NJ: Fire Engineering Books and Videos, 1999).
15. Ronald R. Spadafora, *Sustainable Green Design and Firefighting: A Fire Chief's Perspective* (Clifton Park, NY: Delmar Cengage Learning, 2013).
16. Stephen Kerber, *Evaluating Positive Pressure Ventilation in Large Structures: School Pressure and Fire Experiments* (Washington, DC: National Institute of Standards and Technology, 2008).
17. International Code Council (ICC), *International Building Code* (Washington, DC: ICC, 2012).
18. Ibid.

Additional References

Curtis S. D. Massey, "Stack Effect—What Is It and What Are Its Implications on High-Rise Fires?" *Firehouse* 37, no. 4 (2008).

G. D. Lougheed, "Basic Principles of Smoke Management for Atriums," *Construction Technology Update* 47 (2000).

James Mason, "Size-Up Considerations for High-Rise Fires," *Fire Engineering* 158, no. 4, (2005).

INDEX

A

Abraham Lincoln Presidential Library and Museum, 151
Accelerators, 59
Acidic, 155
Aircraft hangars, 111–113
Alarms. *See* Fire alarms and detection systems
Alcohol resistance, 96
Alcohol-resistant aqueous film-forming foam (AR-AFFF), 97
Alcohol-resistant film-forming fluoroprotein (AR-FFFP), 96
Alkali metals, 183–185
Alkaline, 155
Alkaline earth, transitional and other combustible metals, 183, 185
Allotropes, 178
Alloy, 171
Aluminum, 174–175
American National Standards Institute (ANSI)
    711: *Rating and Fire Testing of Fire Extinguishers*, 234
    864: *Standard for Control Units and Accessories for Fire Alarm Systems*, 240
American Water Works Association (AWWA)
    C-502 *Standard for Dry-Barrel Fire Hydrants*, 33
    C-503 *Standard for Wet-Barrel Fire Hydrants*, 33
    M17 *Standard: Installation, Field Testing, and Maintenance of Fire Hydrants*, 32
    M31 *Standard: Distribution System Requirements for Fire Protection*, 29
Antifreeze solutions, 55–56
Application specific, 225
Aqueous film-forming foam (AFFF), 96–97
Argon, 140–142
Asphyxiation, 121
Atmospheric lifetime (ATL), 119
Atria, 268–271
Automatic fire detectors, 239

B

Backdraft, defined and signs of, 2
Baghouse, 125
Bell Telephone Co., 118
Beryllium, 175–176
Bimetallic element, 69
Blackwater, 32
Bladder tanks, 107
BLEVE (boiling liquid expanding vapor explosion), 3
Boilover, 93
Borax (sodium borate), 155
Burnback resistance, 95

C

Caesium (cesium), 176
Calcium, 176
Carbonation of soft drinks, 121–122
Carbon dioxide
    benefits of, 117–118
    critical point, 119, 120
    critical temperature, 119, 120
    dangers/risks of, 121, 122, 127–132
    extended discharge system, 125
    fire extinguishers, 122, 127–128
    guidelines for first responders, 130–132
    hand hose-line applications, 125
    liquid state, 121
    local applications, 123
    minimum concentrations for extinguishment, 125, 126
    oxygen deficiency, symptoms of, 129
    phases, 119–122
    properties, 118–119
    solid state, 120
    storage of, 132–135
    total-flooding applications, 123, 124–125
    triple point, 119, 120
    vapor state, 122
Carbon monoxide, health effects, 255, 261
Check valves, 86
Chemical chain reaction, 11
Chemical explosion, defined, 3
Chemical foam, 94–95
Chemical Safety and Hazard Investigation Board (CSB), 146, 170, 171
Chips, 176
Clapper valves, 58
Class A fires, 13
Class A foams, 102
Class B fires, 13
Class B foams, 105–106
Class C fires, 13–14
Class D fires, 14
Class K fires, 14
Class I standpipe system, 78–80
Class II standpipe system, 80–81
Class III standpipe system, 81
Clean agents
    deep-seated smoldering fires, 216
    FE-13, 214
    FE-25, 214, 215
    FE-36, 214
    fire extinguishers, 221
    FM-200, 211–212, 213, 214, 215
    halocarbons, 211–213
    how they work, 208–209
    Novec 1230, 211, 212, 213, 215
    overview of, 209
    powdered aerosols, 216–220
    raised floors and, 215–216
    Significant New Alternatives Policy (SNAP), 209–211
    standards, 210–211
    substitute for Halon 1211, 201–202
    substitute for Halon 1301, 205, 210
    total-flooding applications, 215
Cleanroom, 260
Cold-weather valves, 55
Combination system, 23
Combustible metals
    See also Explosions
    alkali metals, 183–185
    alkaline earth, transitional and other combustible metals, 183, 185

Combustible metals (*continued*)
　aluminum, 174–175
　beryllium, 175–176
　caesium (cesium), 176
　calcium, 176
　categories, 183
　defined, 171
　dust explosion pentagon, 172–174
　dust explosions, 170–171
　emergency procedures, 191–192
　fire extinguishing agent applications, 182–183, 184–185
　hafnium, 176
　iron, 176
　Kst value, 172–173
　lithium, 177
　magnesium, 177–178
　manganese, 177
　overview of, 171–172
　phosphorous, 178–179
　plutonium, 179
　potassium, 179
　rubidium, 179
　silicon, 179
　sodium, 179
　strontium, 179
　tantalum, 180
　thallium, 180
　thorium, 180
　tin, 180
　titanium, 180–181
　transitional, 183
　uranium, 182
　zinc, 182
　zirconium, 182
Combustion
　defined, 2
　spontaneous, 5
Compressed-air foam system (CAFS), 102–105
Conduction, 10
Convection, 10
Covalent bonds, 38–39, 209
Critical point, 119, 120
Critical temperature, 119, 120
Cryogenic liquid, 121, 142
CSB. *See* Chemical Safety and Hazard Investigation Board

D

Dead end mains, 31
Decay phase, 13
Decommissioning, 196, 205–206
Decrepitate, 158
Deep-seated fires, 125
Deflagration, defined, 3, 172, 186
Deionized, 225
Deluge mist systems, 232
Deluge sprinklers, 54, 60
Detonation, defined, 3, 186
Dewars, 121
Differential dry pipe valve design, 58
Dip tanks, 123, 166
Distribution mains, 29
Drainboards, 122

Drain valves, 87
Drop-in replacement, 214
Dry chemical agents
　application techniques, 165–166
　fire extinguishers, 164–165
　hand hose-line applications, 164
　local applications, 162, 163
　MET-L-KYL, 156–157
　monoammonium phosphate, 157, 159
　overview of, 44, 154–155
　physical properties, 155–156
　potassium bicarbonate (Purple-K), 157–158
　potassium chloride (Super-K), 158
　properties, 158–159
　sodium bicarbonate, 154, 155, 156
　total-flooding applications, 162–163
　urea-potassium bicarbonate (Monnex), 158
　use of term, 155
Dry chemical extinguishing systems
　operational design and sequence, 161–162
　Purple-K units, 167
　system components, 159–161
Dry hydrants, 26–27
Dry ice, 120, 121
Dry pipe mist systems, 232
Dry pipe sprinklers, 54, 57–59
Dry powders, 155
Dumpster fire fatality, 174
Dust explosion pentagon, 172–174
Dust explosions, 170–171

E

Early-suppression, fast-response (ESFR) sprinklers, 72
Educators, 107–108
Elevator shaft pressurization, 266, 268
Encapsulating agents, 41–42
Endothermic reaction, 3
Entrainment, 11
Environmental Protection Agency (EPA), 209–210
Eutectic, 250
Exothermic reaction, 3
Explosions
　defined, 3, 170, 186
　detection devices and fire alarm control units, 188
　dust, 170–171
　emergency procedures, 191–192
　prevention and protection systems, 186–187, 189–191
Extinguishing agents. *See* Fire extinguishing agents

F

Factory Mutual Insurance Company, 14
Feeder mains, 30
FE-13, 214
FE-25, 214, 215
FE-36, 214
Film-forming fluoroprotein (FFFP), 96
Fines, 176
Fire
　classification of, 13–14
　defined, 2
　phases of, 11–13
　terminology, 2–5

Fire alarm control units (FACUs), 188
   defined, 240
   operational functions, 239
   signals initiated by, 240–241
Fire alarms and detection systems
   addressable, 241
   analog addressable, 241–242
   annunciator panel, 241
   automatic, 239
   components, 239
   connection stations, 256–257
   conventional, 241
   devices, 87
   devices, initiating, 242
   flame (radiant) detectors, 253–254
   gas-sensing detectors, 254–255
   heat detectors, 249–252
   ISO evaluation of, 28
   manual (pull), 242–243
   notification appliances, 256
   power supplies, 242
   pressure-sensing detectors, 255–256
   purpose of, 238–239
   smoke detectors, 243–249
Fire brigade, 78
Fire dampers, 275
Fire department connection (FDC)
   sprinklers and, 59, 64
   standpipe systems and, 81, 86
Fire departments, ISO evaluation of, 28
Fire extinguishers
   carbon dioxide, 122, 127–128
   clean agent, 221
   dry chemical agents, 164–165
   fire rating of, 15–16
   Halon 1211, 200–201
   hydrostatic testing of, 18
   inspection of, 17–18
   labeling and marking system, 14, 15
   maintenance of, 17
   nonmagnetic, 15
   obsolete, 18
   type, size, and placement of, 16–17
   water mist, 234–235
   wet portable, 49
   when to use, 18
Fire extinguishing agents
   dry chemical agents, 42–44, 154–168
   encapsulating agents, 41–42
   saponification, 42–43
   water, 37, 38–40
   wet chemical agents, 38, 42, 43–44
   wetting agents, 38, 40–41
Fire extinguishing systems, 45–49
   See also Sprinklers; Water mist systems
Fire hazard categories, 16
Fire hydrants
   capacity, 34
   color coding, 34
   dry-barrel, 32–33
   flow testing, 35
   gate valves, 33–34
   inspection and maintenance, 35
   signage, 34
   spacing, 34
   standards, 32
   wet-barrel, 33
Fire lines, 29
Fire point of liquids, 7
Fire proximity suit, 11
Fire pumps, 62–64, 82–83
Fire Suppression Rating Schedule (FSRS), 28
Fire Suppression Systems Association (FSSA), 206
Fire Testing of Fire Extinguishing Systems for Protection of Restaurant Cooking Areas (UL-300 standard), 44
Fire tests, 164
Fire tetrahedron, defined, 3, 4
Fire triangle, defined, 3–4
First Interstate Bank fire, 82
Fixed foam systems, 106–113
Flame (radiant) detectors, 253–254
Flammable gases, 7–8
Flashover, defined, 4, 9
Flash point, of liquids, 6–7
Fluoroprotein (FP) foam, 95–96
FM-200, 211–212, 213, 214, 215
Foam
   bladder tanks, 107
   chambers, 98, 108
   chemical, 94–95
   Class A, 102
   Class B, 105–106
   concentrate, 95, 107
   defined, 92
   educators, 107–108
   generators, 109
   high-expansion, 99–101
   how it works, 94
   low-expansion, 98–99
   makers, 99, 108
   mechanical, 95
   monitors, 98, 109
   physical characteristics, 93
   properties, 98
   proportioners, 107–108
   protein, 95–96
   pump skids, 108
   solution, 95
   sprinkler foam-water heads and nozzles, 109
   synthetic, 96–97
   trough, 98
Foam extinguishing systems
   compressed-air, 102–105
   fixed, 106–113
   general principals of, 93–94
Frangible bulb, 69
Frangible pellet, 69
Free radicals, 156
Friction loss, 31
Fuel resistance, 95
Fuels
   gases, 7–8
   heat of combustion of some common, 9
   liquids, 6–7
   solids, 5–6

Fully Developed phase, 13
Fusible link, 69

G

Gases
   See also Inert gases
   chemical properties of, 7–8
   greenhouse, 119
   toxic, 8, 262–263
Gas-sensing detectors, 254–255
Gate valves, 33–34, 86
Global warming potential (GWP), 119, 142
Gravity system, 23, 24
Gravity tanks, 83–84
Great Lakes Chemical Corp., 211
Greenhouse gas, 119
Greywater, 32
Growth phase, 12–13

H

Hafnium, 176
Halocarbons, 211–213
Halogens, 183
Halon, 147
   decommissioning, 196, 205–206
   numbering system, 198, 199
   properties, 197–198
   pros and cons of, 195–196
Halon 1211
   applications, 201
   clean agent substitute for, 201–202
   fire extinguishers, 200–201
   local applications, 200
   overview of, 199–200
   properties, 197
   structural formula, 198
   system components, 200
Halon 1301
   clean agent substitute for, 205, 210
   emergency procedures, 205
   overview of, 202
   properties, 198
   structural formula, 198
   system components, 203–204
   total-flooding applications, 202–205
Halotron I, 201–202
Halotron II, 205
Heat (thermal energy)
   of combustion of some common fuels, 9
   difference between temperature and, 8
   flux, 9
   resistance, 95
   specific, 39
   transfer of, 10–11
   transmission of, 8
   units, 8–9
Heat detectors
   classifications of operation, 249
   coaxial conductor line, 251–252
   paired-wire line, 252
   spot fixed-temperature, 250

   spot rate-compensation, 251
   spot rate-of-rise, 250–251
Heating, ventilation, and air-conditioning (HVAC), smoke control systems, 263–264
Heat release rate (HRR), 9, 11, 13
High-expansion foams, 99–101
Hoeganaes powdered metal plant explosion, 176
Hose stations, 87
Hydrogen cyanide, 261–262
Hydrophilic, 41
Hydrophobic, 41
Hypercapnia, 148
Hypoxia, 148
Hypoxia atmosphere, 235

I

Imperial Sugar factory explosion, 171
Incipient phase, 12
INERGEN, 147–151
Inert gases (IG), 8
   argon, 140–142
   dangers of, in confined spaces, 143, 145–146
   defined, 138
   environmental issues, 141–142
   guidelines for first responders, 143, 146–147, 151
   nitrogen, 143–146
   storage of, 139–140
Ingots, 185
In-rack sprinklers, 71
Insurance Services Office (ISO), 27–28
International Maritime Organization (IMO), 224
International Paper Plant, 42
Iron, 176

J

Jockey pump, 63

K

Knockdown, 96
Kst value, 172–173

L

Lethal concentration in air, 262
Light-emitting diodes (LEDs), 240
Liquid argon, 142
Liquids, characteristics of, 6–7
Liquid state of carbon dioxide, 121
Listed sprinklers, 57
Lithium, 177
Loading, 56
LOAEL (Lowest Observable Adverse Effect Level), 210
Low-expansion foams, 98–99

M

Magnesium, 177–178
Manganese, 177
McDonald's, 122
Mechanical foam, 95
Metals, combustible, 6
MET-L-KYL, 156–157
MGM Grand Hotel and Casino fire, 264

Micelle, 41
Microprocessors, 240
Model Building and Fire Codes, 75
Monnex (urea-potassium bicarbonate), 158
Monoammonium phosphate, 157, 159
Montreal Protocol, 196–197

N

National Association of Fire Equipment Distributors (NAFED), 44
National Fallen Firefighters Foundation (NFFF), 221
National Fire Incident Reporting System (NFIRS), 52
National Fire Protection Association (NFPA)
  carbon dioxide and, 118
  NFPA 10: *Standard for Portable Fire Extinguishers*, 14, 16, 17
  NFPA 11: *Standard for Low-, Medium-, and High-Expansion Foam*, 98, 111
  NFPA 12: *Standard on Carbon Dioxide Extinguishing Systems*, 123, 125, 129
  NFPA 13: *Standard for the Installation of Sprinkler Systems*, 52
  NFPA 13D: *Standard for the Installation of Sprinkler Systems in One- and Two-Family Dwellings and Manufactured Homes*, 56
  NFPA 13R: *Standard for the Installation of Sprinkler Systems in Low-Rise Residential Occupancies*, 56
  NFPA 14: *Standard for the Installation of Standpipes and Hose Systems*, 75–76, 111
  NFPA 16: *Standard for the Installation of Foam-Water Sprinkler and Foam-Water Spray Systems*, 111
  NFPA 17: *Standard for Dry Chemical Extinguishing Systems*, 44, 155
  NFPA 17A: *Standard for Wet Chemical Extinguishing Systems*, 44, 48
  NFPA 18: *Standard on Wetting Agents*, 41
  NFPA 20: *Standard for the Installation of Stationary Pumps for Fire Protection*, 64, 83
  NFPA 25: *Standard for the Inspection, Testing, and Maintenance of Water-Based Fire Protection Systems*, 62, 64, 76, 89
  NFPA 33: *Standard for Pre-Engineered Dry Chemical Extinguishing System Units*, 155
  NFPA 69: *Standard on Explosion Prevention Systems*, 186
  NFPA 70: *National Electrical Code*, 204
  NFPA 72: *National Fire Alarm and Signaling Code*, 204, 240
  NFPA 92: *Standard for Smoke Control Systems*, 261
  NFPA 92B: *Standard for Smoke Management Systems in Malls, Atria, and Large Spaces*, 270
  NFPA 96: *Standard for Ventilation Control and Fire Protection of Commercial Cooking Operations*, 44
  NFPA 291: *Recommended Practice for Fire Flow Testing and Marking of Hydrants*, 34
  NFPA 409: *Standard on Aircraft Hangars*, 111
  NFPA 484: *Standard for Combustible Metals*, 174
  NFPA 704: *Standard System for the Identification of the Hazards of Materials for Emergency Response*, 183
  NFPA 750: *Standard for the Installation of Water Mist Fire Protection Systems*, 225, 228, 230
  NFPA 1142: *Standard on Water Supplies for Suburban and Rural Fire Fighting*, 25
  NFPA 2001: *Standard on Clean Agent Fire Extinguishing Systems*, 142, 210–211
  performance statistics for sprinklers, 52
  smoke detectors and, 246
National Foam Inc., 95

National Institute of Standards and Technology (NIST), 272
Neutral pressure, 273
Nitrogen, 143–146
NOAEL (No Observable Adverse Effect Level), 210
Nonpotable water, 32
Notification appliances, 256
Novec 1230, 211, 212, 213, 215

O

One Meridian Plaza fire, 52–53, 81
Outside screw and yolk (OS&Y) valves, 65–66, 87
Overhaul, 103
Oxidation, 11
Oxidizer, 8, 11
Oxygen deficiency, symptoms of, 129
Ozone-depletion potential (ODP), 119, 142, 197

P

Phase diagram, 119
Phosphorous, 178–179
Plastics, 6
Plutonium, 179
Polymers, 97
Positive-pressure ventilation (PPV) fans, 272–273
Post indicator valves (PIVs), 67
Potable water, 32
Potassium, 179
Potassium bicarbonate (Purple-K), 157–158
Potassium chloride (Super-K), 158
Powdered aerosols, 216–220
PPC (Public Protection Classification) program, 28
Preaction mist systems, 232
Preaction sprinklers, 54, 61
Pressure-control valves, 78
Pressure-reducing devices, 78, 79
Pressure-reducing valves (PRVs), 78, 81
Pressure-restricting devices (PRDs), 78, 88
Pressure-sensing detectors, 255–256
Pressure tanks, 84–85
Propelled extinguishing agent technology (PEAT), 218–219
Proportioners, 107–108
Protein foams, 95–96
Providence, Rhode Island, fires, 263
Pumped system, 23, 24
Pump skids, 108
Purple-K (potassium bicarbonate), 157–158
Purple-K units, 167
Pyrolysis, defined, 4–5
Pyrophoric, 156

Q

Quick-opening devices (QODs), 58
Quick-response sprinklers (QRS), 57

R

Radiation, 10–11
Rapid intervention teams (RITs), 271
Reactive gases, 8
Reservoirs, 24
Residential sprinkler systems, 56
Residual pressure, 80

Ribbon, 177
Roof liquid storage tanks
   floating, 111
   permanent, 110
Rubidium, 179

S

Saponification, 42–43
Scandinavian Star ferry disaster, 226–227
Section valves, 66
Shavings, 177
Sidewall sprinklers, 70–71
Significant New Alternatives Policy (SNAP), 209–211
Silicon, 179
Smoke
   dangers of, 261
   toxic gases, 262–263
Smoke alarms, difference between smoke detectors and, 244
Smoke control systems
   atria, 268–271
   dedicated, 265–271
   elevator shaft pressurization, 266, 268
   nondedicated, 263–264
   passive, 273–276
   principles of, 271–273
   purpose of, 260
   smoke shaft exhaust, 268
   stairwell pressurization, 265–266, 267
   standards, 261
Smoke curtains, 275–276
Smoke dampers, 274, 275
Smoke detectors
   air-aspirating, 246–248
   difference between smoke alarms and, 244
   linear-beam, 248
   role of, 243
   spot ionization, 245–246
   spot photoelectric light-obscuration, 244, 245
   spot photoelectric light-scattering, 244, 245
   video-imaging, 248–249
Smoke inhalation, 261
Smoke-proof enclosure, 276
Smoke shaft exhaust, 268
Sodium, 179
Sodium bicarbonate, 154, 155, 156
Sodium borate (borax), 155
Solids, 5–6
Solvents, 97
Specific heat, 39
Spectrum Naturals, 143
Spontaneous combustion, defined, 5
Sprinkler Identification Number (SIN), 68
Sprinklers
   deluge, 54, 60
   density and area approach, 52
   dry pipe, 54, 57–59
   early-suppression, fast-response, 72
   fire department connection and, 59, 64
   fire pumps, 62–64
   heads, 68–70
   in-rack, 71
   inspection, testing, and maintenance of, 62

   jockey pump, 63
   performance statistics, 52
   piping arrangements, 67–68
   preaction, 54, 61
   quick-response, 57
   residential, 56
   sidewall, 70–71
   system components, 62–71
   valves, 64–67
   wet pipe, 54–56
Stack effect, 271–273, 274
Stairwell pressurization, 265–266, 267
Standard Time/Temperature Curve, 11
Standpipe kit, 87, 88
Standpipe systems
   alarm devices, 87
   automatic dry, 77
   automatic wet, 76
   classes of, 78–81
   combined (dual), 76
   defined, 75
   fire department connection, 81, 86
   hose stations, 87
   inspection, testing, and maintenance of, 88–89
   installation requirements, 75–76
   manual dry, 77
   manual wet, 77
   semiautomatic dry, 78
   supervisory (tamper) switches, 87–88
   valves, 78, 81, 86–87
   water flow alarms, 87
   water supply and operations, 81–85
Static pressure, 265
Station nightclub fire, 263
Stratification, 248
Stratosphere, 196
Strontium, 179
Sublimation, 120
Super-K (potassium chloride), 158
Supervisory (tamper) switches, 87–88
Surface fires, 125
Surfactants, 96
Swarfs, 182
Synthetic fiber, 6

T

Tamper switches, 87–88
Tantalum, 180
Temperature
   difference between heat and, 8
   units (scales), 10
Thallium, 180
Thermal energy. *See* Heat
Thermal imaging camera, 191–192
Thermal lag, 249
Thermistor, 251
Thermocouples, 104, 251
Thorium, 180
3M, 211
Tin, 180
Titanium, 180–181
Toxic gases, 8, 262–263
Transmission mains, 29

Triple point, 119, 120
Truck loading racks, 113–114
Turnings, 176

U

Underwriters Laboratories (UL), 14, 41, 44, 56, 57, 155
   711: *Rating and Fire Testing of Fire Extinguishers,* 234
   864: *Standard for Control Units and Accessories for Fire Alarm Systems,* 240
Union Carbide, Taft/Star plant tragedy, 145
Uranium, 182
Urea-potassium bicarbonate (Monnex), 158
US Army Corps of Engineers, 198, 199
US Fire Administration (USFA), 52
US Naval Research Laboratory, 157

V

Valves
   sprinkler, 64–67
   standpipe systems, 78, 81, 86–87
Vapor density, 7
Venturi effect, 107

W

Wall post indicator valve (WPIV), 67
Water
   advantages of, 39–40
   chemical properties, 38–39
   curtain, 40
   disadvantages of, 40
   as a fire extinguishing agent, 37, 38–40
   physical properties, 39
Water distribution systems
   dual, 32
   network, 29
   types of, 30–32
Water flow alarms, 87
Water mist systems
   advantages and disadvantages of, 227
   deluge, 232
   droplet size, 225–226
   dry pipe, 232
   effectiveness of, 225
   fire detection and alarm, 231–232
   fire extinguishers, 234–235
   foam additives and, 235
   local applications, 234
   nozzles, 230–231
   overview of, 224–225
   piping, 230
   preaction, 232
   pressure classifications, 228–229
   properties, 225
   pumps/cylinders, 230
   safety issues, 235
   system components, 229–232
   total-flooding applications, 234
   water supply methods, 232–234
   wet pie, 232
   zone applications, 234
Water supply systems
   dry hydrants, 26–27
   ISO evaluation of, 28
   standards, 23, 25
   standpipe systems and, 81–85
   types of, 23
   water sources and design criteria, 23–26
Water tender/tanker, 25, 26, 40
Water towers, 24–25
Water volumes and flow rates, 25
Wet chemical agents, 38, 42, 43–44
Wet chemical fire extinguishing systems, 45–49
Wet pipe mist systems, 232
Wet pipe sprinklers, 54–56
Wetting agents, 38, 40–41
Wildland, 4
Wildland/urban interface fires, 102
Wood and wood products, 5–6

Z

Zinc, 182
Zirconium, 182